T0259126

Computer Modelling in Inorganic Crystallography

Computer Modelling in Inorganic Crystallography

edited by

C.R.A. Catlow
The Royal Institution of Great Britain,
21 Albemarle Street, London W1X 4BS, UK.

ACADEMIC PRESS
San Diego · London · Boston
New York · Sydney · Tokyo · Toronto

This book is printed on acid-free paper.

Academic Press, Inc.
525 B Street, Suite 1900, San Diego, California 92101-4495, USA
http://www.apnet.com

Academic Press Limited
24–28 Oval Road, London NW1 7DX, UK
http://www.hbuk.co.uk/ap/

ISBN 0-12-164135-X

A catalogue record for this book is available from the British Library

Typeset by Paston Press Ltd, Loddon, Norfolk

Printed and bound in the United Kingdom
Transfered to Digital Printing, 2011

Preface

Computer modelling techniques are now having a major impact on several key areas of contemporary science. Moreover, the range and scope of the techniques are constantly expanding due to development in methodologies and algorithms and to the continuing and dramatic expansion in the capabilities of the hardware. In atomistic modelling, one of the most basic and enduring class of applications concerns the modelling (and increasingly the prediction) of structures. Such methods may be used as an aid to analysis of experimental data, and more ambitiously as design tools in the development of new materials. The present book focuses on applications to inorganic materials. Its theme is the use of modelling methods as tools in the development of atomistic structural models for the bulk, defect and surface structures of crystalline inorganic solids, although discussion of molecular and amorphous materials is included. The aim is to show the applicability of these techniques to real, complex problems of importance in current experimental studies of inorganic materials.

The book has profited from input and advice from many groups and individuals. But I am especially grateful to Professor A.M. Stoneham, Professor Sir John Meurig Thomas and Professor A.K. Cheetham for discussion during many years on the applications of modelling methods in the science of inorganic materials.

C.R.A. Catlow
April, 1996

Contributors

N.L. Allan School of Chemistry, University of Bristol, Cantock's Close, Bristol BS8 1TS, UK

R.G. Bell The Royal Institution of Great Britain, 21 Albemarle Street, London W1X 4BS, UK

I.D. Brown Brockhouse Institute for Material Research, McMaster University, Hamilton, Ontario, Canada L8S 4M1

C.R.A. Catlow The Royal Institution of Great Britain, 21 Albemarle Street, London W1X 4BS, UK

C.M. Freeman Molecular Simulations Inc., 9685 Scranton Road, San Diego, CA 92121-3752, USA

A.M. Gorman Molecular Simulations Inc., 9685 Scranton Road, San Diego, CA 92121-3752, USA

J.H. Harding Department of Physics and Astronomy, University College London, Gower Street, London WC1E 6BT, UK

N.M. Harrison CLRC, Daresbury Laboratory, Warrington, Cheshire WA4 4AD, UK

R.A. Jackson Department of Chemistry, University of Keele, Keele, Staffs ST5 5BG, UK

P.W.M. Jacobs Department of Chemistry, The University of Western Ontario, London, Canada N6A 5B7

W.C. Mackrodt School of Chemistry, University of St Andrews, St Andrews, Fife KY16 9ST, UK

R.L. McGreevy Studsvik Neutron Research Laboratory, Uppsala University, S-61182 Nyköping, Sweden

J.M. Newsam Molecular Simulations Inc., 9685 Scranton Road, San Diego, CA 92121-3752, USA

S.C. Parker School of Chemistry, University of Bath, Claverton Down, Bath BA2 7AY, UK

G.D. Price Department of Geological Sciences, University College London, Gower Street, London WC1E 6BT, UK

S.L. Price Department of Chemistry, University College London, 20 Gordon Street, London WC1H 0AJ, UK

Z.A. Rycerz Department of Chemistry, The University of Western Ontario, London, Canada N6A 5B7

P. Tschaufeser School of Chemistry, University of Bath, Claverton Down, Bath BA2 7AY, UK

B. Vessal 1830 Redwing Street, San Marcos, CA 92069, USA

A. Wall School of Chemistry, University of Bath, Claverton Down, Bath BA2 7AY, UK

G.W. Watson School of Chemistry, University of Bath, Claverton Down, Bath BA2 7AY, UK

1

Need and Scope of Modelling Techniques

C.R.A. Catlow

1 INTRODUCTION

Atomistic computer modelling techniques are now used extensively in a wide range of fields in both biological and physical sciences (see e.g. van Gunsteren and Berendsen, 1990; Allen and Tildesley, 1987; Catlow *et al.*, 1989). Using these methods we are increasingly able to develop and explore models of the structures, energetics and dynamics of complex molecules and materials; and indeed computer modelling methods are now applied together with experiment in constructing models and solving structural problems. There have been many successes in the applications of these techniques to biological systems; an important and relevant class of examples is the concerted use in protein structural studies of molecular modelling and least squares refinements in solving protein structures (e.g. Brunger and Rice, 1995). The incentive for their application to the structural chemistry of inorganic systems is equally strong, as the study of these materials is increasingly concerned with complex structures, for example zeolites (and other microporous solids), high temperature superconductors and other ternary and quaternary oxides. As this book will show, computational methods now play a central rôle in modelling these and other topical inorganic materials.

To what specific types of problem do we wish to apply computational methods in structural inorganic chemistry? The first and most obvious are in *modelling crystal structures*. Computational techniques are now able to refine and reproduce accurately the crystal structures of a diverse range of inorganic systems; examples will be given in Chapters 3, 5, 9 and 10. Such methods are being used to assist the refinement of crystal structure data; the real challenge in this field is now to develop procedures for predicting structures. Considerable success has been enjoyed for several years by simple but effective procedures based on Pauling's rules for generating approximate structural models of crystals. Such approaches will be described in Chapter 2. Computational methods for generating structures with little prior knowledge are developing fast and will be reviewed in Chapter 5; and the goal of automated prediction of inorganic crystal structures is now in sight.

COMPUTER MODELLING IN INORGANIC CRYSTALLOGRAPHY
ISBN 0-12-164135-X

There is perhaps an even greater incentive for the development of methods for *modelling amorphous structures* owing to the well-known difficulties in the determination of accurate and unambiguous atomistic structures for non-crystalline solids from experimental data alone. Chapter 12 will concentrate on this aspect of structural modelling. *Inorganic surface chemistry* is a field of growing importance and activity. Simulation techniques again have a unique rôle in developing models of structures and energies as discussed in Chapter 7, where we will stress the application of these methods to complex systems including ternary oxides.

One of the oldest and most successful fields of application of simulation techniques in solid state science concerns the study of *defects in solids* which, of course, influence many key properties including transport, reactivity and mechanical behaviour. Simulation techniques have been used with success for over 20 years in calculating formulation and migration energies of defects in solids. An account of the present state of the art is given in Chapter 7.

Dynamical processes in solids, including vibrational properties and atomic diffusion are of obvious interest and importance, not only in a crystallographic context but in broader aspects of materials science. Chapter 4 therefore focuses on the application of molecular dynamics simulation techniques in the explicit modelling of the atomic dynamics of solids.

One of the most exciting recent developments in the fields of computational chemistry and physics has been the increasing scope and accuracy of *electronic structure techniques*. Thus Chapter 8 reviews the current methodologies employed in electronic structure studies of solids with special reference to the qualitative insight which they give into understanding bonding in solids and the quantitative information on charge and spin densities.

Modelling techniques are applicable to a broad range of solids. The present book focuses on non-metallic inorganic systems; detailed discussions of modelling metals can be found elsewhere (e.g. Vitek and Srolovitz, 1989). Particular attention will be paid to oxide, silicate and related systems which include many of the most intensively investigated and important of the inorganic materials. However, the description in Chapter 11 of recent studies of *molecular crystals* allows us to place the modelling of inorganic materials into a broader context.

Our discussions now continue with an account of the basic aspects of modelling methodologies, which we follow with illustrative examples of recent applications.

2 TYPES OF METHODOLOGY

There are two broad classes of method used in current computational chemistry and physics. The first we will denote as '*force field methods*'. In this approach, knowledge of electronic structures is subsumed into effective *interatomic potentials* (V), which describe in analytical or numerical form the variation in the energy of the molecule or solid as a function of the nuclear coordinates ($r_1 ... r_N$) of the N atoms present in the molecule or solid. The total potential is generally broken down into 'pair', 'three-body', 'four-body' and higher order terms:

$$V(r_1 \cdots r_N) = \sum_{i,j=1}' v^{(2)}(r_{ij}) + \sum_{i,j,k=1}' v^{(3)}(\boldsymbol{r}_i, \boldsymbol{r}_j, \boldsymbol{r}_k) + \sum_{i,j,k,l=1}' v^{(4)}(\boldsymbol{r}_i, \boldsymbol{r}_j, \boldsymbol{r}_k, \boldsymbol{r}_l) + \cdots \tag{1}$$

(and as discussed in Chapter 7, an initial term $v^{(0)}$ may be included; the primes indicate that in the summations a given term is only included once). V is often approximated by the leading, pair potential term $v^{(2)}$ which depends simply on the distance between the nuclei i and j; and indeed for systems such as ionic crystals in which interactions between the atoms are non-directional, this is generally a good approximation. It is less acceptable for covalent solids in which there is normally strong directionality to the bonding. The pair potential term itself is usually decomposed into Coulombic and non-Coulombic terms (ϕ):

$$v^{(2)}(r_{ij}) = \frac{q_i q_j}{r_{ij}} + \phi(r_{ij}) \tag{2}$$

where the q_i and q_j are the *effective charges* of the atoms. A number of standard analytical functions are available for ϕ, e.g. the Buckingham and Lennard–Jones forms, discussed in Chapters 3 and 11, which are more appropriate for non-bonded interactions. However, as noted, numerical functions may also be used. The types of three-body and higher order terms used in currently available force fields are considered in detail in Chapter 3, while the crucial question of the methods used in summations of the Coulomb interactions is reviewed in Chapter 4.

One of the major thrusts in current computational studies of molecules and solids is the development of high quality force fields. To develop force fields we may use 'empirical fitting' procedures in which parameters in the analytical function describing the potential are adjusted via a least squares fitting procedure in order to reproduce as accurately as possible the observed properties of a molecule or crystal. The range of properties must include the structure, while information on vibrational, elastic and dielectric properties is also of great value as discussed in Chapter 3. The second strategy, which has been used for many years, but which is being increasingly refined is to calculate the potentials directly using quantum mechanical methods of the type discussed later. Thus the energy of a molecule or of a pair or cluster of non-bonded atoms is calculated for a variety of geometries and the resulting energy variation is again fitted to an interatomic potential function. The method may also be applied to periodic structures (Gale *et al.*, 1992). Further discussion will follow in Chapter 3.

Having developed a suitable interatomic potential, there is a range of computational techniques which may be employed for calculating structural properties and the dynamical behaviour of molecules and solids, the computational basis of which is now described; subsequent chapters will give more details of the techniques.

3 ENERGY MINIMIZATION

In this simplest of approaches we determine the minimum energy configuration of the molecule or crystal corresponding to the specified interatomic potential. These methods require the specification of an initial configuration or 'starting point'; the system is then driven (by an iterative procedure) down in energy to the nearest minimum. A wide variety of algorithms are available, which are classified according to the order of the derivative of the total energy function that is employed in the calculation. In general, methods employing second derivatives of the energy with respect to atomic positions are more rapidly convergent than first derivative techniques; and the latter are considerably more efficient than 'zero derivative' or search procedures.

Modern minimization methods can be applied effectively and rapidly to large complex systems and, as we will see, are making an increasingly important contribution to real crystallographic problems. The approach is, however, limited for several reasons, of which the following three are the most important:

1. *The need for a starting point.* This is inherent to this method which, as noted, locates the nearest minimum to the initially specified structure. The first or 'trial' structure may be devised by chemical intuition or analogy; approximate experimental information may also be used. There are also a number of more automated procedures. *Simulated annealing* methods (which employ Monte Carlo algorithms as discussed below) allow very approximate, including random, configurations to be refined to generate a range of candidates for starting points. These methods necessarily employ a *cost function* which is normally simpler to evaluate than the energy of the system and which measures the quality of the structure. The method allows configurational space to be explored, but in a way that biases the search systematically towards structures with low cost functions. An alternative approach based on *genetic algorithms* uses evolutionary programming techniques to 'breed' successive structures whose probability of survival is dependent on the value of the cost function. Again, random or near random initial configurations can be used to generate plausible starting points. We return to the simulation annealing method below in the context of the Monte Carlo simulation method and Chapter 5 will discuss these techniques and the question of the choice of cost functions in greater detail; an example of the

successful use of the genetic algorithm approach will be given later in this chapter.

2. *The local minimum problem.* This is intimately related to the issue just discussed. Minimization methods find, of course, the *nearest* minimum to the starting point. All but the simplest molecules and materials have multiple minima and there is no guarantee that the lowest energy structure has been located; indeed as the complexity of the system increases, this becomes an increasingly remote eventuality. The only real solution to the problem is to sample large numbers of starting points in order to ensure that all low energy minima have been identified. The procedures outlined in the previous paragraph are often used in selecting a range of plausible starting points; a number of sophisticated mathematical procedures have been developed with the aim of simplifying the overall potential energy surface and 'washing out' local minima (e.g. Kostrowicki and Sheraga, 1992), but there can never be a guarantee that the 'global' minimum—the minimum of lowest energy—has been identified.
3. *Omission of dynamics.* Minimization identifies the static configuration of lowest energy and there is no representation of the vibrational or other dynamical properties of the system. In formal terms, these are 'zero Kelvin' calculations, with zero point motion omitted. It is, however, relatively straightforward to add a treatment of the vibrational properties of the system within the harmonic (or quasi-harmonic) approximations. Such methods will be discussed in Chapter 3.

Despite these limitations, minimization techniques are straightforward, robust and readily applicable, and can be applied to systems with large and complex unit cells. They are being used increasingly to refine approximate crystal structures and to calculate the relative energies of different polymorphs, as discussed later in this chapter and in Chapters 3 and 5. Highly complex structures may, moreover, now be modelled routinely as will be apparent in the discussion of silicates in Chapter 9 and high temperature superconductors in Chapter 10.

4 MOLECULAR DYNAMICS (MD)

The basis of these very widely used techniques is again conceptually simple. We simulate the system studied by an ensemble of particles (usually representing atoms, but occasionally groups of atoms) which are contained in a simulation box. Periodic boundary conditions (pbcs) are commonly applied, i.e. the basic simulation box is replicated infinitely in three directions (although 2-D pbcs may be used in surface simulations); in simulations of crystals, the simulation box will comprise the unit cell or, more commonly, a supercell. As the simulations are 'dynamical', the atoms are given velocities that are chosen with a target temperature in mind. The simulation proceeds by solving the classical (Newtonian) equations of motion in an iterative manner, in which each iteration (x_i) and velocity (v_i) of the particles in the simulation box are updated after the lapse of a 'time step', Δt, for an infinitesimal value of which we would have the simple update expressions:

$$x_i(t + \Delta t) = x_i(t) + v_i(t)\Delta t \tag{3a}$$

$$v_i(t + \Delta t) = v_i(t) + \frac{f_i}{m_i}(t) \cdot \Delta t \tag{3b}$$

where m_i and f_i are respectively the mass and the total force acting upon the ith particle; the forces are, of course, calculated from the derivatives of the interatomic potentials. Since, in practice, a finite value of Δt must be used, all update algorithms in effect include terms with higher order powers of Δt. The choice of Δt is of critical importance in MD simulations. It must be smaller than the time scale of any important dynamical processes at the atomic or molecular level; thus it must be at least an order of magnitude smaller than the typical period of atomic vibrations (10^{-12}–10^{-13} s); and the shorter the time step, the more accurate is the numerical integration of the equations of motion. But of course short time steps mean more computational effort for the same amount of 'real time' simulation. Values in the range 10^{-14}–10^{-15} s normally represent the best compromise as discussed in detail by Allen and Tildesley (1987) and by Jacobs and Rycerz in Chapter 4 of this book. These references also review the range of update algorithms used in contemporary MD studies and describe the more sophisticated variants of the technique.

MD simulations therefore proceed by successive applications of the time step, which allows us to develop a dynamical record of the time evolution of the system over the period of simulation. During the early stages of the simulation, the system 'equilibrates', i.e. it achieves equipartition of energy and a Maxwellian distribution of velocities. Once equilibrium is achieved, a production run follows, from the analysis of which details of the structure and dynamics of the system may be calculated. Modern simulations involve typically several thousand particles, sampling 100 ps to 1 ns of 'real time' (i.e. 10^5–10^6 time steps).

MD methods give the most detailed and rich information of any atomistic simulation method. In particular, they have the major advantage of incorporating dynamical (including anharmonic) effects explicitly. They are, however, computationally far more intensive than energy (and free energy) minimization methods. Moreover, the amount of 'real time' sampled is limited to the order of nanoseconds even with the largest, most powerful contemporary computers. Dynamical processes with longer time scales cannot therefore be sampled. The real strength of the method in the context of the present book is seen first in the simulation of solids containing mobile atoms or ions, notably the so-called superionic or fast-ion conductors discussed later in this chapter and in Chapter 4: the same chapter will also describe the successes in modelling certain classes of phase transition; while in Chapter 12, Vessal discusses the important application to the generation of models for vitreous materials by directly simulating the melt–quench procedure used in the preparation of glasses.

One rôle of MD is in counteracting the 'local minimum problem' in attempts to identify global minima. Since thermal effects are included in the simulation, it is possible to surmount energy barriers between different minima. But it should be emphasized that an MD simulation is unlikely to overcome energy barriers which are much greater than thermal energies; and indeed, the simulation cannot be expected to escape from regions of the potential hypersurface that are bounded by barriers $\gg kT$. An alternative strategy for exploring different regions of the energy surface is the simulated annealing strategy, referred to above, which is based on the Monte Carlo methodologies discussed in the next section.

5 MONTE CARLO TECHNIQUES

Monte Carlo (MC) resembles MD in that successive configurations of the simulated system are generated, but there is no relationship in time between such configurations. Indeed successive configurations are generated by random 'moves' which can be an atomic translation or molecular rotation, or the insertion or deletion of an atom or molecule. In this way, ensembles of configurations are accumulated, from which statistical averages may be calculated. In generating such ensembles, it is essential to formulate an 'acceptance' criterion, i.e. a procedure that determines whether a new configuration created by a move will be accepted within the ensemble. The most commonly used approach is the Metropolis method, which proceeds as follows:

1. The energy change, ΔE, associated with the move is calculated; from this the Boltzmann factor $B = \exp(-\Delta E/kT)$ is calculated if ΔE is positive. We note that MC simulations are most commonly run in the NVT (or canonical) ensemble in which both volume and temperature are fixed.
2. If ΔE is negative, the configuration is accepted and we proceed with another random move. If not, we continue with the next stage.
3. A random number (P) in the range 0–1 is generated.
4. If $P < B$, the move is accepted; if not, it is rejected. In either case, we continue with the generation of a new move.

The effect of this procedure is to generate an ensemble of configuration that is consistent with Boltzmann statistics.

Like MD, Monte Carlo simulations are typically performed on ensembles containing several thousand particles, to which periodic boundary conditions are applied in the case of the simulation of solids. The simulation again starts with an equilibration phase during which the system is 'thermalized', followed by a 'production run' in which, typically, several million configurations are generated. We refer once more to the monograph of Allen and Tildesley (1987) for more details of the methodologies employed.

In modelling crystalline solids, the MC technique is of particular value in three distinct fields. The first concerns studies of *sorbed systems*, e.g. microporous solids loaded with organic molecules. MC techniques are particularly suitable for studying the variation with temperature of the distribution of sorbed molecules in such systems (Yashonath *et al.*, 1988). Secondly, the method has been fruitfully applied to the study of atomic diffusion. In this case the 'moves' are atomic jumps of defined frequencies. In complex solids (including, e.g., alloys and ionic conductors), use of the MC technique allows accurate sampling of all the different jump mechanisms contributing to the diffusion, as shown in several studies of Murch and coworkers (e.g. Murch, 1982).

The third class of applications referred to above uses the MC method to explore complex energy surfaces and to find low energy regions, from which minima can subsequently be located by standard minimization procedures. As noted earlier and as discussed in greater detail in Chapter 5, *simulated annealing* methods have been used with great effect in generating crystal structures from initial random configurations of atoms or ions. We note that the 'energy term' in simulating annealing calculations may take a variety of forms. It is commonly, as noted earlier, a simple cost function based on coordination numbers and connectivity. Lattice energies, calculated from Born model potentials may, of course, be employed, but this procedure becomes computationally expensive. It should, however, be noted that simulated annealing procedures employing energies calculated by electronic structure techniques are becoming feasible, although in practice it would seem to be more computationally efficient to use simpler procedures to calculate the energy (or cost function), and to refine the approximate configuration generated by simulated annealing by electronic structure methods when the use of such methods is needed.

The 'temperature' used in a simulated annealing study is normally a parameter with no real physical significance. Higher temperatures will result in the acceptance of an increasing number of high energy configurations, allowing the exploration of a wider range of the energy surface but, of course, at increased computational cost. In practice, the simulation usually starts with a high temperature which is then reduced as the lower energy regions of the surface are identified, as further discussed in Chapter 5. Simulated annealing, like MC methods, may also be adapted to the analysis of experimental data. In this case the 'energy term' becomes the deviation between calculated data (e.g. the calculated X-ray intensity versus scattering angle for a structural model) and experiment. Such techniques are described as 'Reverse Monte Carlo' (RMC) and will be considered in detail in Chapter 6.

The account given above of the basic simulation methodologies will, of course, be extended in subsequent chapters. Further details and discussion may be found in Allen and Tildesley (1987), Catlow *et al.* (1989) and Catlow (1992). The remainder of this chapter aims to give an impression of the range and scope of simulation studies of inorganic materials by highlighting a number of recent applications.

6 CASE STUDIES

The applications we now discuss relate to the modelling and prediction of crystal structures, to the development of atomistic models for amorphous materials, to the modelling of surfaces of inorganic solids, to the simulation of the dynamical and defect properties of solids and to the explicit calculation of the electronic structure of crystals. They will foreshadow the much more detailed accounts that follow in later chapters.

6.1 Modelling and prediction of crystal structures

Since the 1970s, minimization techniques have been used to model crystal structures, and early work on both molecular and ionic crystals concentrated on reproducing observed crystal structures using suitable interatomic potentials (see Catlow *et al.*, 1993, for a review). Some of the greatest challenges here have been posed by microporous crystal structures, typical examples of which are shown in Fig. 1.1. These materials are attracting growing attention owing to their importance in catalysis, gas separation and ion exchange (see Thomas, 1992) as well as to the fundamental fascination of their crystal architectures. Moreover, their chemical range and diversity continues to expand: to the original aluminosilicate *zeolites* (constructed from SiO_4 and AlO_4 tetrahedra linked by corner sharing) were added microporous aluminophosphates (ALPOs) in the early 1980s (Wilson *et al.*, 1982); and substitution of framework tetrahedral (T) atoms by B, Ga, Ge and a wide range of transition metals has become widespread in both zeolites and ALPOs.

The work of Sanders *et al.* (1984), Jackson and Catlow (1988) and Tomlinson *et al.* (1990) showed that with Born model potentials it is possible accurately to reproduce the crystal structures of silicates including zeolites. A typical example is shown in Plate I which illustrates the calculated and experimental structures of a purely siliceous (i.e. pure SiO_2 polymorph) zeolite, silicalite. (A closely related, isostructural material, ZSM-5, which contains a low concentration of Al is an effective isomerization and hydrocarbon synthesis catalyst.) The agreement between theory and experiment is evidently good; more discussion follows in Chapter 9.

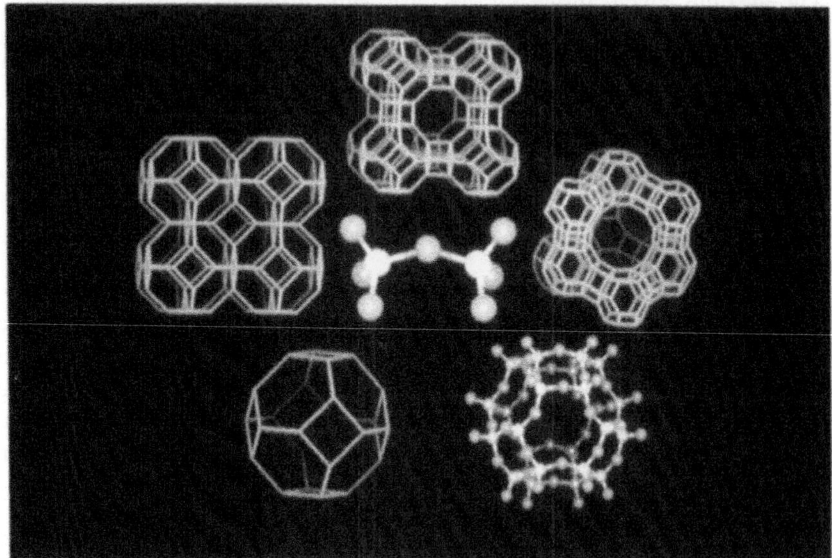

Figure 1.1 Microporous (zeolitic) crystal structures showing the sodalite cage (bottom), zeolite A structure (top), sodalite structure (left), and zeolite Y (right).

More recently, increased use has been made of minimization techniques in *refining* approximately known structures, which is of particular value in the study of microporous systems as these materials are commonly only available as powders and unassisted refinement by the widely used Rietveld techniques of the powder data for structures of this complexity may be difficult. Several examples illustrate the value of such calculations, as discussed in greater detail in Chapter 9. The first relates to a zeolite, Nu87, synthesized in the late 1980s at the ICI Laboratories in Wilton, UK. The material has useful catalytic properties. Its structure, however, could not be solved, even from high resolution X-ray powder diffraction data collected using Synchrotron Radiation. An approximate structural model was proposed, which was, however, unable to account for all the Bragg peaks in the diffraction pattern of the material, as shown in Fig. 1.2. This model was then used as the starting point of an energy minimization calculation; the resulting minimum, which was of lower symmetry than the initial structure, is shown in Fig. 1.3. This lower symmetry, minimized structure now fully accounts for all the reflections in the diffraction pattern; and indeed, when input into a Rietveld refinement of the powder diffraction data, rapidly yields a low R factor. With the assistance of the energy minimization calculation, the structure has been solved (Shannon *et al.*, 1991).

Several other examples in the recent crystallographic literature of microporous materials illustrate the power of this combination of powder diffraction

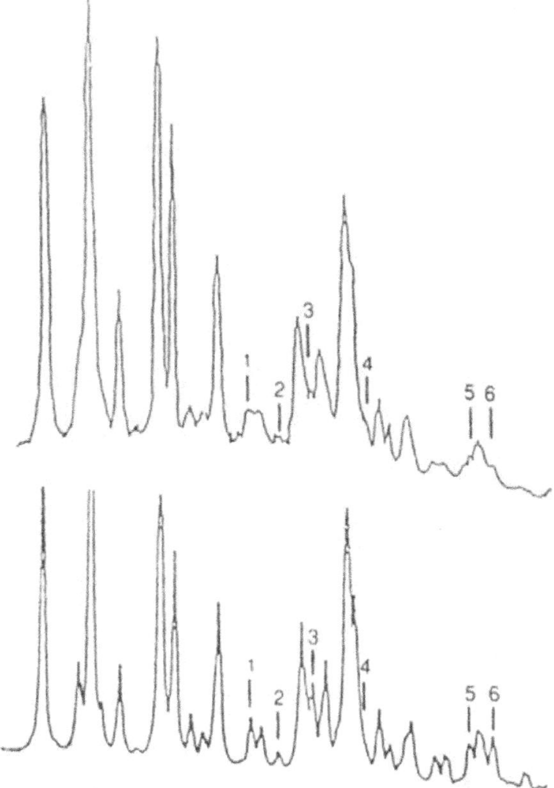

Figure 1.2 Part of the diffraction pattern for Nu 87. Upper diagram shows experimental data; peaks unassigned on the basis of the model prior to energy minimization are highlighted. Lower diagram is calculated pattern for energy minimized structure. Reproduced by kind permission of Dr P. Cox.

and energy minimization techniques. For example, the structures of two recently synthesized aluminophosphates—MAPO36 and DAF1—(Wright *et al.*, 1992, 1993) have been investigated by both powder diffraction and computational methods. The complex structure of the latter, with its two separate and non-communicating channel structures, is shown in Fig. 1.4. Futher examples of the modelling of these systems will be given in Chapter 9.

As discussed earlier, the greatest challenge in this field is now the *prediction* of crystal structures, an exercise in global optimization which as we have noted becomes, in practice, a question of identifying a suitable range of 'starting points' for subsequent energy minimization. Chapter 5 will describe in greater detail how the challenge of generating plausible initial structures from random starting points is met by the use of the simulated annealing and genetic algorithm procedures referred to earlier. The power of this approach can be illustrated by one example: the recent study of Bush *et al.* (1995) who solved the

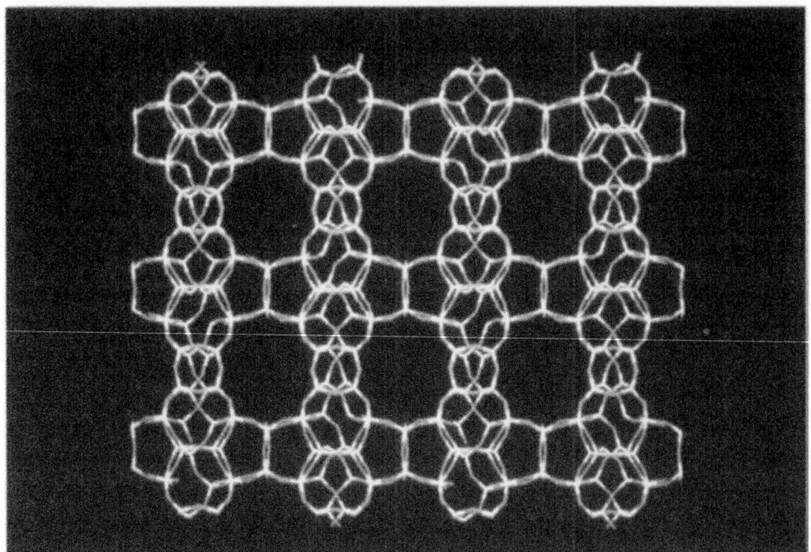

Figure 1.3 Energy minimized structure for Nu 87. Reproduced by kind permission of Dr P. Cox.

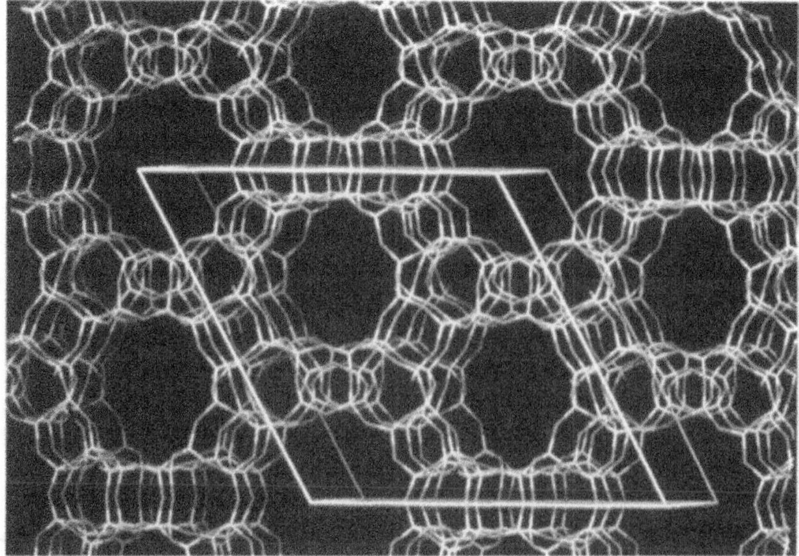

Figure 1.4 Crystal structure of DAF 1, solved from high resolution powder diffraction data and energy minimization.

structure of a complex ternary oxide, Li_3RuO_4, which had previously eluded structure determination. Their approach was based on a combination of genetic algorithm techniques (using a simple cost function of the type discussed by Brown in Chapter 2) and energy minimization techniques. The resulting structure (which leads to an acceptable Rietveld refinement of powder X-ray data) is illustrated in Plate II. As will be discussed in greater detail in Chapter 5, there is now a growing number of cases where unknown structures of this order of complexity have been solved by computer modelling techniques.

6.2 Structures of amorphous solids

There is an even greater incentive for the development of reliable computational procedures for modelling the atomistic structures of amorphous solids than for the case of crystalline systems. The development of such models from experimental data is difficult, as the interpretation of Radial Distribution Functions (RDFs) obtained from diffraction data is often ambiguous beyond the first few shells of neighbours. Local structural information produced by Nuclear Magnetic Resonance (NMR) and Extended X-ray Absorption Fine Structure (EXAFS) is of great value, but the information is again largely confined to the near neighbour environment of particular atom types. However, the most intriguing aspect of glass structure is possibly the intermediate range order (often characterized by the distribution of ring sizes) which has to a large extent been the subject of conjecture.

As discussed in Chapter 12, modelling of glass structures has ranged from the construction of simple physical models (the importance of which in all aspects of structural chemistry is not removed by the availability of computer graphics) to computational procedures based on MC and MD techniques. For glasses (which are defined as that subset of amorphous materials which show a distinct glass transition temperature, T_g, and which are produced by quenching from the melt), the MD approach appears to be particularly appropriate, as we can simulate the melting of the crystalline phases and its subsequent cooling below T_g, although, as will be discussed in detail in Chapter 12, the speed of the computational quench is necessarily orders of magnitude greater than that of any real physical quench. Nevertheless the atomistic models generated for glassy materials by this procedure are commonly consistent with experimental data and provide valuable physical insights into the structures of glassy materials.

A good example of the rôle of such calculations is provided by the recent work of Vessal *et al.* (amplified in Chapter 12) who have developed by MD quenching procedures, models for both silica and silicate glasses. Plate III shows their model of vitreous $Na_2Si_2O_5$, an intriguing feature of which is the clear indication of a microsegregation of the framework modifying Na^+ cations into loosely defined channels in the silicate matrix. Such behaviour is in accordance with the 'Modified Random Network' (MRN) model of silicate systems advanced by Greaves (1985). A detailed analysis by Vessal *et al.* (1992) showed clearly how such computer generated models for alkali silicates were consistent with experimental EXAFS and other structural data.

We also note that the RMC method discussed in Chapter 6 by McGreevy has made a substantial contribution to the development of models for classes of materials.

6.3 Simulation of surface structures

Again, the difficulty of developing unambiguous models of surface structures from experimental data provides a powerful impetus for the application of computational techniques. For inorganic materials, the most appropriate tool is straightforward energy minimizations applied to an infinite 2-D periodic model of the surface structure, embedded in a rigid representation of the remainder of the lattice—procedures that are discussed in detail by Harding in Chapter 7. As shown in that chapter, relaxation in the surface region can be a major effect, with simulations predicting large changes in the *d*-spacings of the surface layers. Such relaxations will have substantial effects on the physical and chemical properties of the surfaces.

A good illustration of the complexity of the surface structure that can now be simulated by these techniques is provided by the recent work of Gay and Rohl (1995). Their recent studies have generated models for the surfaces of

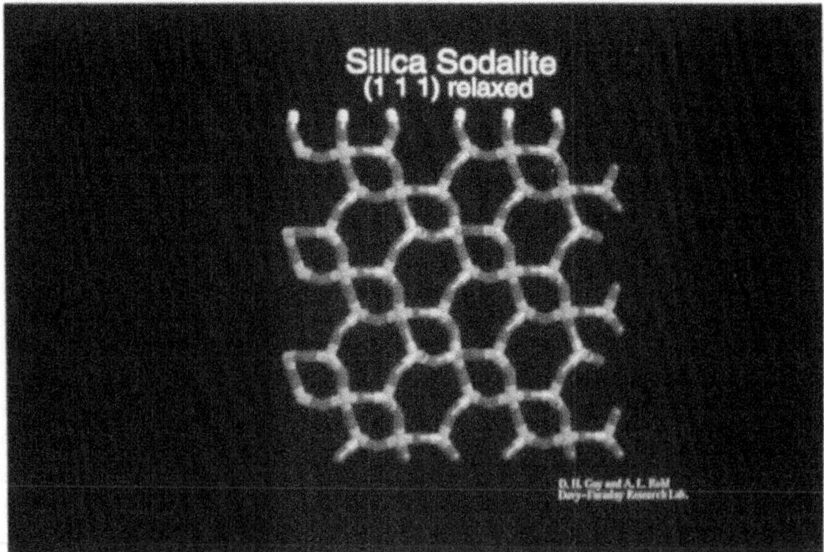

Figure 1.5 Relaxed surface structure of silicalite. Surface OH groups indicated. (Taken from Gay and Rohl, to be published.)

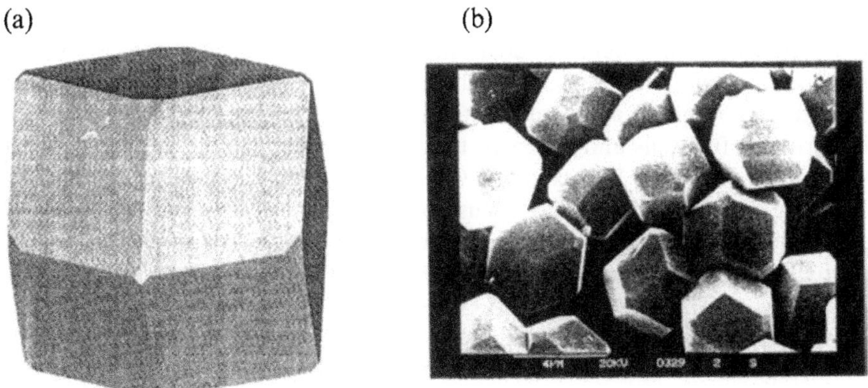

Figure 1.6 Calculated (a) and experimental (b) morphology of sodalite.

microporous materials. Thus in Fig. 1.5 we illustrate their simulation of the relaxed surface structure of siliceous sodalite. They find that in common with several other silica and silicate systems, the surfaces are only stable when hydrogenated, i.e. when surface oxygen atoms are replaced by surface hydroxyl species—a finding that is consistent with known properties of silica surfaces. Appreciable relaxation of the surface layer is also apparent in the energy minimized structure shown in Fig. 1.5.

From calculated surface energies, it is also possible to produce crystal morphologies using a procedure due originally to Gibbs which essentially assumes equilibrium control of the morphology (which may not be a valid assumption in many cases). Using this procedure, Loades *et al.* (1994) were able to generate the calculated morphology shown in Fig. 1.6(a); it shows a close correspondence to the experimentally observed morphologies illustrated in Fig. 1.6(b). Such agreement gives us further confidence in the validity of the interatomic potentials and simulation procedures employed.

6.4 Dynamical processes in solids

As will be apparent in Chapter 4, MD simulations have made major contributions to our understanding of the structures and dynamical properties of solids. Problems and processes simulated include phase transitions (e.g. Impey *et al.*, 1985; Meyer and Ciccotti, 1985), orientation dynamics in molecular crystals (e.g. Dove and Pawley, 1984) and ionic and diffusion (e.g. Gillan and Dixon, 1980; Vashishta and Rahman, 1978). In addition, they have been used to study melting and, as noted above, to prepare glass structures by a simulated melt–quench cycle as described by Vessal in Chapter 12.

The focus of this book is the use of simulations in predicting and interpreting structural properties. In this context, the most relevant applications of MD techniques are in developing models for the structures of dynamically disordered solids. Some of the best examples are to be found in the study of fast ion conductors, considered in detail in Chapter 4 which, as noted, are solids showing ionic mobility typically of those found in a melt. Such materials have an extensive range of actual and potential applications in high energy density batteries, in fuel cells and in sensor technology. Many simple inorganic solids show this behaviour. For example, CaF_2, $SrCl_2$, PbF_2 and indeed all fluorite structured compounds show a diffuse phase transition within a few hundred degrees of the melting point above which there is high anion mobility and a partially disordered anion sublattice within an ordered cation matrix (e.g. Hutchings *et al.*, 1984; Dixon and Gillan, 1980). In contrast, extensive cation disordering occurs in silver iodide which shows a first order phase transition occurring at 147°C, above which silver ions are highly mobile; the anion sublattice remains ordered.

These materials pose interesting structural problems, that of disorder on *one* sublattice; the extent and nature of this disorder was widely debated and investigated in the 1970s and 1980s. MD simulations made a unique and important contribution to the understanding of these problems in both the classes of materials referred to above. Thus Gillan and coworkers (Gillan and Dixon, 1980; Dixon and Gillan, 1980; Gillan, 1995) in a series of studies of fluorite structured materials, clearly demonstrated the partial nature of the disorder on the anion sublattice. The structural models developed by their MD simulations of CaF_2 are illustrated in Fig. 1.7. An extensive range of neutron scattering data—both Bragg, diffuse and quasi-elastic—were collected on these systems, most notably by Hutchings *et al.* (1984). The results are fully compatible with models for the anion disorder found by the simulations. For example, a comparison of calculated (from the averaged MD structure) and experimental $S(Q)$ obtained for neutron scattering studies of high temperature $SrCl_2$ gave excellent agreement with experiment.

The power of MD techniques in the study of dynamically disordered systems is also apparent in a more recent study of Cox *et al.* (1994) on the fluorite structured mixed cation system $RbBiF_4$. This intriguing material shows a disordered distribution of the two types of cation over the regular fluorite lattice sites. The work of Réau and coworkers (Matar *et al.*, 1980) shows that the material has a high ionic conductivity at modest temperatures (~500°C) owing to high anionic mobility. Neutron diffraction studies indicated extensive anion disorder. The material is an ideal candidate for EXAFS spectroscopy which yields local structural information on individual atom types. EXAFS data collected on both the Rb and Bi edges revealed that the disorder was far more extensive around the Rb ion. The MD simulations of Cox *et al.* (1994) found large levels of anionic disorder consistent with experiment. They also showed different extents of disorder around the two types of cation in good agreement with the EXAFS data. Perhaps of greatest interest was the information gained from the simulations on the mechanisms of ion migration. They show that the migrating fluoride ions move by an 'interstitialcy' mechanism in

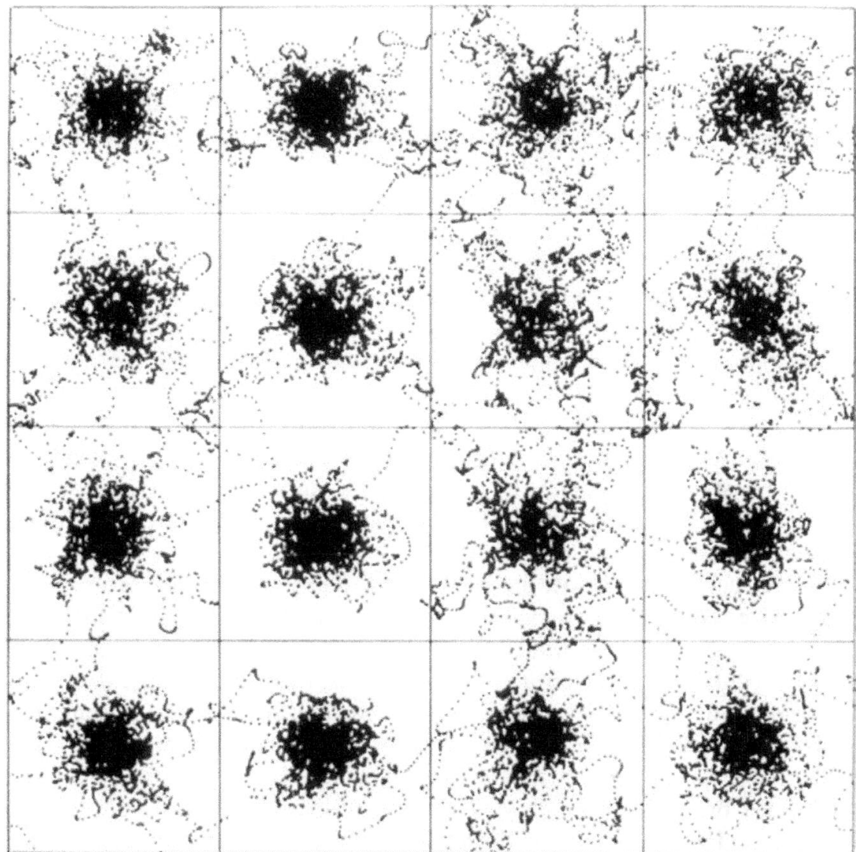

Figure 1.7 Anion trajectories for simulated CaF_2 at 1657 K over an exposure time of 3 ps. Cations are not shown, but square grid indicates cation sublattice. (Reproduced by kind permission of Professor M.J. Gillan.)

which the migrating interstitial ion displaces lattice ions into an interstitial site and itself occupies the vacated lattice sites. Moreover, such motions are highly correlated with several such motions occurring simultaneously as shown in Plate IV, which illustrates the trajectories of several ions as revealed by the simulation. MD simulations have revealed this kind of correlated behaviour in other fast ion conducting systems, e.g. Li_3N (Wolf and Catlow, 1984).

Similar valuable information on the structures and dynamics of the mobile sublattice in fast ion conducting silver components have been observed by Vashishta and coworkers in a series of MD studies which have included simulations of AgI and Ag_2Se (Vashishta and Rahman, 1978; Vashishta, 1986; Ray *et al.*, 1989). Again the simulations have proved of great value in analysing and interpreting experimental scattering data.

6.5 Defect structures

The modelling of the structures and energies of defects in ionic materials was one of the earlier successes of the solid state simulation field. As will be amplified in Chapter 7, straightforward methods based on energy minimization using interatomic potentials have proved enormously fruitful in deriving structures and energies of defects and defect aggregates that are consistent with experiment. The experimental determination of the detailed structures of defects and impurity states is difficult. Diffraction is inherently an averaging technique and the structural properties of defects can only be inferred from their effects on the average unit cell contents. Such information is necessarily indirect and defects will, moreover, only have a significant effect on Bragg intensities when they are present in appreciable concentrations. Moreover, the interpretation of diffuse scattering which contains information on short range order is often difficult. It is interesting to note that the interpretation of some of the most successful diffraction studies of defective compounds (such as non-stoichiometric $Fe_{1-x}O$ (Cheetham *et al.*, 1971b; Catlow and Fender, 1975) has relied heavily on the results of computer simulations.

Local structural techniques are clearly needed in structural characterizations of defective compounds. High Resolution Energy Microscopy (HREM), as it approaches atomic resolution, allows defects and defect aggregates to be investigated directly in favourable circumstances; but even with the best available resolution, structural details cannot be obtained. Spin resonance techniques (both NMR and ESR) yield valuable information on site symmetries and can yield more detailed structural information. Possibly the most powerful technique is provided by EXAFS, which yields details of local structures (coordination numbers of radii of surrounding shells of atoms) around specific impurities, since we may tune the X-ray wavelength to the absorption edge appropriate to that atom type. The method, of course, yields little information on intrinsic defects, e.g. vacancies, unless they are present in high concentrations.

Even with the best local and crystallographic data on defective systems, assistance in the interpretation of computer modelling studies is often vital. A good case study is provided by the widely studied yttria stabilized zirconia (Y/ZrO_2) system; the high oxygen ion mobility of the material (which can accommodate high concentrations of low valence dopants) has led to its widespread use as a solid electrolyte in fuel cells and sensors. The basic defect structure is simple: Y replaces Zr in the host which adopts the fluorite structure when doped with low valence ions, with the lower valence of the dopant ion resulting in the creation of charge compensating oxygen vacancies, which have low migration energies and hence high mobilities. The crucial question in this and many defective compounds concerns the nature of the interaction between the dopants and the defects. These species have opposite 'effective charges' (i.e. charges relative to those at the relevant sites in the perfect lattice). Coulomb interactions will therefore tend to drive aggregation. The simplest type of

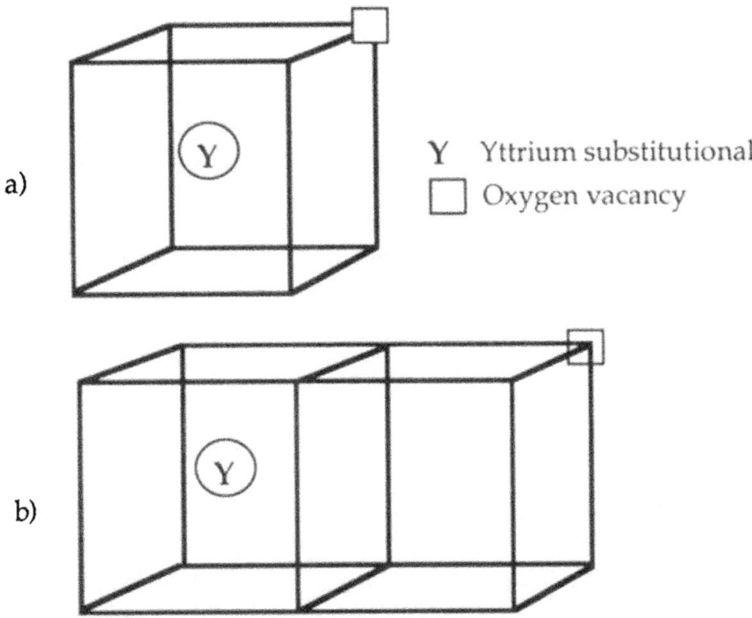

Figure 1.8 Nearest neighbour (a) and next nearest neighbour (b) dopant–vacancy configurations in fluorite structured Y/ZrO_2. Note oxygen ions at one corner of cubes.

aggregation in the present system would involve the yttrium dopant capturing an oxygen vacancy to form the complex shown in Fig. 1.8(a). The formation of such a complex effectively reduces the coordination number of the yttrium from 8 to 7. Interestingly, no support for the formation of such complexes is found in EXAFS (Catlow *et al.*, 1986a) studies of the material. The EXAFS indicate *higher* coordination numbers for yttrium than for zirconium, but more significantly they find on analysis of Debye–Waller factors that there is a more disordered environment around Zr than Y. These findings were reinforced by an anomalous dispersion study of the material (Moroney *et al.*, 1988). By collecting data close to the absorption edges of both the Y and the Zr, partial radial distribution functions can be obtained. (We note that the availability of the tuneable X-ray sources in synchrotron radiation is vital for the viability of such experiments.) Analysis of the data reveals higher Debye–Waller factors (which in this context almost certainly indicates more static disorder) in pair correlation functions involving Zr than those involving Y.

How are we to interpret these results? They certainly do not support the kind of model indicated by Fig. 1.8(a). It would be surprising, however, if there were no interaction between dopants and defects in this material. A simple resolution of this problem is provided by the simulations (Catlow *et al.*, 1986b) which show that Y dopants and vacancies do indeed interact, but that the vacancy occupies the next nearest rather than the nearest neighbour site with respect to the impurity, as shown in Fig. 1.8(b)—behaviour that is found in many other doped ionic materials. If such a complex predominates, then the Y will have a higher, more regular coordination number than the Zr, which nicely explains the experimental findings.

Several other examples of the rôle of simulations in explaining defect structures are to be found in the solid state chemistry of ionic materials. For example, the solid solutions formed by dissolving rare earth fluorides in alkaline earth fluoride hosts give rise to a complex and intriguing range of defect clusters, the nature of which was elucidated in the 1970s and 1980s by a combination of diffraction, EXAFS and computer modelling studies (Cheetham *et al.*, 1971a; Catlow, 1973; Catlow *et al.*, 1984).

6.6 Electronic structure studies

The large and growing topic of *ab initio*, electronic structure studies of solids will be reviewed in Chapter 8. Here we wish simply to emphasize the ability of such methods to give accurate information on the electron density distribution of increasingly complex materials. Recent illustrations are provided by the work of Nada *et al.* (1990, 1992) who have studied the electronic structure of silica and silicates using *ab initio* Hartree–Fock techniques (employing the CRYSTAL (Pisani *et al.*, 1988) code). Their work on SiO_2 and $MgSiO_3$ yielded minimum energy structures close to those observed experimentally (i.e. with calculated structural parameters within $\sim 1\%$ of those observed experimentally). Moreover, the calculations reveal a wealth of detail on the distribution of electron chemistry in the material. Thus Plate V displays a calculated *difference* electron density map through a section of the quartz structure; it therefore shows the redistribution of electron density due to bonding. Transfer of density from Si to O is clearly apparent, in line with the well-known partial ionicity of the material; but a concentration of density between the Si and O, i.e. partial covalent bonding, is also evident. The difference density map for the high pressure ilmenite structure of $MgSiO_3$ also revealed interesting features: the distribution of density around the magnesium ion is indicative of ionic bonding for this species and indeed there appears to be a higher ionicity in the silicate sublattice. High quality calculations of this type clearly illuminate our understanding of bonding in solids. They also provide information on electron density distributions, which is of direct value to the crystallographer.

7 SUMMARY AND CONCLUSIONS

The brief survey presented in this chapter has aimed to demonstrate the central rôle that computational techniques now play in structural studies of inorganic materials. In many cases the techniques we have discussed are standard tools, for which efficient and general purpose computer codes are available (and which are increasingly being interfaced to high quality graphics, the importance of which in advancing our understanding and intuition concerning complex structural problems should be emphasized). In other areas, techniques are actively being developed and refined. Moreover, the horizon of the field continues to expand with the apparently unending exponential growth in computer power. We hope the stage is now set for the detailed discussions of techniques and applications to be presented in subsequent chapters of this book.

REFERENCES

Allen, M.P. and Tildesley, D.J. (1987) *Computer Simulation of Liquids*, Oxford University Press.

Brunger, A.T. and Rice, L.M. (1995) In *Adaptation of Simulated Annealing to Chemical Optimisation Problems* (ed. J.H. Kalivas) p. 259, Elsevier.

Bush, T.S., Catlow, C.R.A. and Battle P.D. (1995) *J. Mater. Chem.*, **5**, 1269.

Catlow, C.R.A. (1973) *J. Phys. C: Solid State Phys.*, **6**, L64.

Catlow, C.R.A. (1989) In *Superionic Solids and Solid Electrolytes* (eds A.L. Laskar and S. Chandra), Academic Press, San Diego, p. 339.

Catlow, C.R.A. (ed.) (1992) *Modelling of Structure and Reactivity in Zeolites*, Academic Press, London.

Catlow, C.R.A., and Fender, B.E.F. (1975) *J. Phys. C*, **8**, 3267.

Catlow, C.R.A., Chadwick, A.V., Greaves, G.N. and Moroney, L.M. (1984) *Nature (Lond.)*, **312**, 601.

Catlow, C.R.A., Chadwick, A.V., Greaves, G.N. and Moroney, L.M. (1986a) *J. Am. Ceram. Soc.*, **69**, 272.

Catlow, C.R.A., Chadwick, A.V., Cormack, A.N., Greaves, G.N., Leslie, M. and Moroney, L.M. (1986b) *Materials Research Society Symposium Proceedings*, Vol. 60: *Defect Prop. Process. High-Technology Nonmetallic Materials*, p. 173.

Catlow, C.R.A., Parker, S.C. and Allen, N.M.P. (eds) (1989) *Computer Modelling of Fluids, Polymers and Solids*, Vol. 293, NATO Series, Kluwer Academic Publishers, Dordrecht.

Catlow, C.R.A., Thomas, J.M., Freeman, C.M., Wright, P.A. and Bell, R.G. (1993) *Proc. R. Soc. Lond. A*, **442**, 85.

Cheetham, A.K., Fender, B.E.F. and Cooper, M.J. (1971a) *J. Phys. C: Condens. Matter*, **4**, 3107.

Cheetham, A.K., Fender, B.E.F. and Taylor, R.I. (1971b) *J. Phys. C*, **8**, L34.

Cox, P.A., Catlow, C.R.A. and Chadwick, A.V. (1994) *J. Mater. Sci.*, **29**, 2725.

Dixon, M. and Gillan, M.J. (1980) *J. Phys. C*, **13**, 1919.

Dove, M.T. and Pawley, G.S. (1984) *J. Phys. C*, **17**, 6581.

Funke, K. (1976) *Prog. Solid State Chem.*, **11**, 345.

Gale, J.D., Catlow, C.R.A. and Mackrodt, W.C. (1992) *Modelling Simul. Mater. Sci. Engng*, **1**, 73.

Gay, D.H. and Rohl, A.L. (1995) *J. Chem. Soc., Faraday Trans.*, **91**, 925.

Gillan, M.J. (1985) *Physica B*, **131**, 157.

Gillan, M.J. and Dixon, M. (1980) *J. Phys. C*, **13**, 1901.

Greaves, G.N. (1985) *J. Non-Cryst. Solids*, **71**, 203.

Hutchings, M.T., Clausen, K., Dickens, M.H., Hayes, W., Kjems, K.J., Schnabel, P.G. and Smith, C. (1984) *J. Phys. C*, **17**, 3903.
Impey, R.W., Sprik, M. and Klein, M.L. (1985) *J. Chem. Phys.*, **83**, 3638.
Jackson, R.A. and Catlow, C.R.A. (1988) *Mol. Simul.*, **1**, 207.
Keen, D.A. and McGreevy, R.L. (1988) *Nature, Lond.*, **344**, 423.
Kostrowicki, J. and Sheraga, H.A. (1992) *J. Phys. Chem.*, **96**, 7442.
Loades, S.D., Carr, S.W., Gay, D.H. and Rohl, A.L. (1994) *J. Chem. Soc. Chem. Comm.*, **11**, 1369.
Matar, S., Réau, J.M., Demazeau, G., Leuat, J. and Hagenmuller, P. (1980) *Mater. Res. Bull.*, **15**, 1295.
Meyer, M. and Ciccotti, G. (1985) *Mol. Phys.*, **56**, 1235.
Moroney, L.M., Thompson, P. and Cox, D.E. (1988) *J. Appl. Crystallogr.*, **21**, 206.
Murch, G.E. (1982) *Phil. Mag.*, **A46**, 575.
Nada, R., Catlow, C.R.A., Dovesi, R. and Pisani, C. (1990) *Phys. Chem. Minerals*, **17**, 353.
Nada, R., Catlow, C.R.A., Dovesi, R. and Saunders, V.R. (1992) *Proc. R. Soc. Lond. A*, **436**, 499.
Pisani, C., Dovesi, R. and Roettie, C. (1988) *Lecture Notes in Chemistry*, Vol. 48, Springer, Heidelberg.
Ray, J.R., Rahman, A. and Vashishta, P. (1989) In *Superionic Solids and Solid Electrolytes* (eds A.L. Laskar and S. Chandra), Academic Press, San Diego.
Sanders, M.J., Leslie, M. and Catlow, C.R.A. (1984) *J. Chem. Soc. Chem. Comm.*, 1271.
Shannon, M.D., Casci, J.L., Cox, P.A. and Andrews, A. (1991) *Nature*, **253**, 417.
Thomas, J.M. (1992) *Scientific American*, **4**, 112.
Tomlinson, S.M., Jackson, R.A. and Catlow, C.R.A. (1990) *J. Chem. Soc., Chem. Comm.*, 813.
van Gunsteren, W.F. and Berendsen, H.J.C. (1990) *Angewandte Chemie Int.*, **29**, 992.
Vashishta, P. (1986) *Solid State Ionics*, **18/19**, 3.
Vashishta, P. and Rahman, A. (1978) *Phys. Rev. Lett.*, **40**, 1337.
Vessal, B., Greaves, G.N., Marten, P.T. Chadwick, A.V., Mole, R. and Houde-Watet, S. (1992) *Nature*, **356**, 504.
Vitek, V. and Srolovitz, D.J. (1989) *Atomic Simulation of Materials*, Plenum, Press, New York.
Wilson, S.T., Lok, B.M., Messina, C.A., Cannon, T.R. and Flanigen, E.M. (1982) *J. Am. Chem. Soc.*, **104**, 1146.
Wolf, M.L. and Catlow, C.R.A. (1984) *J. Phys. C*, **17**, 6635.
Wright, P.A., Natarajan, S., Thomas, J.M., Bell, R.G., Gai-Boyes, P.L., Jones, R.H. and Chen, J. (1992) *Angew. Chem. Int. Ed. Engl.*, **31**, 1472.
Wright, P.A., Jones, R.H., Natarajan, S., Chen, J., Bell, R.G., Hursthouse, M.B. and Thomas, J.M. (1993) *J. Chem. Soc. Chem. Comm.*, 633.
Yashonath, S., Thomas, J.M., Nowak, A.K. and Cheetham, A.K. (1988) *Nature*, **331**, 601.

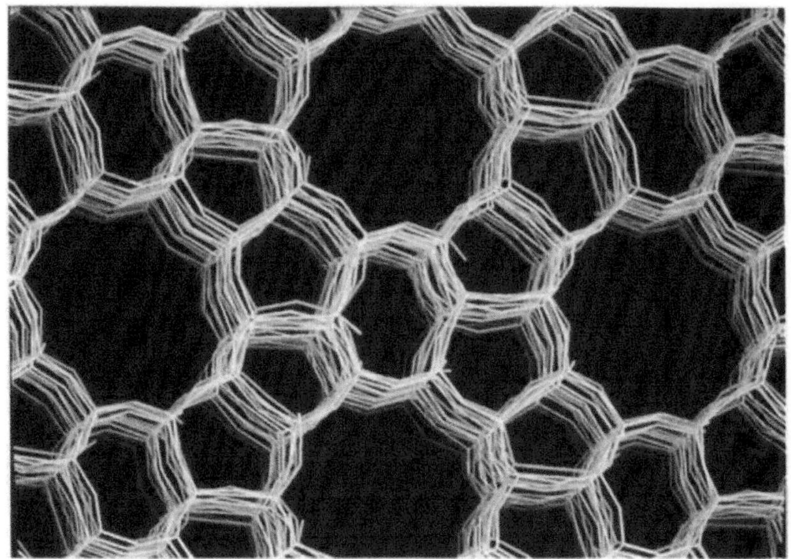

PLATE I Calculated (yellow) and experimental (blue) structures for silicate (from Bell *et al.*, 1990).

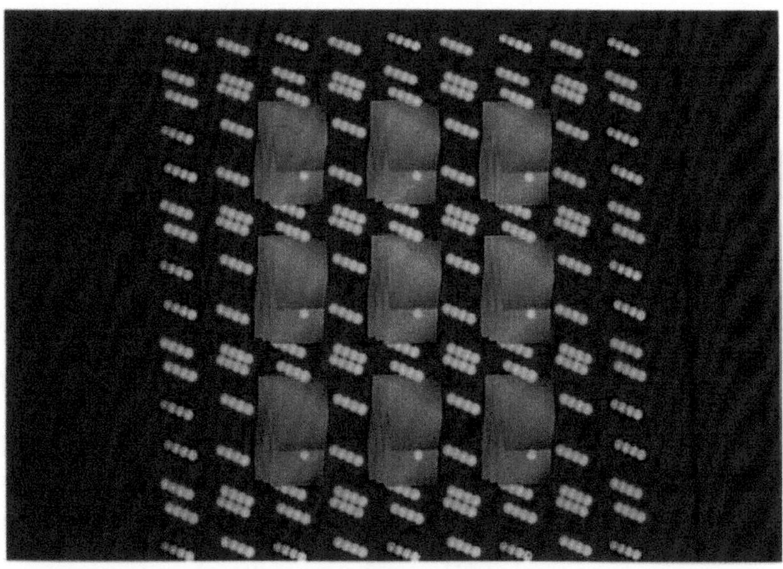

PLATE II Stucture of Li_3RuO_4, solved by genetic algorithms/energy minimzation techniques. Purple polyhedra indicate chains of edge-sharing RuO_6 octahedra.

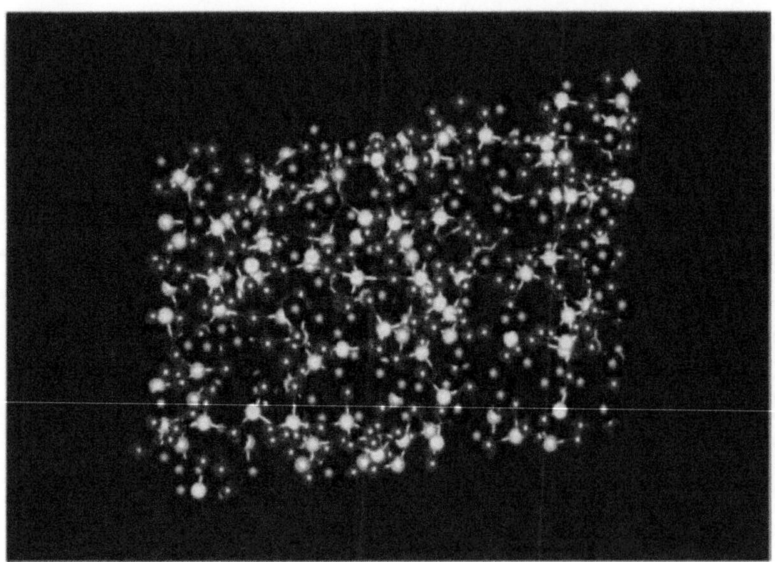

PLATE III Structure of vitreous $Na_2Si_2O_5$ obtained by MD simulations. Blue spheres indicate Na^+ ions; red indicate oxygen; yellow indicate silicon.

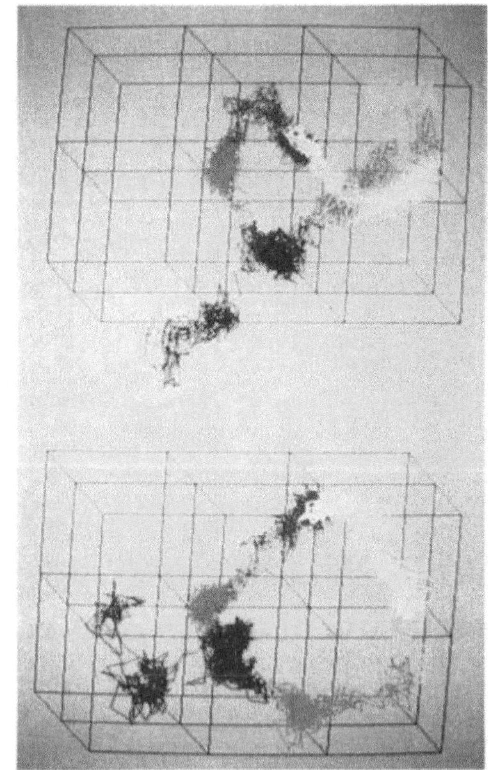

PLATE IV Cooperative interstitialcy motion of F^- ions in $RbBiF_4$. Each colour indicates a different ion trajectory.

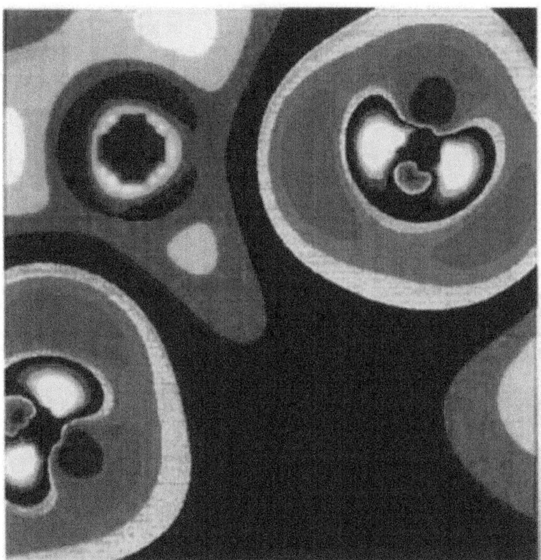

PLATE V Calculated *difference* electron density maps in SiO_2. Differences are relative to atomic wavefunctions. Red colours indicate depletion of electron density; blue colours indicate enhancement.

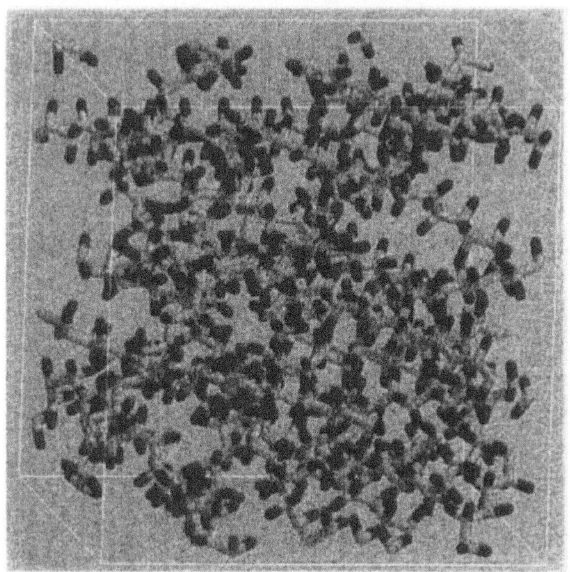

PLATE VI RMC model of molten PTFE. White sections show the main chain and green sections the side groups.

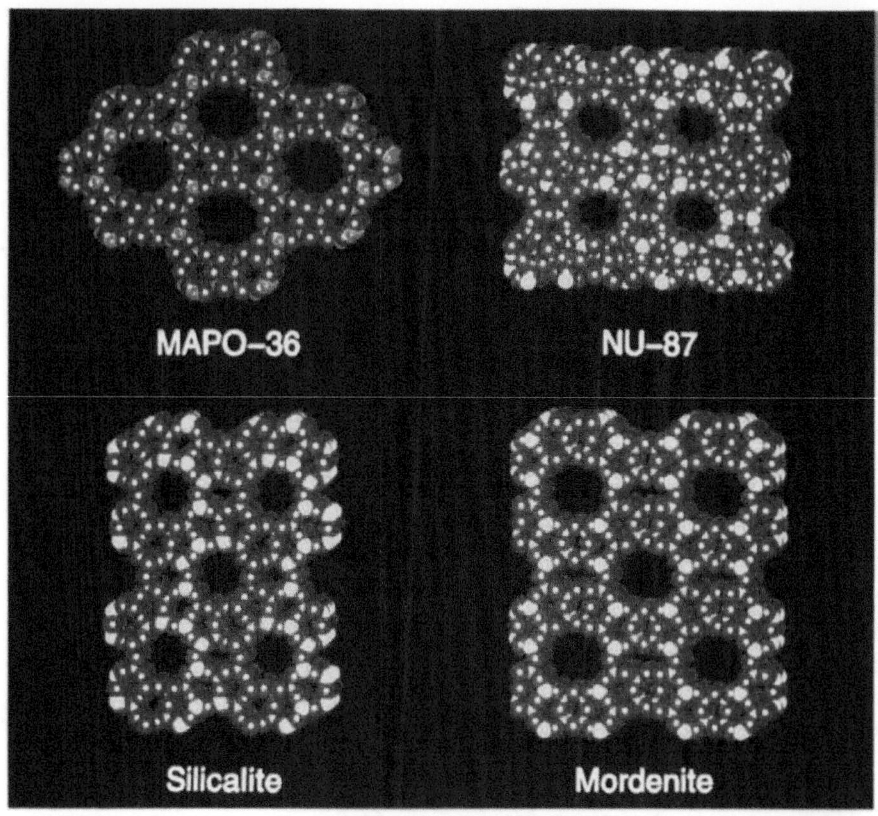

PLATE VII Molecular graphic representation of the framework structures of MAPO-36, NU-87, silicate and mordenite. Colour code: yellow, Si; Magenta, Al; Green, P; Red, O.

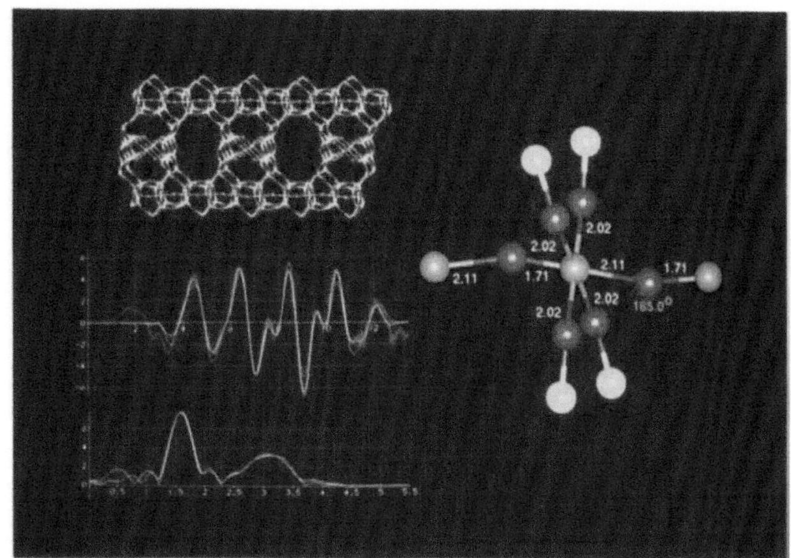

PLATE VIII Top left – framework structure of ETS-10; right – local structure around framework Ti in ETS-10, as determined from EXAFS and computer modelling; bottom left – comparison between experimental and simulated EXAFS data for ETS-10.

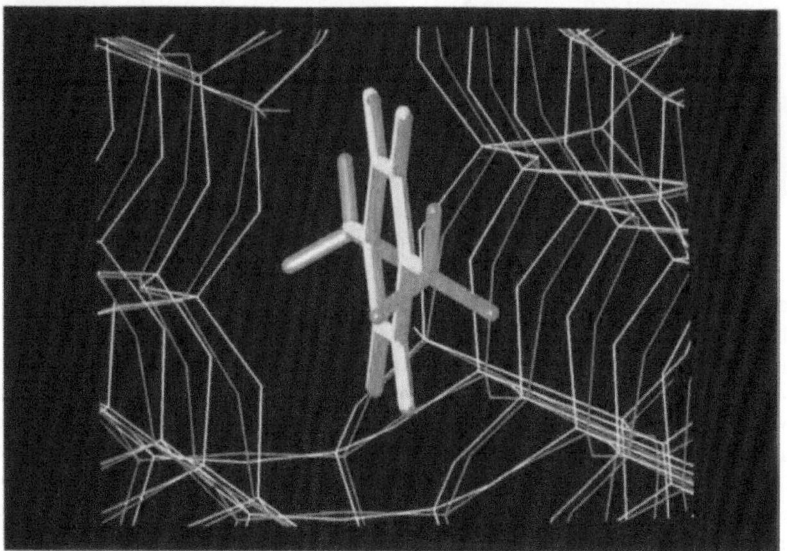

PLATE IX Comparison between experimental (yellow) and calculated (blue) sites for *p*-xylene in ferrierite. Methyl hydrogens are omitted from the experimental structure, since their positions are not uniquely defined.

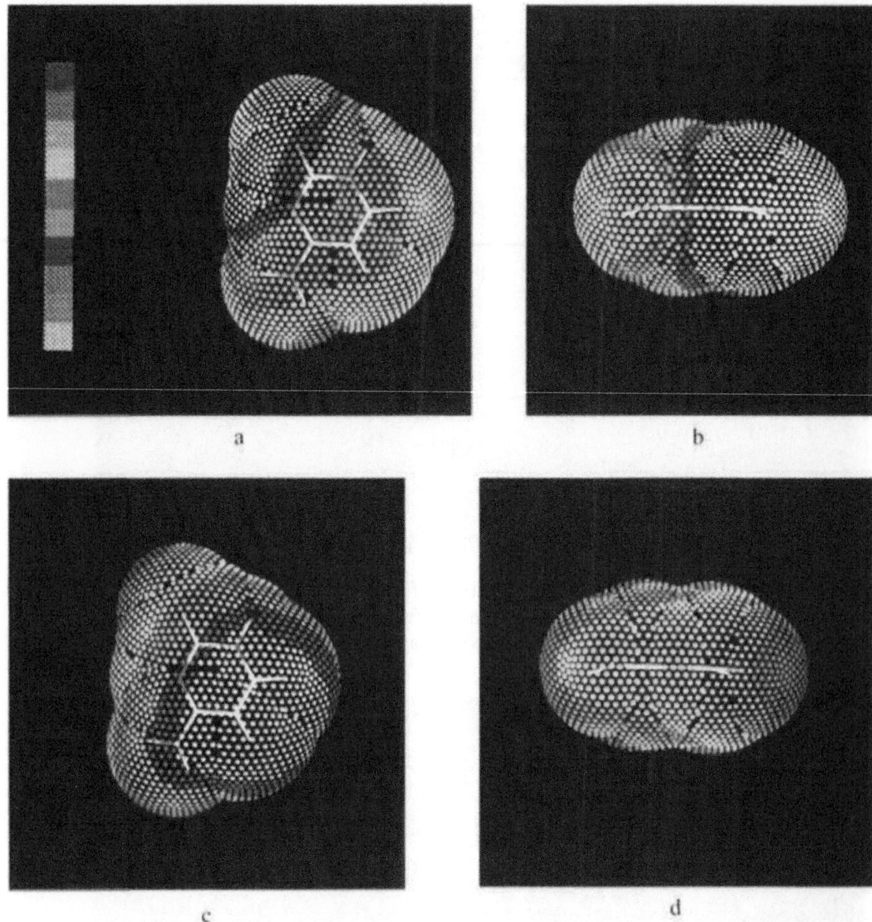

PLATE X The electrostatic potential, V, calculated 1Å from the van der Waals surface of cytosine, and the contribution of the anisotropic atomic multipoles, shown from a viewpoint above the aromatic ring and from the NH_2–C–N–C=O side. (a, b) The potential, V (kJ/mol), as calculated from the DMA of the 6-31G**SCF wavefunction, colour coded: white < -160 < grey < -120 < magenta < -80 < blue < -40 < cyan < 0 < green < 40 < yellow < 80 < orange < 120 < brown < 160 < red. The potential varies from -190 to 166 kJ/mol. (c,d) Errors in the potential, $\Delta V = V - V(Q)$ (kJ/mol), where $V(Q)$ is calculated using just the charge component of the DMA, colour coded: white < -60 < grey < -45 < magenta < -30 < blue < -15 < cyan < 0 < green < 15 < yellow < 30 < orange < 45 < brown < 60 < red. The potential due to the charge varies from -246 to 163 kJ/mol, with the error due to the neglect of the anistropic atomic multipoles varying from -59 to 71 kJ/mol.

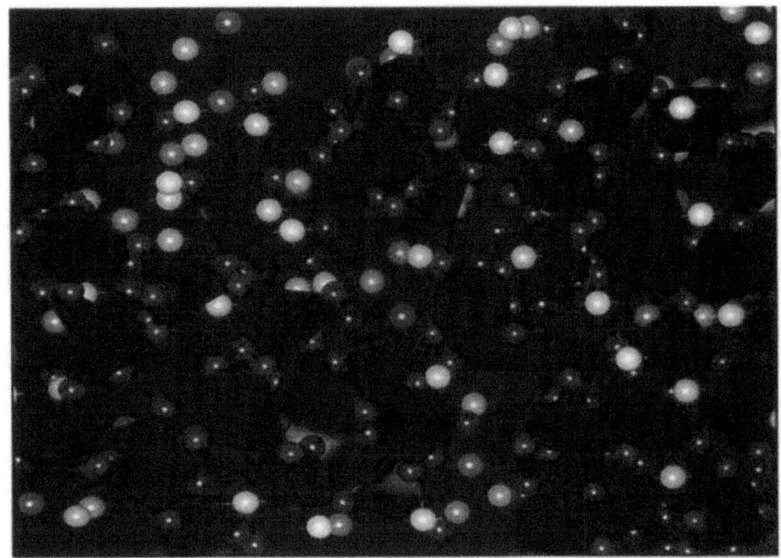

PLATE XI Structure of vitreous $Na_2Si_2O_5$ generated using BIOSYM software: SiO_4 tetrahedra are shown in purple, oxygens are shown as red spheres, and Na^+ ions as white spheres.

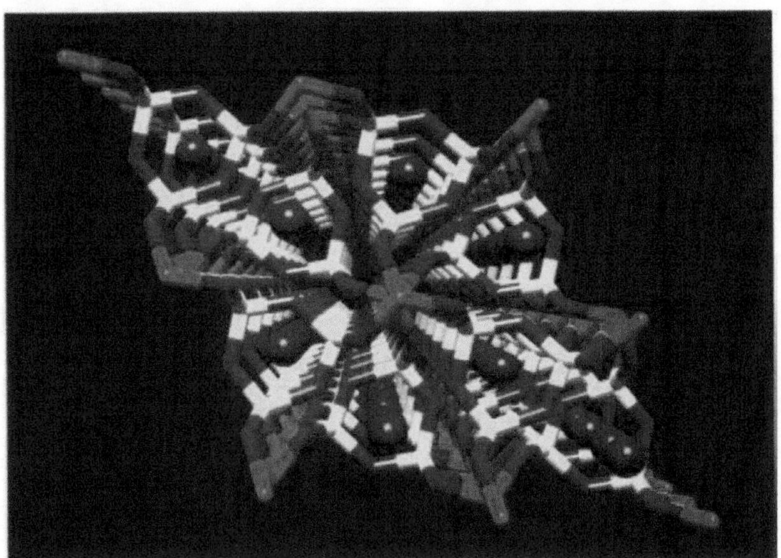

PLATE XII Molecular graphics representation of structure of Ti-wadeite crystal. Si is yellow, O is red, Ti is blue, Potassium ions are shown as purple spheres.

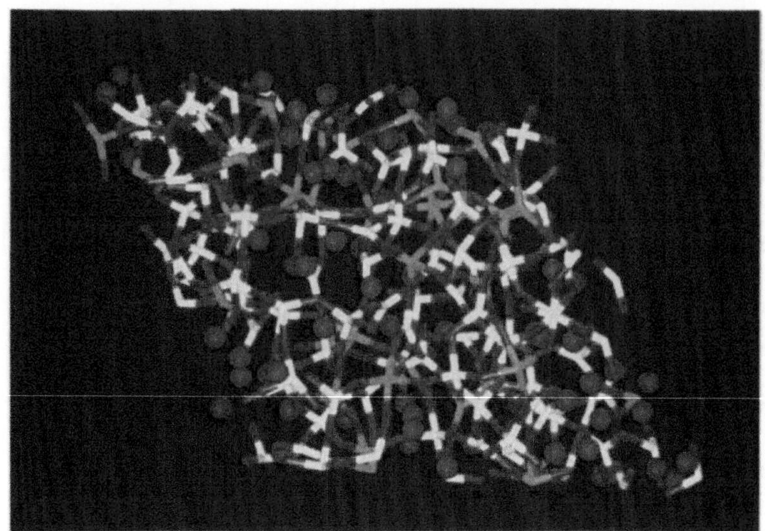

PLATE XIII Molecular graphics representation of structure of simulated Ti-wadeite glass. The colours for atoms are the same as those in Plate XII.

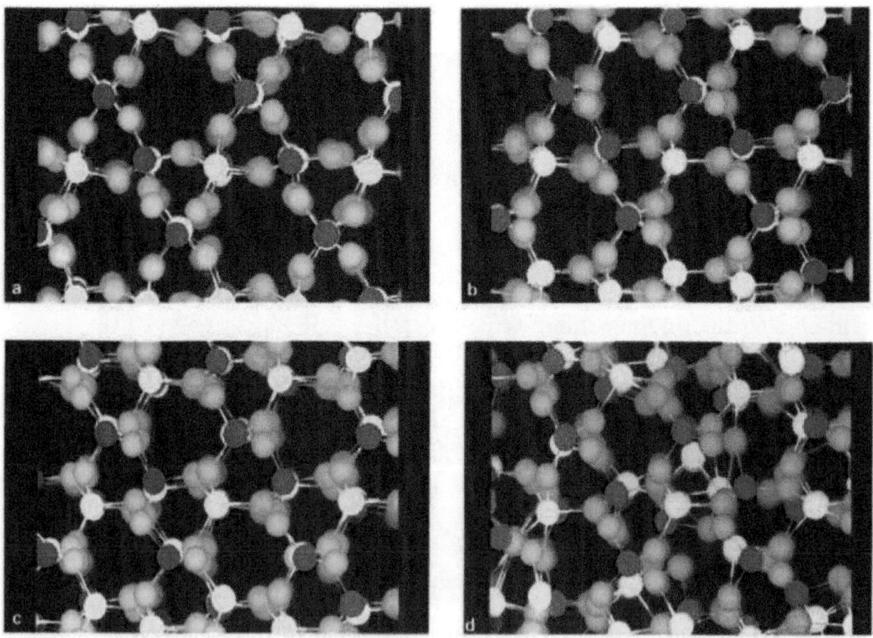

PLATE XIV Structure of simulated $AlPO_4$ at a pressure of (a) 0GPa, (b) 13GPa, (c) 29GPa, and (d) 33GPa. Yellow spheres are aluminium atoms, purple spheres are phosphorus atoms and blue spheres are oxygen atoms.

SECTION ONE: METHODOLOGIES

2

Bond Valence Methods

I.D. Brown

1 INTRODUCTION

Since the middle of the nineteenth century, chemists have been using the chemical bond model to predict the structure and properties of organic molecules and, surprisingly, the events that followed the discovery of the electron in 1898 made remarkably little difference to the way this model was formulated and applied. Quantum mechanics has provided insight into the nature of chemical bonding, but the bond model itself was essentially complete by the end of the nineteenth century. The only significant addition to the model in this century has been the determination of bond lengths and angles. In spite of the great advances in our understanding of the physics of the atom, the chemical bond model remains essentially empirical. The idea of a bond does not arise naturally from quantum mechanics and the only justification for the model lies in its success in predicting the existence, structure and properties of organic molecules.

Until recently it was generally thought that this model could not be used to describe the structure of inorganic solids because in inorganic materials the bonds clearly do not have integral bond orders. Consequently, in the early years of the twentieth century two other models were proposed. The ionic model was developed to describe inorganic salts in which the bonding electrons are incorporated into the valence shell of the anion, and the covalent model was developed for those compounds where the bonding electrons were more equally shared between the bonded atoms. While the early versions of these models have proved to be pedagogically useful, they were not able to give a quantitative prediction of structure except for the simplest compounds. Recently the ionic model has been transformed into the very successful interatomic potential model which is discussed in detail elsewhere in this book (see Chapters 3 and 7), but it has also become apparent that the chemical bond model itself, with only minor modifications, can also be adapted to inorganic compounds. It has shown itself as powerful in describing the structure and properties of inorganic solids as it has in describing those of organic molecules. This chapter will describe the modifications needed to adapt the model to inorganic solids and it will show how the model can be applied.

COMPUTER MODELLING IN INORGANIC CRYSTALLOGRAPHY
ISBN 0-12-164135-X

2 THE CHEMICAL BOND MODEL FOR INORGANIC SOLIDS

2.1 Differences between the organic and inorganic bond models

How must the rules of the chemical bond model be modified when used to describe inorganic solids? In both versions of the model, each atom is assumed to have an *atomic valence* which corresponds to the number of electrons (positive valence) or holes (negative valence) that are available in the valence shell of the neutral atom. In the original bond model the valence simply represented the number of bonds that the atom formed, i.e. its coordination number, but compounds were later discovered that contained double and triple bonds. In these compounds the coordination number was different from the atomic valence and each bond was associated with more than one unit of valence. However, in all cases the sum of the valences (or strengths) associated with the bonds around each atom was found to be equal to the valence of the atom. This is the principal rule of the chemical bond model and is known as the *valence sum rule*:

> The sum of the *bond valences* around any atom is equal to the valence of the atom.

The term 'bond valence', which is used here for precision, is essentially synonymous with the more widely used but less well-defined term 'bond strength'. The valence sum rule holds in both the organic and inorganic versions of the model.

The first important difference between the two versions of the model is that in inorganic compounds the bonds are not restricted to integral (or semi-integral) valences. The valence of each atom is distributed as uniformly as possible between the bonds that it forms (*the equal valence rule* discussed in more detail below) so that the valences of the bonds may have non-integral as well as integral values. In inorganic compounds, bonds with integral valences are the exception rather than the rule. Thus in crystals of NaCl, each of the six bonds formed by the Na atom has an equal share of its atomic valence of +1, giving them each a bond valence of 1/6 valence units (v.u.). Similarly, in the SO_4^{2-} ion each of the four bonds formed by S shares equally in its atomic valence of +6, giving each a bond valence of 1.5 v.u.

The second difference between the two models is that in inorganic solids the sign of the atomic valence becomes important. It is therefore convenient to label atoms with positive valence as *cations* and those with negative valence as *anions* but without in any way implying that the model is restricted to those compounds traditionally thought of as ionic; the model works equally well for both ionic and covalent compounds and it is no longer necessary to distinguish between them. However, the model is mathematically restricted to compounds in which all the bonds have a cation at one end and an anion at the other.

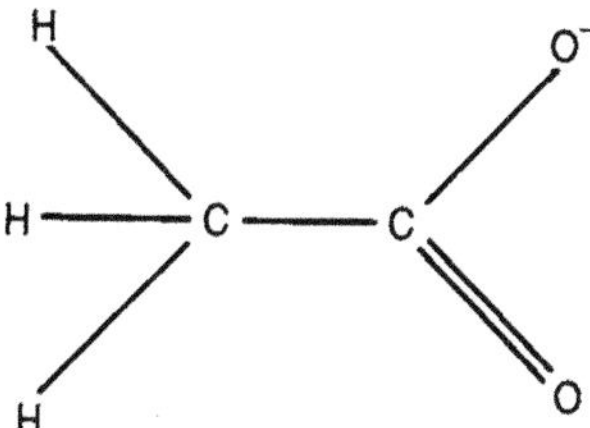

Figure 2.1 Bond graph of acetate ion showing that it also gives a good representation of the molecular geometry.

Although this appears to be a serious restriction, it is obeyed by almost all metal oxides and halides and those that do not conform can be treated by extensions to the model described by Brown (1992a).

A third difference lies in the way the structures are represented graphically. Organic molecules can be represented by a two-dimensional bond diagram which shows the topology of the molecule, i.e. the way in which the atoms are connected together. If carefully drawn, this diagram can be made to look like a projection of the three-dimensional structure of the molecule (Fig. 2.1).

Unfortunately, the corresponding bond diagram for inorganic compounds usually extends to infinity in one or more directions, making the same kind of simple two-dimensional representation impossible. However, for a single formula unit it is possible to draw a finite bond graph that retains the essential topological properties of the infinite bond network, though such a diagram does not give the same pictorial impression of the three-dimensional structure (Fig. 2.2). In order to use the bond model for inorganic solids it is necessary to know how to draw and interpret this bond graph.

2.2 Bond networks and bond graphs

Figure 2.2 illustrates how the finite bond graph is related to the infinite bond network. A single formula unit of the compound is isolated and extracted from the network. This necessarily involves breaking a number of bonds but these can be reconnected to other broken bonds within the formula unit in such a way that the nearest neighbour bonding topology is preserved. The resulting diagram differs in an important way from the traditional organic bond diagram, namely that the presence of two or more lines connecting a pair of atoms does not indicate a bond of double or triple strength; rather it indicates that each of the atoms is bonded to two or more spatially separated but chemically similar atoms. The bond graph for NaCl (Fig.2.2(b)) contains only two atoms, Na and Cl, but these are connected by 6 lines, indicating that each Na atom bonds to 6 different Cl atoms and vice versa. Similarly, the bond graph of SiO_2 (Fig. 2.2(d)) contains three atoms, with Si forming four bonds to different O atoms and each of the two O atoms forming bonds to two different Si atoms. The number of

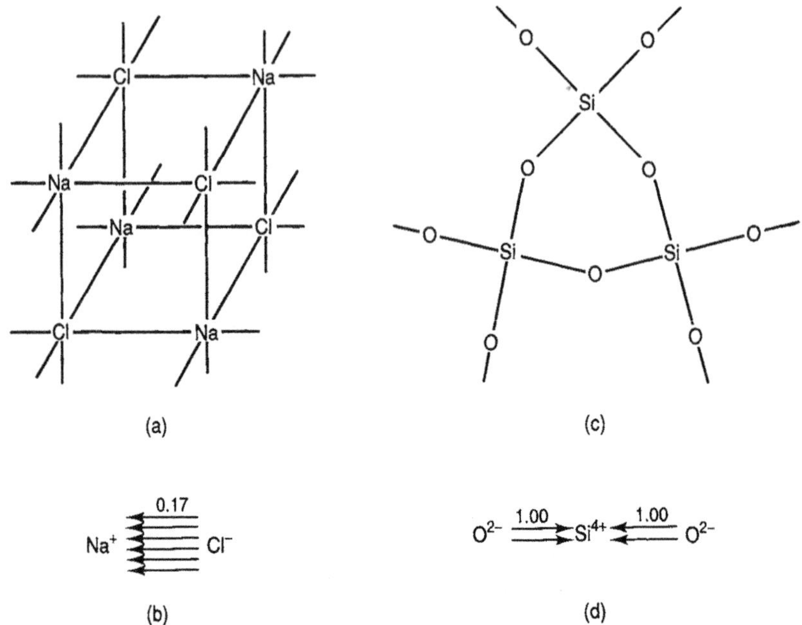

Figure 2.2 Extraction of bond graph from bond network: (a) and (b) NaCl, (c) and (d) silica; the two-dimensional geometry is shown in (a) and (c) and the bond graph in (b) and (d).

lines terminating at an atom is its coordination number. For inorganic compounds it makes no sense to use single and double lines to represent the bond valence because in general the bonds do not have integral valences. Where necessary the bond valence can be shown by means of a number written alongside the bond.

Graphs which only have bonds linking cations to anions are called bipartite graphs, a property that can be indicated by arrows showing the bond directed from the anion to the cation. All the loops in a bipartite graph contain even numbers of bonds.

2.3 Prediction of bond valences

The valence sum rule and the equal valence rule can be expressed in mathematical form by equations (1) and (2), respectively (Brown, 1992a):

$$\sum_j S_{ij} = V_i \tag{1}$$

$$\sum_{\text{loop}} S_{ij} = 0 \tag{2}$$

where S_{ij} is the valence of the bond between atoms i and j and V_i is the valence of atom i. The analogy is with Kirchhoff's equations for the analysis of electrical networks where the electric current is assumed to distribute itself as equally as possible between the various branches of the network. The meaning of equation (2) is that the sum of bond valences around any loop in the network (taking account of the direction of the bonds) is zero and this condition is fulfilled when the valences are most uniformly distributed (Brown, 1992b, appendix). Equations (1) and (2) are called the *network equations* and are sufficient to determine the ideal distribution of the valence between the bonds once the bond graph and atomic valences are known.

2.4 Correlation between bond valence and bond length

The usefulness of bond valence lies in its correlation with bond length. For each pair of ions that form a bond, there is a correlation between S_{ij} and the bond length, R_{ij}, similar to that shown in Fig. 2.3. Weaker bonds, i.e. those with smaller valence, are longer and those with higher valence are shorter. Thus a knowledge of the valence of a bond leads to a prediction of its length and vice versa.

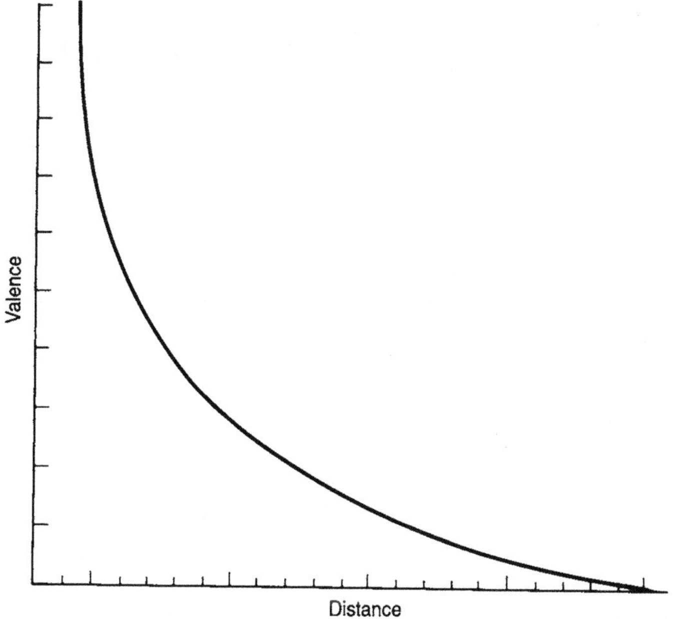

Figure 2.3 Typical example of a bond valence–bond length correlation.

Although the relationship shown in Fig. 2.3 does not have a simple analytical form, over the small range of bond lengths and valences encountered for most kinds of bond it is possible to represent the relationship using equation (3):

$$S_{ij} = \exp[(R_o - R_{ij})/B] \quad (3)$$

where R_o and B are two constants that must be fitted for each type of bond. In practice, B, which defines the slope of the curve shown in Fig. 2.3, changes very little between one bond type and another and can usually be set to 37 pm. R_o represents the length of a bond of unit valence. Empirically determined values of R_o and B have been tabulated by Brown and Altermatt (1985) for most common types of bond. O'Keeffe and Brese (1991) have proposed a way of estimating R_o when experimental values are not available.

2.5 The distortion theorem

The concave shape of the curve in Fig. 2.3 leads to an important theorem known as *the distortion theorem* which states that:

> If the environment of an atom is made less symmetric by increasing the length of some bonds and decreasing the lengths of others so as to leave the average bond length unaltered, the average bond valence, and hence the bond valence sum at the atom, will increase.

There is a corollary to this theorem that states that:

> If an atom is placed in a cavity that is too large for it, its bonds will tend to distort.

This can be understood as follows. Because the cavity is too large, the bonds will be longer than predicted by equations (1)–(3). As a result, the bond valences calculated from their lengths will be too small and the valence sum around the atom will be too low. By distorting its environment so that some bonds become longer and others shorter, the atom will increase its bond valence sum without changing the average bond length, since the shortened bond will gain more valence than the longer one loses. If the distortion is sufficiently large, the bond valence sum can be made equal to the atomic valence. The distortion theorem therefore predicts that whenever an atom is in a cavity that is too large, it will tend to distort its environment by changing the bond lengths, typically by moving off-centre.

The theorem is illustrated by the perovskites $BaTiO_3$, $SrTiO_3$ and $CaTiO_3$, in which the size of the cubic cell is defined by the TiO_3 framework (Fig. 2.4(b)). The Sr^{2+} ion fits almost exactly into the central cavity of the cube but, when it is replaced by the larger Ba^{2+} ion, the framework expands, leaving the Ti^{4+} ion in a cavity that is too large for it. The structure distorts by Ti^{4+} moving away from the centre of its octahedron (Fig. 2.4(a)). Ca^{2+} is, however, smaller than the

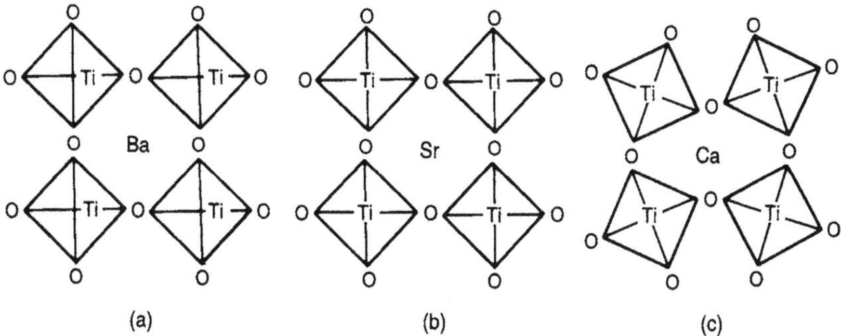

Figure 2.4 Perovskite structures: (a) $BaTiO_3$ with Ti^{4+} off-centre, (b) $SrTiO_3$ (cubic), (c) $CaTiO_3$ with a collapsed TiO_3 framework.

central cavity, so that in $CaTiO_3$, it is the environment of Ca^{2+} that distorts by a collapse of the cubic TiO_3 framework around it (Fig. 2.4(c)).

2.6 Coordination number

In modelling inorganic structures it is often useful to know what coordination number might be expected around a given cation. The factors determining the coordination number are complex, involving both anion–anion repulsion and the relative numbers of cations and anions in the compound (Brown, 1988, 1995). For the present purpose, it is sufficient to recognize that the coordination numbers found for a particular cation tend to cluster close to their average value which can therefore be taken as a typical coordination number. Average Observed Coordination Numbers (AOCN) have been reported for many cations (Brown, 1988) and a selection of values is given in Table 2.1. They provide a strong constraint in structure modelling.

2.7 Bonding strength and the valence matching principle

If the AOCN represents the typical coordination number of a cation, then from the valence sum rule the valence, s_c, of a typical bond formed by this cation is given by equation (4):

$$s_c = V_c/\text{AOCN} \tag{4}$$

where V_c is the atomic valence of the cation. The quantity s_c is called the *bonding strength* of the cation and represents the strength of the bonds that the cation might be expected to form.

Table 2.1 Selected cation and anion bonding strengths and Average Observed Coordination Numbers (AOCN) to oxygen (taken from Brown, 1988, 1992a).

	AOCN	Bonding strength
Cations		
Cs^+	8.8	0.113
Rb^+	8.0	0.124
K^+	7.9	0.126
Na^+	6.4	0.156
Ba^{2+}	10.2	0.195
Li^+	4.9	0.205
Sr^{2+}	8.6	0.233
Ca^{2+}	7.3	0.274
Mg^{2+}	6.0	0.334
Cr^{3+}	6.0	0.500
Be^{2+}	4.0	0.501
Al^{3+}	5.3	0.57
B^{3+}	3.5	0.87
Si^{4+}	4.3	1.00
P^{5+}	4.0	1.25
C^{4+}	3.0	1.35
S^{6+}	4.0	1.5
N^{5+}	3.0	1.67
Cl^{7+}	4.0	1.75
Anions		
ClO_4^-		0.08
NO_3^-		0.11
Cl^-		0.17
SO_4^{2-}		0.17
PO_4^{3-}		0.25
BO_3^{3-}		0.33
SiO_4^{4-}		0.33
O^{2-}		0.50

One can also define an anion bonding strength, s_a, in a similar way, taking the coordination number of O^{2-} and F^- to be 4 and of S^{2-} and the other halogens to be 6. A selection of cation and anion bonding strengths is given in Table 2.1.

Since the bonding strength of both the cation and the anion is an estimate of the valence of the bond formed between them, a cation will preferentially bond to an anion that has the same bonding strength. The *valence matching principle*, which is useful in determining which cations and anions will bond to each other, states that:

The bond between a cation and an anion will be most stable when they both have the same bonding strength.

The valence matching principle is used in the modelling of structures by ensuring that strong cations bond to strong anions and weak cations bond to weak anions, as discussed below.

3 DETERMINING THE STRUCTURE

3.1 Introduction

The determination of the crystal structure of a compound is conveniently divided into two stages: the determination of the three-dimensional arrangement of the atoms relative to each other (the *structure* or *topology* of the compound) and the determination of the bond lengths and angles (the *geometry* of the compound). While there are many reliable ways to determine the geometry once the structure is known, much less progress has been made in the problem of determining the structure *ab initio*. The two problems are quite distinct and will be discussed separately.

Although there are still no reliable ways of predicting the structure, this has not been for want of trying. There are currently two approaches that involve the use of bond valences. In the first, bond valences are used to assess the quality of, and improve, randomly generated trial structures; in the second the structure is assembled directly using the crystal chemical rules of the bond valence model.

3.2 Random structure methods

In these methods, trial structures are generated by randomly placing the atoms of one or more formula units into a box which, through the use of cyclic boundary conditions, represents the unit cell. It is therefore necessary to know, at least approximately, the unit cell of the expected structure. The randomly generated structures can be initially screened to eliminate obviously impossible structures such as those that have atoms too close together. The quality of a structure that passes this screening is evaluated using a cost function—the closer the cost function is to zero, the closer the structure conforms to the expected crystal chemical rules. The criteria for choosing a cost function are that it should discriminate between likely and unlikely structures, and should be easy to calculate so that many different structures can be rapidly tested. The bond valence model offers a number of parameters suitable for this purpose, the most obvious being the difference between the atomic valence and bond valence sum. Other possible parameters are the difference between the AOCN for the cation and the coordination found in the model, the difference between the bond valences and the bonding strengths of the cations, and the sum of the bond valences around loops. It is usually also necessary to include a term that becomes larger when non-bonding atoms approach too closely.

Since it is not feasible to generate every possible arrangement of atoms, it is necessary to have some means of improving good candidate structures. Straightforward procedures, such as simplex or least squares, refine to the nearest local minimum which, in general, is not the desired global minimum. Procedures are needed that allow the refinement to explore different parts of configuration space in order to locate the true minimum. Pannetier *et al.* (1990) used the simulated annealing technique referred to in Chapter 1. The trial structure is changed in a random manner, but if the change results in an increase in the cost function, the new structure may not be accepted. Initially the chance that a structure with an increased cost function is rejected is small, corresponding to keeping the crystal at a high temperature, but as the calculation proceeds the probability of rejection is made larger, corresponding to a gradual lowering of the temperature. Eventually a minimum in the cost function is reached. Pannetier *et al.* report that in some, but not all, cases this minimum corresponded to the observed structure, but that different structures could be obtained by using different cooling rates. The approach shows promise, but much more work is needed to explore the best cooling conditions if it is to be made reliable. Further discussion follows in Chapter 5.

As noted in Chapter 1, Bush *et al.* (1995) have described a different approach using a genetic algorithm. Several trial structures are generated and evaluated. These are then combined to produce progeny bearing characteristics taken from both parents. Structures with low cost functions are allowed to produce more progeny than those with high cost functions and a low rate of random mutations is introduced into the population to maintain genetic diversity. After several thousand generations, a series of structures with low cost functions is obtained. Bush *et al.* describe how, when applied to Li_3RuO_4 whose structure was previously unknown, the final population contained several individuals with the correct structure.

While these approaches show much promise, they are not strictly *ab initio* methods as they require an *a priori* knowledge of the unit cell. However, they may prove useful in the structure solving packages used to interpret powder diffraction patterns where the unit cell is obtained early in the analysis but where it is not possible to use traditional structure determination methods.

3.3 Crystal chemical methods

3.3.1 Introduction

The crystal chemical methods comprise a variety of techniques that attempt to predict the structure directly by applying crystal chemical principles at each stage of its development. These methods were successfully used in the early days of X-ray diffraction to propose structures for many minerals. In the hands of experienced crystal chemists such as Pauling and Bragg they achieved remarkable success, but each structure required a different technique, making it difficult to find a general approach to structure prediction.

The modern goal of crystal chemical modelling is to produce a computer program that can reliably predict the structure of any compound given only the chemical formula. Needless to say, this goal is difficult to achieve and may remain so, since many compounds exist in more than one polymorph and, to distinguish between them, the program would have to account for changes in temperature and different methods of synthesis. Nevertheless, the exercise of looking for a universal modelling algorithm is useful in drawing attention to the range of validity of each of the crystal chemical rules that are used.

Since no general algorithm has yet been shown to exist, the best that can be done in this section is to describe some of the components that might eventually be incorporated into a general structure generating program.

A guiding principle in the following analysis is the *principle of maximum symmetry* which states that:

> At each stage in the generation of a structure, one selects the highest symmetry structure that is consistent with the crystal chemical constraints.

For example, in NaCl the Na atoms are clearly not equivalent to the Cl atoms, but all the Na atoms are equivalent to each other, as are all the Cl atoms, and all the Na–Cl bonds are equivalent. The cubic space group of solid NaCl maintains the full symmetry of the formula unit. In other examples, some of which are discussed below, there are good crystal chemical reasons why such high symmetry cannot be maintained and structures with monoclinic or lower symmetry are found.

3.3.2 Determination of the bond topology

In some cases it is possible to predict the bond graph using only a knowledge of the atoms in the formula unit. The atoms are divided into cations and anions and are arranged in descending order of their bonding strengths. In accordance with the valence matching principle, bonds are then drawn between the cations and the anions with the highest bonding strength, assigning to each bond a valence comparable to the cation and anion bonding strengths. The process is continued until the valence of the cation is satisfied. In general, there are several ways in which this may be done, but the principle of maximum symmetry requires that the graph selected must be the one with the highest symmetry.

This procedure results in the strongest cation surrounding itself with anions to form a complex anion which will have its own bonding strength. The atoms forming the complex are then removed from the list and the complex is inserted at the place appropriate to its bonding strength. The process is repeated until all

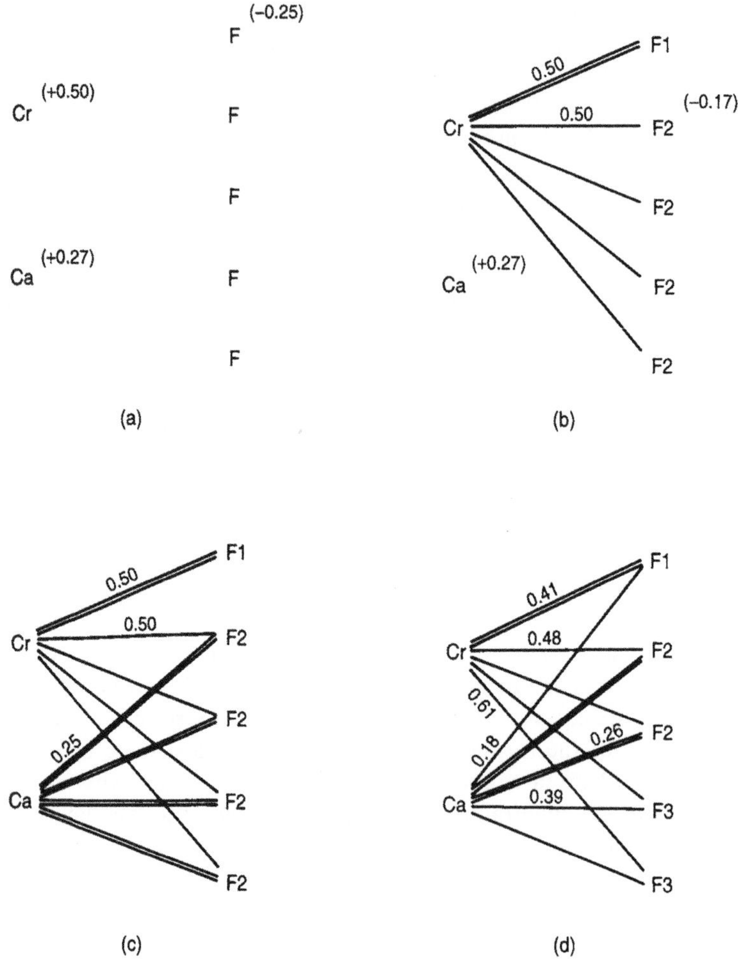

Figure 2.5 Creating the bond graph for $CaCrF_5$: (a) ions ordered by bonding strength (given in parentheses), (b) after the formation of the CrF_5^{2-} complex, bond valences shown on the bonds, (c) the complete predicted bond graph, (d) bond graph observed for the monoclinic structure with the predicted bond valences.

bonds stronger than about 0.2 v.u. have been assigned. In general it is not possible to make an *a priori* assignment of bonds weaker than 0.2 v.u. since there are often so many bonds that several graphs with the same high symmetry are possible. In practice, the weak bonds often violate the principle of maximum symmetry because their arrangement is strongly affected by the requirements of crystal packing.

The method is illustrated in Fig. 2.5 for $CaCrF_5$. The Cr^{3+} ions have a bonding strength of 0.5 v.u. since they are usually six-coordinate. The bonding strength of Ca^{2+}, taken from Table 2.1, is 0.27 v.u. The Cr–F bonds are therefore formed first and, since Cr^{3+} will form six bonds, one of the F atoms must be bonded to two Cr^{3+} ions. At this stage (Fig. 2.5(b)) the five F ions are no longer equivalent. There is a single F1 atom (bonded to two Cr^{3+}) and four equivalent F2 atoms are bonded to a single Cr^{3+}. The CrF_5^{2-} complex will form bonds only through the four F2 atoms, the valence of F1 being already saturated, and since the coordination number for F is expected to be 4, the anion bonding strength of the complex is 2/12 = 0.17 v.u. The Ca^{2+} binding strength is a little larger, 0.27 v.u., but symmetry can be preserved if each F2 ion forms two bonds to Ca^{2+} as shown in Fig. 2.5(c).

Once the bond graph (or part of it) has been established, the two network equations (equations (1) and (2)) can be used to predict the bond valences, and from these the expected bond lengths can be calculated from equation (3) for use in later stages of modelling. In the case of the bond graph of $CaCrF_5$ shown in Fig. 2.5(c), the bond valences can be assigned by inspection. All the Cr–F bonds have valences of 0.50 v.u. and all the Ca–F bonds have valences of 0.25 v.u., corresponding to lengths of 191.3 pm and 235.7 pm, respectively.

3.3.3 Determination of the space group and cell

The expansion of the bond graph into a three-dimensional network proves to be the most difficult step because there are a very large number of ways in which this can be done; however, a method recently proposed by Galliulin and Khachaturov (1994) shows promise of being able to solve this problem in many cases. If the infinite bond network is to preserve as much of the symmetry of the bond graph as possible, it will adopt the space group with the highest symmetry that is compatible with the symmetry of the bond graph. Galliulin and Khachaturov show that this space group can be readily found by comparing the number of special positions of a given multiplicity required by the chemical formula with those available in the different space groups. They define the *spectrum of special positions* of a space group as the sequence of ten numbers corresponding to the number of special positions with multiplicities of 1, 2, 3, 4, 6, 8, 12, 16, 24 and 48, respectively, in the primitive unit cell. For example, the spectrum of *Fm3m* is (2100***0**) where * indicates that the special position contains a free parameter and can be occupied by as many atoms as necessary. The spectrum of a bond graph can be written in the same form. In the case of $CaCrF_5$, this spectrum is (30010000) since there are three atoms (Cr^{3+}, Ca^{2+} and F1) that occur only once in the formula unit and one atom (F2) that appears four times. The compatible space groups are the ones whose spectrum contains digits that are as high as or higher than those in the spectrum of $CaCrF_5$. Clearly *Fm3m* does not meet this criterion since it only has two, not three, positions of multiplicity 1. Since the multiplicity of the general position in the primitive cell is a convenient measure of the amount of symmetry, the space groups with a primitive general multiplicity of 48 are searched first, then those with

Table 2.2 Spectra of special positions for selected space groups.

Space group	Spectrum	Multiplicity
Fd3m	02020**0**	48
Fm3m	2100***0**	48
Pm3m	2020***0**	48
I$\bar{4}$3m	101**0*0*0	24
F$\bar{4}$3m	400**0*0*0	24
I4$_1$/amd	020*0*0*00	16
I4/mmm	2*0*0*0*00	16
P4/mmm	4*0*0*0*00	16

multiplicities of 24, 16, etc., in decreasing order. The two best matches for $CaCrF_5$ are *F$\bar{4}$3m* with a spectrum of (400**0*0*0) and general multiplicity of 24, and *P4/mmm* with a spectrum of (4*0*0*0*00) and a general multiplicity of 16. All other matches occur with space groups of multiplicities of 12 or less. The spectra of special positions for a selection of space groups is given in Table 2.2.

The higher symmetry *F$\bar{4}$3m* can be ruled out because the special positions have the wrong point symmetry. The Cr^{3+} ion is expected to be octahedrally coordinated and the only possible arrangement that leaves the four F2 atoms equivalent is the one in which the F1 atoms are *trans* to each other. The F2 atoms must therefore lie in the equatorial plane. In *F$\bar{4}$3m* the F2 atoms would occupy the 16e (*x*,*x*,*x*) sites which are arranged tetrahedrally around the Cr^{3+} ion at 4a (0,0,0) and cannot be made to lie in a plane. *F$\bar{4}$3m* can therefore be eliminated from considerations since it does not provide the right chemical environment for Cr^{3+}. This leaves *P4/mmm* as the best candidate to use for a trial structure.

Once a space group has been selected, the atoms can be placed at the appropriate special positions. The structure of $CaCrF_5$ in *P4/mmm* is shown in Fig. 2.6. The Ca atoms occupy site 1d (1/2,1/2,1/2) and the Cr atoms 1a (0,0,0). F1 is found between the Cr atoms along the *c* axis at 1b (0,0,1/2) and F2 at 4j (x,x,0). The *c* axis length is 382.6 pm (2 × Cr–F1) and the *a* axis length can be calculated in a similar way from the predicted bond lengths and angles to be 464.8 pm. From this mapping, therefore, it is possible to predict both the space group and the unit cell dimensions as well as the atomic coordinates.

However, this structure turns out not to be the one that is observed, though it approximates to the observed structure in that the overall arrangement of the atoms in space is correct.

The problems with the tetragonal proto-structure of $CaCrF_5$ (Fig. 2.6) are illustrated in Figs 2.7(a) and 2.7(c) where, as required by the observed structure, two of the four F2 atoms are labelled F3. The overlap between F2 and F3 (Fig. 2.7(c)) and the large cavities close to Ca^{2+} (Fig. 2.7(a)) are stresses that prevent

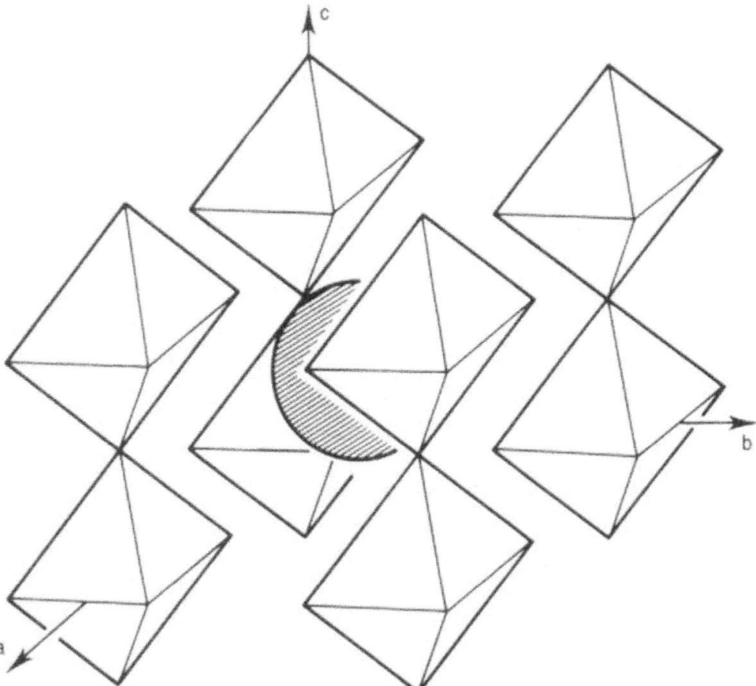

Figure 2.6 Tetragonal (*P4/mmm*) crystal structure corresponding to graph shown in Fig. 2.5(c). The CrF_5 chains are shown as octahedra linked along the *c* axis, Ca^{2+} is shown as a shaded circle. The F atoms lie at the corners of the octahedra with F1 linking the octahedra along the *c* axis.

the tetragonal structure from existing. They are removed by buckling the chain of CrF_6 octahedra (Fig. 2.7(b)) and shearing the tetragonal (010) plane of the structure along *c* (Fig. 2.7(d)). In the process, the *c* cell edge is doubled, F1 is brought within bonding range of Ca^{2+} (Fig. 2.7(b)) and one of the bonds from Ca to F3 is broken. In addition to reducing the symmetry of the crystal from *P4/mmm* to *C2/c*, the relaxation alters the bond graph to that shown in Fig. 2.5(d). The bond valences calculated for this bond graph lead to bond lengths and lattice parameters that correspond closely to those observed (Table 2.3).

For some compounds the high-symmetry proto-structure can be observed at high temperature but, as the temperature is reduced, the crystal chemical stresses are gradually relaxed by distortions that lead to a sequence of phase transitions to lower symmetry structures. The origins of these stresses will become apparent as we discuss the problem of refining the geometry in the next section.

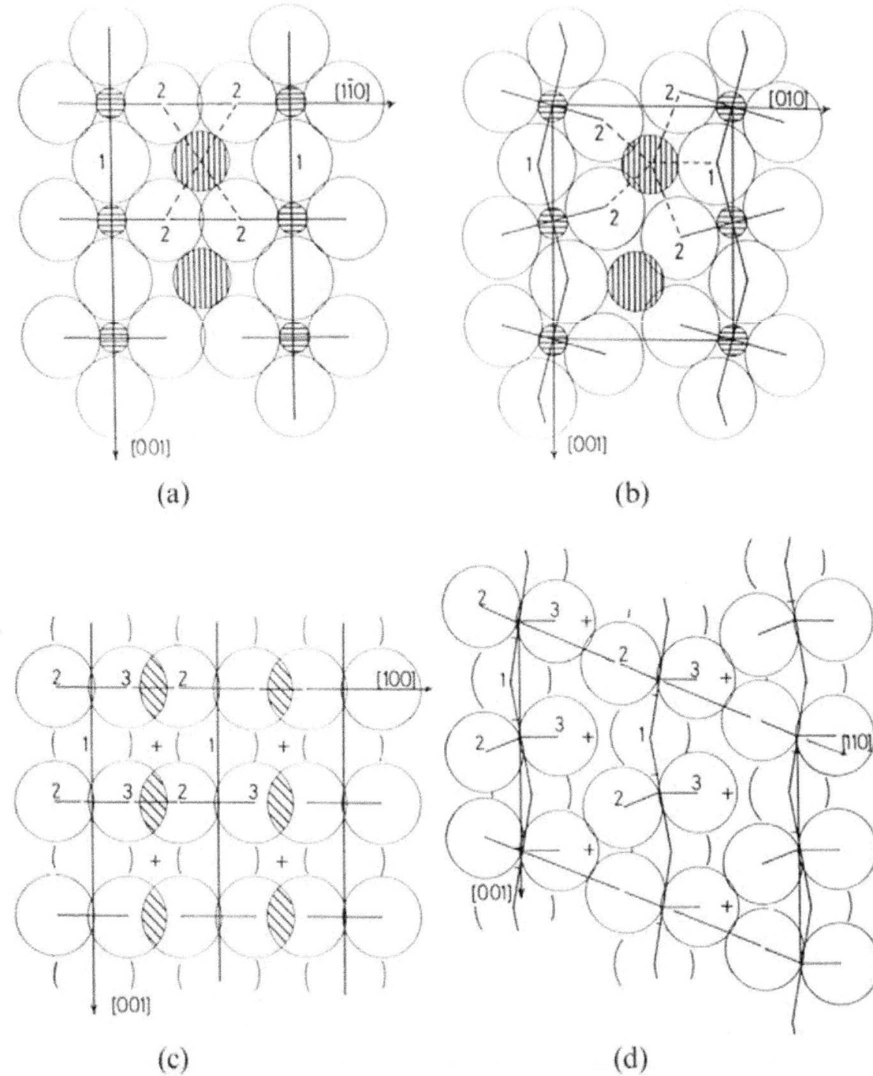

Figure 2.7 (a), (b) [110] projection of $CaCrF_5$; (c), (d) [010] projection: (a) and (c) show the predicted tetragonal structure, (b) and (d) the relaxed monoclinic structure. The Cr^{3+} ions are shown with horizontal shading, Ca^{2+} with vertical shading, F^- unshaded. The overlap of the F atoms is shown in (c) with diagonal shading. The positions of the Ca^{2+} ions in the planes behind those illustrated in (c) and (d) are indicated with crosses.

Table 2.3 Predicted and observed bond lengths and lattice parameters in $CaCrF_5$ in pm and degrees (Brown, 1992a).

	P4/mmm predicted	*C2/c*	
		Predicted	Observed
a (pm)	465	952	900
b (pm)		657	647
c (pm)	383	765	753
β (degrees)		118	116
Cr-F(1)(pm)	191	199	194
Cr-F(2)(pm)		193	192
Cr-F(3)(pm)		184	185
Ca-F(1)(pm)	236	248	250
Ca-F(2)(pm)		234	229, 239
Ca-F(3)(pm)		219	221

4 REFINING THE GEOMETRY

4.1 Introduction

The previous section described a possible strategy for the determination of the structure, i.e. the basic arrangement of the atoms in space. Refining the geometry means locating the exact positions of the atoms within the unit cell, subject to the symmetry restrictions of the selected space group. The geometry is refined by moving the atoms until the distances between them correspond to the bond lengths predicted using the network equations (1) and (2). In practice, it is usually also necessary to include some constraint to avoid non-bonded atoms from approaching too closely. Only if the number of free parameters (the cell dimensions and free atomic coordinates) is equal to the number of constraints (predicted bond lengths and non-bonded distances) will there be a unique solution. More often there are either too few or too many free parameters.

If the selected space group of the starting structure has high symmetry, the number of free parameters will be smaller than the number of constraints and it is not possible for all the predicted distances to be realized. If the deviations are small, the structure may be stable, but if they are large, the structure will relax or, in extreme cases, be so unstable that it cannot be prepared. Relaxation may involve only a small adjustment to the bond lengths so that the valence sum rule continues to be obeyed (at the expense of the equal valence rule), or it may involve a reduction in the symmetry as is found in the case of $CaCrF_5$ described above. If the symmetry is reduced, the number of free variables is increased and the atomic coordinates may be under-determined. In this case, the constraints on the sizes of non-bonding distances become important.

In straightforward cases the atomic coordinates can be refined by least squares using one of a number of minimization functions. Distance Least Squares (DLS, Meier and Villiger, 1969) minimizes the difference between the distances in the refined model and target distances such as the bond lengths predicted by the network equations (equations (1)–(3)). It is usually necessary to add a few non-bonded distances but the target values assigned to these are not critical since they are given much less weight than the bonds. Alternatively, the refinement can minimize the deviation between the atomic valences and bond valence sums of the model, though additional constraints corresponding to the equal valence rule must be introduced in some form or other. These approaches have been incorporated in the refinement programs STRUMO (Brown, 1989) and DVLS (Kroll *et al.*, 1992).

These refinement procedures work well in those cases where the network equations give a good prediction of the bond lengths, but in many cases the observed bond lengths differ from those predicted as a result of crystal chemical effects not properly described by the bond valence model. In some structures non-bonded atoms cannot be brought as close as the model predicts (steric effects), in some the bonds are strained in order to maintain translational symmetry (lattice effects) and in some distortions are induced by the electronic structure of the atoms (electronic effects). Quite often, several of these effects occur together in a mutually supporting role. Since these crystal chemical effects have an important bearing on structure and geometry, they are discussed separately.

4.2 Steric effects

4.2.1 Introduction

While the network equations give predictions of the distances between bonded atoms, they say nothing about the distances between non-bonded atoms. These clearly become an important factor in modelling when two anions or two cations are brought too close together. Anion–anion repulsions are important when the two anions are strongly bonded to the same cation; cation–cation repulsions need to be considered when two cation coordination spheres share edges or faces.

4.2.2 Anion–anion repulsion

The most important effect of the repulsion between anions is to limit cation coordination numbers. Strongly bonding cations tend to surround themselves with as many ligands as they can fit into their coordination sphere, but the closeness with which two ligands can approach will depend on the valences of the bonds to their common cation, the stronger these bonds, the closer the approach of the anions. To provide a quantitative estimate of this effect, it is useful to define an effective valence, s', which represents the projection of the cation–anion bond valence, s, on the anion–anion vector (equation (5)):

$$s' = s \cdot \cos \alpha \tag{5}$$

where α is the angle between the anion–anion vector and the bond. Regular coordination environments are only found in oxides if the O–O distances are larger than R_{min} given by equation (6):

$$R_{min} = R_o - B \cdot \ln(s') \tag{6}$$

where R_o = 220 pm and B = 34 pm are fitted constants (Brown, 1995). Equation (6) therefore provides an upper limit on cation coordination numbers. From the point of view of modelling, this restriction is mainly of interest insofar as it limits the coordination number of strongly bonding cations like C^{4+}, N^{5+} and S^{6+} and hence determines their bonding strength (Table 2.1).

However, there is one place where anion–anion repulsion plays a special role, namely in the crystal chemistry of H^+, because the predicted O–O distance is less than R_{min} for all possible regular coordination environments. Even for two-coordination the predicted H–O bond length of 115 pm is less than half of R_{min} = 244 pm. Consequently the H–O bonds will be stretched and, in this situation, the distortion theorem predicts that the environment of H^+ will distort in order to increase the bond valence sum to 1.0. If one of the O–H bonds shortens to 97 pm and the other lengthens to 181 pm, corresponding to bond valences of 0.82 and 0.18 v.u., respectively, the sum of the two O–H distances just equals 278 pm, the value of R_{min} for s' = 0.18 v.u. This is the smallest distortion that can lead to the correct bond valence sums around H^+ without straining the bonds. It is the geometry of a typical hydrogen bond, the geometry found, for example, in ice.

Stronger (more symmetric) hydrogen bonds can be formed by stretching the O–H bonds until the O–O distance is as large as R_{min}. Such a strain can only be accommodated if it relieves stresses in other parts of the structure. In KH_2PO_4, for example, the phosphate ions are linked by hydrogen bonds, as shown schematically in Fig. 2.8. The network equations predict the symmetric structure shown in Fig. 2.8(a) in which all the P–O bonds have valences of 1.25 v.u. and all the H–O bonds have valences of 0.5 v.u. However, as explained above, O–O repulsion is expected to distort the symmetric hydrogen bonds. If the hydrogen bonds adopt the geometry found in ice, they induce an energetically unfavourable distortion of the PO_4^{3-} ion (Fig. 2.8(b)). The compromise is a structure somewhere in between (Fig. 2.8(c)) in which smaller distortions are found around the P^{5+} cation and the H–O bonds are somewhat stretched.

Weak (more distorted) hydrogen bonds occur when the network equations predict a greater distortion than that found for normal hydrogen bonds. This situation is typically found in hydrogen bonds to weakly bonding anions such as

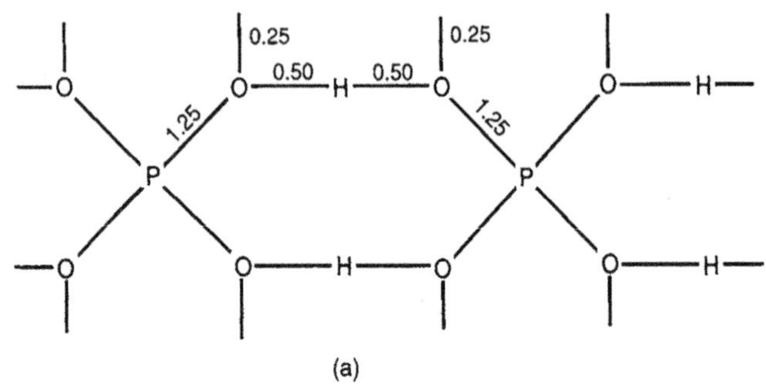

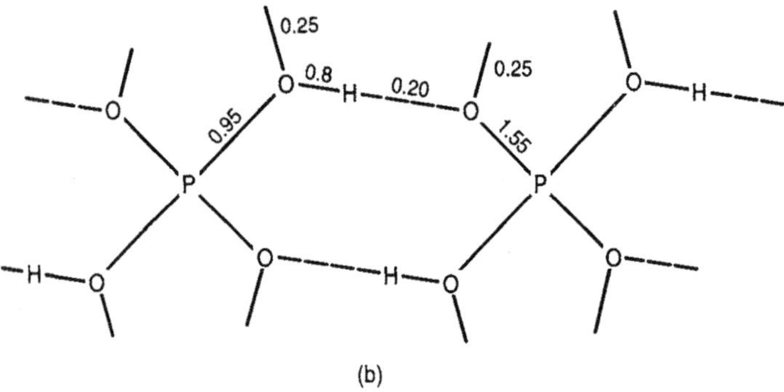

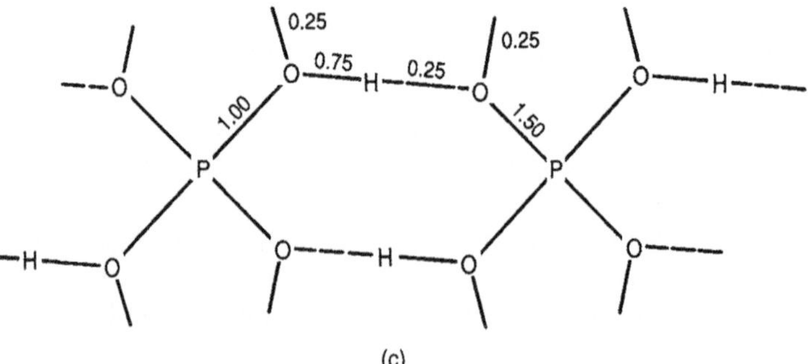

Figure 2.8 Hydrogen bonding in KH_2PO_4: (a) symmetrical arrangement predicted by network equations, (b) arrangement with normal hydrogen bonding, (c) observed arrangement lying between (a) and (b).

ClO_4^- (bonding strength 0.08 v.u.) In this case the O–O distance can be longer than R_{min} and the bonds are free to bend. Weak hydrogen bonds are usually bent with an O–H···O angle corresponding to an O–O distance equal to R_{min}.

Any modelling of compounds that contains H^+ ions needs to take the hydrogen bond distortion into account. The valence sum rule continues to be obeyed around H^+ but the distortion violates the equal valence rule which must be replaced by some other constraint. For strong hydrogen bonds, the cost of stretching the H–O bonds needs to be balanced against a gain in energy in some other part of the crystal. For weak bonds, crystal packing is at least as important as the equal valence rule in determining the bond lengths. It is not yet clear how these constraints can be introduced into the bond valence model to produce a reliable prediction of the hydrogen bond geometry.

4.2.3 Cation–cation repulsion

Cation–cation repulsions appear to be important when cations approach closer than about 310 pm. They usually need to be considered only when the coordination spheres around the two cations share an edge or a face. Unless there is metallic bonding between the cations, they will repel, moving away from the centres of the their coordination spheres. This effect has not been quantitatively studied, but Kunz and Brown (1995) have shown that, in compounds of d^0 transition metals, cation–cation repulsion is important in determining the directions in which electronic distortions occur.

4.3 Lattice induced strain

In some crystals a particular lattice parameter is determined by more than one set of bonds. For example, layer compounds are composed of a sequence of different layers, each of which will have its lattice translations determined by the lengths of the bonds within the layer. In general, the lattice parameters predicted for one layer will be different from those predicted for the others, so some accommodation is needed if the layers are to coexist in the same crystal. There are then three possibilities: (1) the incommensuration between the layers may be so severe that the compound cannot form, (2) each layer may keep its own lattice spacing and so form an incommensurate structure or (3) the bonds in some layers will stretch and in others will compress so as to ensure that the lattice parameters of all layers are the same. The second solution is found in structures such as cannizzarite (Fig. 2.9) where the bonding between the incommensurate layers is weak and the third is found in perovskite-related structures (e.g. La_2NiO_4, Fig. 2.10) where the interlayer bonding is strong.

La_2NiO_4 can be thought of as being composed of NiO_2 layers perpendicular to the unique four-fold axis alternating with two LaO layers. The *a* lattice spacing is constrained by geometry to be $\sqrt{2}R_{La-O}$ and $2R_{Ni-O}$, but the network

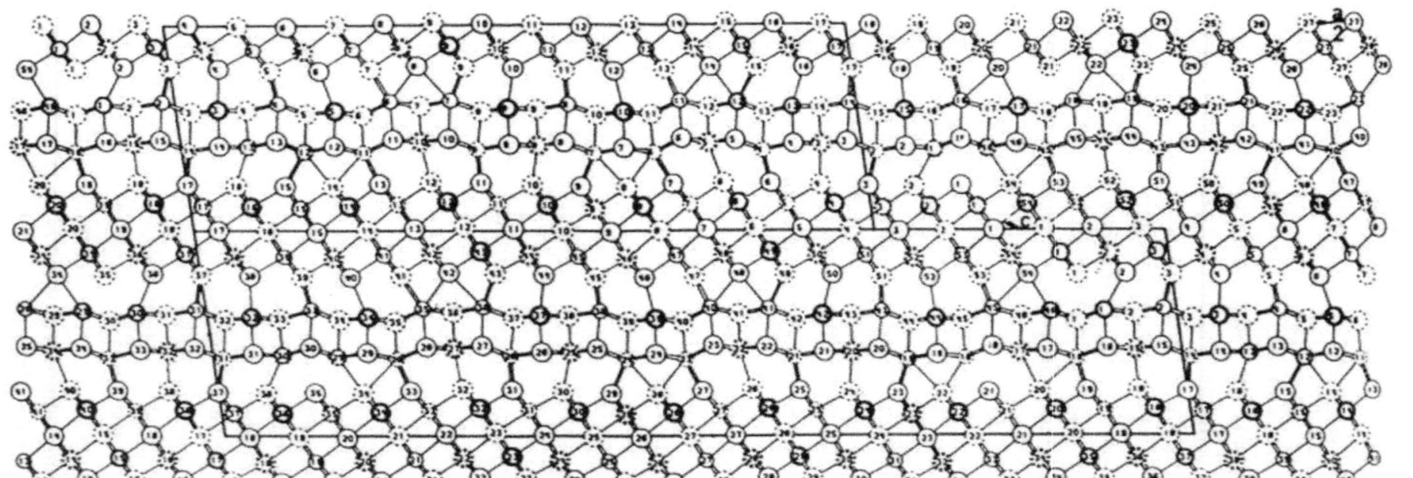

Figure 2.9 Structure of mineral cannizzarite, $Pb_{46}Bi_{54}S_{127}$, showing the interleaving of two layers having incommensurate lattice parameters (from Matzat, 1979).

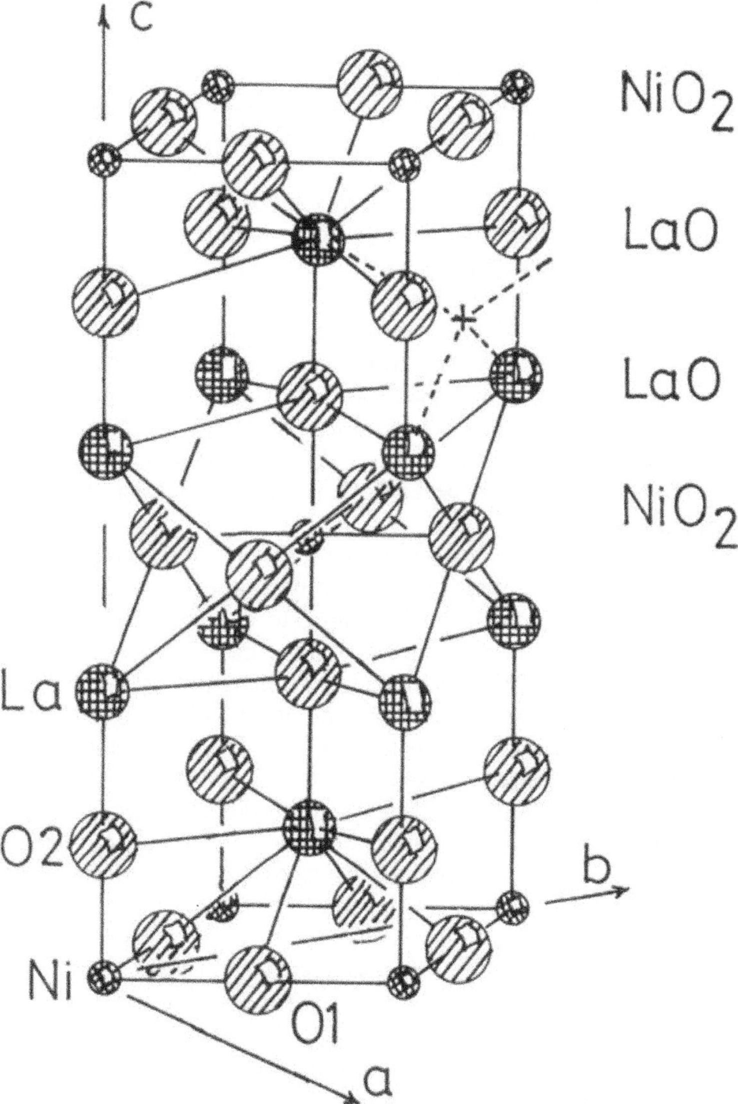

Figure 2.10 Structure of La_2NiO_4 showing the arrangement of LaO and NiO_2 layers. La^{3+} is shown with vertical cross-hatching, Ni^{2+} with diagonal cross-hatching, O^{2-} with diagonal shading. The cross indicates the position of the interstitial O.

equations predict that $R_{La\text{-}O} = 257.6$ pm and $R_{Ni\text{-}O} = 206.0$ pm, corresponding to the lattice spacings $a = 364$ and 412 pm, respectively (Brown, 1992b). If these layers are to form a commensurate structure, the LaO layer must be stretched and the NiO_2 layer compressed until both have the same spacing of 380 pm (average of 2×364 and 412 pm). This requires that the La–O bonds be stretched from 258 to 269 pm and the Ni–O bonds compressed from 206 to 190 pm. Such strains result in a large deviation from the valence sum rule, the sums around La^{3+} and Ni^{2+} both being around 2.5 v.u., and unless there is some way for the structure to relax, it cannot exist. La_2NiO_4 shows several modes of relaxation, each of which helps to restore the valence sum rule. In the first instance the La–O bonds between the layers are shortened and the Ni–O bonds lengthened, but this change is not, by itself, sufficient to make the valence sums equal to the atomic valences. If the compound is prepared in air, interstitial O atoms (shown by a cross in Fig. 2.10) are absorbed between two adjacent LaO layers, increasing the number of La–O bonds, hence increasing the valence sum at La^{3+}. At the same time the oxidation number of Ni^{2+} is increased, which has the effect of shortening the Ni–O bonds. Quantitative calculations using the bond valence model (Brown, 1992b) predict that about 1/6 of an O atom per formula unit is needed to bring the valence sums within an acceptable range of the atomic valences. The observed formula of the air prepared material is found to be $La_2NiO_{4.17}$ with Ni increasing its oxidation state from 2 to 2.34.

If La_2NiO_4 is prepared in the absence of air, the crystals cannot absorb excess oxygen and a different relaxation mechanism is adopted with the structure distorting to a lower symmetry. According to the distortion theorem, the valence sum around La^{3+} can be increased if the La–O bond lengths are made to have different values. This is achieved by buckling the NiO_2 layer, with half the NiO_6 octahedra tilting one way and the other half tilting the other way. Since the apical O atoms of these octahedra lie in the LaO plane, some of the La–O distances are increased and some decreased. In addition, the buckling of the NiO_2 plane allows the Ni–O distances to be slightly longer than $a/2$, thus relieving the strain in the NiO_2 layer as well. On cooling to room temperature La_2NiO_4 transforms from the tetragonal *I4/mmm* high-temperature structure to orthorhombic *Bmab*, and at lower temperatures the symmetry is reduced even further.

Even when all these relaxations have taken place, the bond valence sums are still not equal to the atomic valence, but the average deviation is less than 0.2 v.u. Structures whose valence sums differ on average by more than 0.2 v.u. are generally unstable, and structures reported with larger deviations have been shown to be improperly described (e.g. Armbruster *et al.*, 1990).

In the above example of La_2NiO_4, not only the valence sum rule, but also the equal valence rule is clearly violated and the network equations cannot therefore be used to predict the bond lengths; however, the network equations do give a set of ideal bond lengths, i.e. the lengths that the bonds would have if it were not for the constraints imposed by the translational symmetry of the crystal. This allows us to determine the sizes of the strains that such symmetry imposes, and allows us to understand why certain compounds undergo phase transitions as the temperature is reduced, and why they sometimes adopt unusual oxidation states and stoichiometries.

It should be noted that each of the relaxation mechanisms shown by La_2NiO_4 helps to relax both the tensile strain in the LaO layers and the compressive strain in the NiO_2 layers, a characteristic of the relaxation of many strained systems. Relaxation also leads to complex structures with unusual properties. Lattice strain can result in a lowering of symmetry, which in some cases can lead to ferroelectricity (e.g. $BaTiO_3$), and it can lead to changes in composition and to the stabilization of unusual oxidation states as in $La_2NiO_{4.17}$. La_2CuO_4 has the same structure as La_2NiO_4 but becomes superconducting at low temperatures, a property which appears only when Cu has an average oxidation state of +2.3. The appearance of superconductivity is thus a direct consequence of the lattice strain.

4.4 Electronic induced distortions

4.4.1 Introduction

A number of the distortions found in cation coordination environments are the result of electronic instabilities in the atoms themselves. Such distortions are typically found among the transition metal cations and the main group elements in low oxidation states. The origins of these instabilities are different and they are discussed separately.

4.4.2 Transition metal cations

Much has been written about the effects of partially filled d shells on the environments of transition metal cations, but these discussions tend to be qualitative and do not claim to predict the quantitative changes found in the bonding geometry (e.g. Orgel, 1960; Burdett, 1980). The chemical literature interprets these distortions as a consequence of the Jahn–Teller theorem (Jahn and Teller, 1937) which states that systems in a degenerate electronic state will spontaneously distort in such a way as to remove the degeneracy. The most frequently quoted example is the tetragonally distorted octahedral coordination typically found around Cu^{2+}, a cation with a d^9 configuration in which the single hole in the $d_{x^2-y^2}$ orbital pointing along two of the octahedral axes leads to the two bonds along the third axis becoming longer and the four equatorial bonds becoming shorter. Similar distortions are predicted for other transition metal cations in degenerate states, but these distortions are much smaller and are often masked by distortions caused by other influences such as the steric and lattice effects described above.

The largest distortions among transition metal cations are not shown by cations with partially filled d shells but by those whose d shells are empty or nearly empty (Kunz and Brown, 1995). There is still an electronic instability caused by degeneracy, but in these cations the degeneracy is accidental, and the distortion is described as resulting from a second-order Jahn–Teller effect (Burdett, 1980) or dynamic covalence (Bilz *et al.*, 1987). It results from the overlap of the empty d orbitals of the transition metal cation (LUMO) and the filled p orbitals of the ligands (HOMO). The size of the distortion increases as the charge on the cation increases and the energy of the LUMO is lowered. Distortions are observed in most of the octahedrally coordinated compounds of Ti^{4+}, Ta^{5+}, Nb^{5+}, W^{6+} and Mo^{6+}, but are particularly pronounced in the octahedral complexes of V^{5+}. The distortion is so great for Cr^{6+} that no octahedral complexes are known.

Kunz and Brown (1995) have shown that it is possible to predict the bond lengths for the d^0 transition metal cations in octahedral coordination using weighted network equations:

$$\sum_j S_{ij} = V_i \tag{7}$$

$$\sum_{\text{loop}} S_{ij}/C_{ij} = 0 \tag{8}$$

The valence sum rule (equation (7)) is the same as equation (1), but the loop rule (equation (8)), has been modified by the addition of bond weights, C_{ij}. The effect of giving a bond a large weight is to increase its predicted valence. The problem then reduces to choosing the weights, C_{ij}, to be assigned to each bond.

If the weights are all set to 1.0, equation (8) reduces to equation (2) and the solution is the same as for the unweighted network equations. When different values are used, two problems have to be addressed. The relative magnitudes of C_{ij} in a given coordination sphere must be determined and these magnitudes must be assigned to particular bonds. Kunz and Brown showed that for d^0 transition metal cations octahedrally coordinated by O^{2-}, it is possible to assign values of C_{ij} that are transferable between all the compounds of a given cation (Table 2.4). Although their study was restricted to a small set of cations, the method appears to be general and it is likely that similar transferable weights can be assigned for other transition metal cations.

Table 2.4 Network equation weights for d^0 cations (Kunz and Brown, 1995).

Cation	C(strong bond)	C(weak bond)
Ti^{4+}	1.3	0.7
Ta^{5+}	1.3	0.7
Nb^{5+}	1.4	0.6
V^{5+}	1.6	0.1[a]
	1.6	0.7[b]

[a]Opposite most distorted bond.
[b]Opposite next most distorted bond.

The problem of which bonds to weight is more difficult. Typically two of the six bonds in the octahedron are given a weight greater than 1.0 (corresponding to the bonds that will be stronger and shorter) and the two bonds *trans* to these are given a weight less than 1.0, but in an isolated octahedron all bonds are equivalent so that there is no basis for weighting one bond more than any other. Only when the octahedron is placed in the crystal is this equivalence removed. Small distortions may be induced by the bond network (reflected in the solution of the unweighted network equations) and by cation–cation repulsions across shared edges or faces. Both of these mechanisms will predispose the cation to move off-centre in a particular direction, and this direction will be picked up by the electronic instability. Kunz and Brown calculated the direction in which the bond network and the cation–cation repulsion move the cation, and assigned large weights to any bond that was within 65° of this direction. In all cases the smallest weight was assigned to bonds *trans* to those that were given the largest weight. All other bonds were given a weight of 1.0. With this procedure, they were able to predict the bond valence to within 0.1 v.u. of the observed value.

4.4.3 Main group cations in low oxidation states

Main group cations in their highest oxidation state use all their valence electrons for bonding. Their electronic structure is symmetric and makes little or no contribution to anisotropies in the bonding. For the large number of compounds that contain such cations, the bond valence model using the unweighted network equations gives satisfactory predictions. Cations in lower oxidation states, however, have electrons in the valence shell that are not used in bonding. Such electrons can influence the bonding geometry in a way that is well predicted by the Valence Shell Electron Pair Repulsion model (VSEPR, Gillespie and Hargittai, 1991). The model assumes that each pair of electrons in the valence shell is localized. Those that are bonding lie between the cation and a ligand, while those that are non-bonding occupy a position with no ligand. Thus, if there are four electron pairs in the cation valence shell and one of them is non-bonding, the cation will form three bonds to ligands at three of the corners of a tetrahedron. The fourth corner has no ligand and is said to be occupied by the lone pair.

The VSEPR model was developed with molecular compounds in mind, and does not take account of the long, secondary bonds (Alcock, 1972) that are frequently observed in solids. While the primary bonds are found in the direction of the bonding electron pairs, the secondary bonds tend to avoid both bonding and non-bonding electron pairs. However, in some compounds primary and secondary bonds have similar lengths and cannot be easily distinguished. Tl^+, which in compounds like Tl_3BO_3 is three-coordinate with the lone pair occupying the fourth corner of a tetrahedron, behaves in TlCl like a perfectly symmetric alkali metal slightly larger in size than Rb^+.

The cations that are found in low oxidation states are those that lie in the periodic table below a line drawn between S and Tl. The further a cation is from that line, the more likely it is to be found in a low oxidation state. Bi is more often found in its +3 state than +5, and Xe can be found with oxidation states +2, +4 and +6. Cations like S^{4+} near the top of the periodic table tend to adopt the ideal VSEPR geometry without any secondary bonds, but the further down the periodic table a cation is found, the more likely it is to form secondary bonds, and the more similar the secondary and primary bonds are to each other. Thus the secondary bonds in compounds of Tl^+ and Pb^{2+} are often quite strong and, as mentioned above, in a few cases are indistinguishable from the primary bonds.

The bond valence model can be used to predict what kind of environment will be found around those cations that, like Tl^+, show coordination geometries in which the lone electron pair may or may not be stereoactive. When it is stereoactive, the Tl^+ ion forms three bonds at three of the corners of a tetrahedron and its bonding strength is therefore 0.33 v.u. This environment will be found with strong anions such as BO_3^{3-} whose bonding strength is also 0.33 v.u. When Tl^+ occurs in a symmetric environment it typically has a coordination number close to 10, leading to a bonding strength of 0.10 v.u. This type of environment will occur with weak anions such as NO_3^- whose bonding strength is only 0.11 v.u.

The distorted environment around cations with lone pairs is clearly not correctly predicted by the unweighted network equations which assume that the bonding is as symmetric as possible. Quantitative predictions of the bond lengths may be possible using the weighted network equations but this has yet to be demonstrated. Clearly for cations like Tl^+ the magnitudes of the weights will depend on the base strength of the counterion.

4.5 Combined effects

It should come as no surprise that in many compounds several different distortion mechanisms are present together as illustrated by compounds in the series $R^{2+}M_2^{5+}O_6$. Compounds with R^{5+} = Sb^{5+}, Ta^{5+}, Nb^{5+} or V^{5+} mostly crystallize in one of three structure types: trirutile, columbite and brannerite. Trirutile has the bond graph that would be expected according to the principle of maximum symmetry (Fig. 2.11(a)), but columbite and brannerite both have the less symmetric graph shown in Fig. 2.11(b). The latter two structures differ in the way in which the chains of MO_6 octahedra share edges to form chains, the arrangement in brannerite leading to larger cation–cation repulsions. Trirutile, in which M is at the centre of a nearly regular octahedron, is adopted by compounds of Sb^{5+} which are expected to show no distortion. Compounds of Ta^{5+} and Nb^{5+}, in which some distortion is expected, crystallize with the columbite structure while compounds of V^{5+}, in which large distortions are

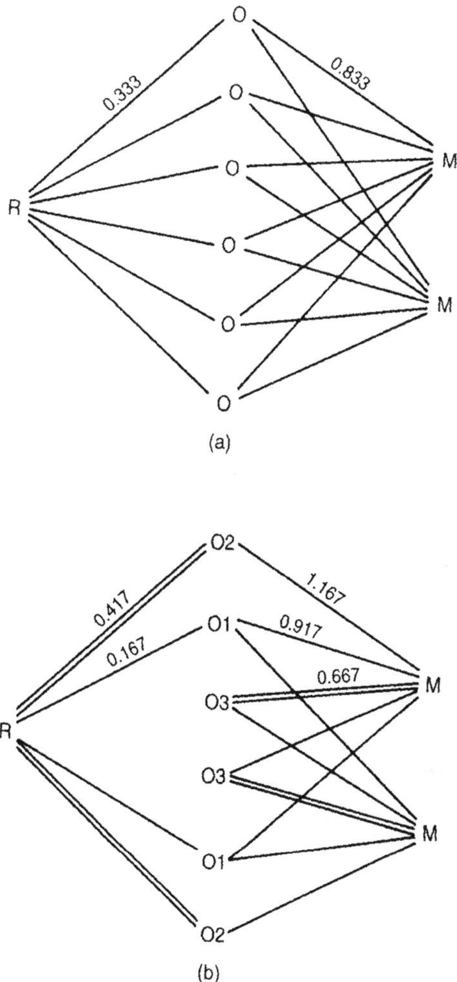

Figure 2.11 Bond graphs of RM_2O_6 structures: (a) trirutile, (b) columbite and brannerite.

expected, crystallize with the brannerite structure. These examples show that the electron instabilities in the transition metal cations result in compounds that crystallize in a structure in which both the bond graph and the cation–cation repulsions support the distortion that lowers the electronic energy. In all these structures there is also a lattice induced strain whose relaxation serves to increase the distortion around M^{5+} and to mitigate the distortion caused by the bond network around R^{2+}. All the various strains thus work together to provide a crystal structure with a low energy. Examples of such synergistic effects are found in many other systems. Even though it may not always be possible to use the bond valence model to provide accurate predictions of the bond lengths in these cases, the model does provide a framework in which the various effects can be discussed.

5 SUMMARY

This chapter has explored the way in which the bond valence model can be used for modelling inorganic crystal structures. As yet there is no algorithmic way in which a structure can be modelled starting with only the chemical formula, and such an algorithm may be impossible to construct given the extent of polymorphism and the small differences in energy between different polymorphs. However, the model does provide chemical insights that can be useful in modelling. It contains a set of simple chemical rules like the network equations which can be used to predict an ideal geometry for the chemical bonding in much the same way as the organic bond model does for organic molecules. Just as the geometry of organic molecules can be modified by steric strain, so the geometry of inorganic solids is influenced by a wide variety of extraneous effects – steric strain resulting from non-bonded repulsions, lattice induced strain resulting from the translational symmetry of the crystal, and electronic induced distortions arising from asymmetries in the electronic structures of many cations.

There is much more work to be done before the model can make quantitative predictions of all of these effects, but in its present form, the bond valence model does provide a means of understanding the structure and predicting the geometry that would be expected in the absence of extraneous effects. This can then be used to examine the relative importance of different types of strain, leading to a better appreciation of the roles of, and interconnections between, the different influences at work in inorganic solids.

REFERENCES

Alcock, N.W. (1972) *Adv. Inorg. Radiochem.*, **15**, 1.
Armbruster, T., Röthlisberger, F. and Seifert, F., (1990) *Am. Min.*, **75**, 847.
Bilz, H., Büttner, H., Bussmann-Holder, A. and Vogl, F. (1987) *Ferroelectrics*, **73**, 493.
Brown, I.D. (1988) *Acta Crystallogr.*, **B44**, 545.
Brown, I.D. (1989) *J. Chem. Inf. Comp. Sci.*, **29**, 266.
Brown, I.D. (1992a) *Acta Crystallogr.*, **B48**, 553.
Brown, I.D. (1992b) *Zeit. Krist.*, **199**, 255.
Brown, I.D. (1995) *Can J. Phys.*, **73**, 676.
Brown, I.D. and Altermatt, D. (1985) *Acta Crystallogr.*, **B41**, 244.
Burdett, J.K. (1980) *Molecular Shapes*, Wiley–Interscience, New York.
Bush, T.S., Catlow, C.R.A. and Battle, P.D. (1995) *J. Mater. Chem.*, **5**, 1269.

Galliulin, R.V. and Khachaturov, V.R. (1994) In *Mathematical Modelling of Complex Objects* (ed. V.R. Khachaturov) Computing Center of the Russian Academy of Sciences, Moscow, p. 28.

Gillespie, R.J. and Hargittai, I. (1991) *The VSEPR Model of Molecular Geometry*, Prentice Hall, New York.

Jahn, H.A. and Teller, E. (1937) *Proc. R. Soc.*, **A161**, 220.

Kroll, H., Maurer, H., Stöckelmann, D., Beckers, W., Fulst, J., Krüsemann, R., Stutenbäumer, Th. and Zingel, A. (1992), *Zeit. Krist.*, **199**, 49.

Kunz, M. and Brown, I.D. (1995), *J. Solid State Chem.*, **115**, 395.

Matzat, E. (1979) *Acta Crystallogr.*, **B35**, 133.

Meier, W.M. and Villiger, H. (1969), *Zeit. Krist.*, **129**, 411.

O'Keeffe M. and Brese N.E. (1991) *J. Am. Chem. Soc.*, **113**, 3226.

Orgel, L.E. (1960) *Introduction to Transition Metal Chemistry Ligand Field Theory*, Methuen & Co. Ltd, London, John Wiley & Sons Inc., New York.

Pannetier, J., Bassas-Alsina, J., Rodriguez-Carvajal, J. and Caignaert, V. (1990) *Nature*, **346**, 343.

3

Lattice Energy and Free Energy Minimization Techniques

G.W. Watson, P. Tschaufeser, A. Wall, R.A. Jackson and S.C. Parker

1 INTRODUCTION

The focus of this chapter is the use of energy minimization techniques to model the crystal structures of inorganic solids. However, it is both desirable and possible to go beyond lattice energy minimization where no account is taken of temperature, to perform free energy minimization which allows us to simulate crystal structures over a wide range of temperatures and pressures. One approach which has long been known to be a powerful technique for modelling the thermal properties of solids is lattice dynamics. The technique is discussed in many reviews (e.g. Born and Huang, 1954; Cochran, 1977; Barron *et al.*, 1980). The approach is to calculate the vibrational properties from the energy surface and use statistical mechanics to obtain the thermodynamic properties, including free energy. Although the theory was well established, this approach was not applied to more complex inorganic solids such as silicate minerals because of the computational requirements of CPU time and memory. However, this problem has been reduced owing to recent developments in computer hardware. Indeed, it is not only possible to calculate the free energy of a solid but the free energy can also be calculated within an iterative cycle to allow the configuration to be adjusted until a free energy minimum is obtained.

The full range of applications of lattice dynamics is very considerable and beyond the scope of this chapter. This chapter therefore focuses on the recent modelling of inorganic crystal structures. Further discussions of recent applications are also to be found in the work of Harding (1986, 1989), Price *et al.* (1987), Cormack and Parker (1990), Allan *et al.* (1992, 1993), Vocadlo *et al.* (1995) and Watson and Parker (1995). We describe, first, the key steps and approximations in calculating and minimizing the energy of the crystal; next we illustrate the scope and limitations of the methodology by describing recent applications to several complex silicate minerals. However, the success of these simulations is limited by the accuracy with which the interatomic potentials

COMPUTER MODELLING IN INORGANIC CRYSTALLOGRAPHY
ISBN 0-12-164135-X

model the forces in the crystal over the pressure and temperature region in which we are interested. So before describing the simulation techniques we describe the potential models.

2 POTENTIAL MODELS

The atomistic approach to modelling the crystal structure and properties involves the definition of interatomic potential functions to simulate the forces acting between ions. As discussed in Chapter 1, interatomic pair potentials can be written as:

$$v_{ij} = \frac{q_i q_j}{r_{ij}} + \phi(r_{ij})$$

where q_i and q_j are the ionic charges, r_{ij} is the interatomic distance and $\phi(r_{ij})$ are the short-range interactions which are attributed to the repulsion between electron charge clouds, van der Waals attraction, bond bending and stretching.

The definition of the charges can be a problem. Formal charges equated to the oxidation state are often used, although it can be argued that such charges for partially covalent systems such as silicates are not realistic. However, problems of retaining charge neutrality are encountered when studying defects or phase transitions if non-formal charges are used, as reviewed by Catlow and Stoneham (1983).

The recent applications described in Section 3.6 use formal charges with short-range potential parameters derived in order to be consistent with these charges. It should be noted that these are 'effective charges' and together with the short-range term form an 'effective potential', although the physical significance of the individual terms may in some cases be uncertain.

The electrostatic part of the potential is slowly convergent when lattice summations are carried out, and this problem is overcome by the use of the Ewald method, details of which are given in contrasting but complementary approaches by Tosi (1964) and by Jackson and Catlow (1988). A detailed discussion follows in Chapter 4.

The short-range interaction combines a number of components including non-bonded interactions (electron repulsion and van der Waals attraction), electronic polarizability and, where relevant, covalent interactions, modelled by bond-bending and bond-stretching terms.

The non-bonded potential can take a number of forms but for ionic or partially ionic materials the most commonly used is as discussed in Chapter 1, the Buckingham form:

$$\phi_{ij} = A_{ij} \exp\left(\frac{-r_{ij}}{\rho_{ij}}\right) - \frac{C_{ij}}{r_{ij}^6}$$

with a repulsive exponential and an attractive term.

We recall from Chapter 1 that for ionic materials, ionic polarizability can be taken into account using the shell model of Dick and Overhauser (1958), which treats each ion as a core and shell, coupled by a harmonic spring. The ion charge is divided between the core and shell such that the sum of their charges is the total ion charge. The free ion polarizability, α, is related to the shell charge, Y, and spring constant, k, by:

$$\alpha = \frac{Y^2}{k}$$

For most predominantly ionic systems the above expressions are sufficient, but where there is a degree of covalency, additional terms are used, in particular the following.

(a) Bond-bending Bond-bending terms introduce an energy penalty for deviation of the bond angle from the equilibrium value, thus inferring directionality on the bonding. They are used in modelling covalent and semi-covalent systems such as silicates and aluminosilicates (e.g. zeolites where oxygen atoms are bonded to a central silicon or aluminium atom with tetrahedral coordination, Jackson and Catlow, 1988) or molecular ions such as carbonates (Parker *et al.*, 1993a) and sulphates (Parker *et al.*, 1993b). The following analytical expression is used:

$$U = \tfrac{1}{2}k_{ijk}(\theta - \theta_0)^2$$

where k_{ijk} is the bond-bending force constant between bonds ij and ik, and θ_0 is the equilibrium bond angle.

(b) Bond-stretching Bond-stretching terms are used in modelling bonded interactions in covalent systems, e.g. the hydroxide ion in ionic systems (e.g. Baram and Parker, 1995). A Morse potential can be used for such interactions:

$$\phi_{ij} = A_{ij}\{1 - \exp[-B_{ij}(r_{ij} - R_{ij})]\}^2 - A_{ij}$$

where A_{ij} is the bond dissociation energy, R_{ij} is the equilibrium bond distance, r_{ij} is the bond length and B_{ij} is a function of the slope of the potential energy well and can be obtained from spectroscopic data. A simple alternative is the widely used harmonic potential (which has been applied recently to molecular ions by Telfer *et al.*, 1995):

$$\phi_{ij} = \tfrac{1}{2}k_{ij}(r_{ij} - R_{ij})^2$$

where k_{ij} is the harmonic force constant.

(c) Torsional terms Four-body potentials (often call torsionals) are used to model systems which have a planar nature due to π bonding, e.g. in modelling calcite (Parker *et al.*, 1993a; Jackson and Price, 1992), or benzene adsorption in zeolites (Titiloye *et al.*, 1992).

Two planes are defined, between the ions, i, j and k and between j, k and l with the potential energy being dependent on n times the angle between these planes, thus giving rise to a minimum when the angle is 0, and multiples of $2\pi/n$ radians. The full potential function is given by:

$$U = K_{ijkl}[1 - \cos(n\theta)]$$

Having defined the form of the interatomic potential, the next step is to determine the variable parameters.

3 PARAMETERIZATION OF POTENTIALS

As noted in Chapter 1, there are two principal methods by which potential parameters may be obtained: (i) empirical fitting and (ii) direct calculation. Empirical fitting will be described in more detail here, as direct calculation by electronic structure methods will be described in Chapter 8. A comparison of the strengths and weaknesses of each method is given following the description of the methods.

3.1 Empirical fitting

Empirical fitting is essentially the reverse of the lattice simulation methods discussed in Section 4, in that structural and lattice properties are used to determine the potential, by requiring that the potential reproduces these properties.

The procedure is as follows: starting with an initial guess of the parameters, they are adjusted systematically, usually via a least squares fitting technique, until the differences between the calculated and experimental structure and properties are minimized. This process is summarized as follows:

1. Make an initial guess of the parameters in the specified expressions.
2. Calculate, at the observed structure:
   (i) The crystal properties.
   (ii) The lattice (bulk) and ion (internal) strains.
3. Minimize, by adjusting the potential parameters:
   (i) The difference between the observed and the calculated properties.
   (ii) The lattice and internal strains.
4. Go back to stage 2 and repeat stages 2 and 3 iteratively until optimal agreement is obtained.

Empirical fitting requires the availability of structural data and preferably other data such as elastic and dielectric constants.

3.2 Direct calculation

Potentials may be calculated directly using methods based on quantum mechanics. These range from electron gas methods (Mackrodt and Stewart, 1979) to *ab initio* calculations using either Hartree–Fock (Gale *et al.*, 1992) or local density methods. In these methods the interaction energy between a pair or periodic array of atoms can be calculated for a range of distances, and the resulting potential energy curve fitted to a suitable functional form. Pairs or clusters of atoms may be embedded in some representation of the surrounding crystal. These methods can, in principle, calculate potentials for any interaction, but may be dependent on the availability of suitable basic sets. More details will be found in Chapter 8.

3.3 Comparison of the two methods

Empirical potentials are only applicable with certainty over the range of interatomic distances used in the fitting procedure, which can lead to problems if the potential is used in a calculation that accesses distances outside this range. This can happen in defect calculations, molecular dynamics simulations or lattice dynamics calculations at high temperature and/or pressure. In addition experimental data is required and thus direct calculation is the only method available when there is no relevant experimental data. It may, of course, be possible to take potentials derived for one system and transfer them to another. This method has been successful with potentials derived for binary oxides (Lewis and Catlow, 1985; Bush *et al.*, 1994) being transferred to ternary systems (Lewis and Catlow, 1985; Price *et al.*, 1987; Cormack *et al.*, 1988; Purton and Catlow, 1990; Bush *et al.*, 1994).

Directly calculated potentials can be obtained for any interatomic distance or relative atomic configuration, but care must be taken that any environmental factors (e.g. Madelung fields) are taken into account correctly. These requirements may impose considerable difficulty in many systems. Where possible, empirical and non-empirical methods should be used in a concerted manner in deriving interatomic potentials.

4 LATTICE ENERGY MINIMIZATION

The first stage of any lattice simulation is to equilibrate the structure, i.e. bring it to a state of mechanical equilibrum. The simplest procedure is to equilibrate under conditions of constant volume, i.e. with invariant cell dimensions. Extensions to the procedure were introduced by Parker (1982, 1983) who introduced the use of constant pressure minimization in the computer code METAPOCS, in which lattice energy minimization was performed with respect to both coordinate position and bulk lattice strains. The following sections will describe some of the commonly used procedures.

4.1 Constant volume minimization

As noted, such calculations are undertaken without variation in cell dimensions. Most modern simulation codes employ second derivative methods, in particular the Newton–Raphson variable matrix method (Norgett and Fletcher, 1970) to minimize the lattice energy, $U(r)$, with respect to the coordinates, r. In this approach $U(r)$ is expanded to second order (Born and Huang 1954):

$$U(r') = U(r) + g^T \cdot \delta r + \tfrac{1}{2}\delta r^T \cdot \mathbf{W} \cdot \delta r$$

where δ is the displacement (or strain) of the ions, g is the first derivative and $\mathbf{W}$ the second derivative of energy with respect to position r.

By assuming the equilibrium condition, i.e. that the change in energy with strain is zero:

$$\frac{\partial U}{\partial \delta r} = 0 = g + \mathbf{W} \cdot \delta r$$

which gives:

$$\delta r = -\mathbf{W}^{-1} \cdot g$$

If the energy of the system was perfectly harmonic in $\mathbf{r}$, this would give rise to the minimum energy of the system in one step. However, the energy of the system is not harmonic, but the displacement normally gives rise to a lower energy configuration. The minimum energy is thus found by iteratively repeating this step. This procedure is normally found to be the most efficient. But if the initial configuration is far from equilibrium, simpler gradient (first derivative) methods may be more appropriate in the early stages of the minimization.

4.2 Constant pressure minimization

The basis of constant pressure minimization is that both lattice vectors and coordinates are adjusted to remove forces on both the atoms and the unit cell as a whole. This is commonly performed simultaneously, using the same approach for the lattice vectors as was used in constant volume minimization for the coordinates (i.e. treating the lattice vectors as additional variables).

The 'bulk strains', i.e. strains on the whole unit cell, are defined using the following relationship:

$$r' = (\mathbf{I} + \boldsymbol{\varepsilon}) \cdot r$$

where **I** is the identity matrix, $\boldsymbol{\varepsilon}$ is the strain on the lattice vectors and r, and r' are the transformed lattice vectors. This can be expressed in vector form using the Voigt notation:

$$\begin{bmatrix} x' \\ y' \\ z' \end{bmatrix} = \begin{bmatrix} 1+\varepsilon_1 & \frac{1}{2}\varepsilon_6 & \frac{1}{2}\varepsilon_5 \\ \frac{1}{2}\varepsilon_6 & 1+\varepsilon_2 & \frac{1}{2}\varepsilon_4 \\ \frac{1}{2}\varepsilon_5 & \frac{1}{2}\varepsilon_4 & 1+\varepsilon_3 \end{bmatrix} \begin{bmatrix} x \\ y \\ z \end{bmatrix}$$

where x, y and z are components of r.

The first derivative of lattice energy with respect to strain, the static pressure, is determined using the chain rule.

$$P_{\text{static}} = \frac{\partial U}{d\varepsilon_j} = \frac{\partial U}{\partial r} \cdot \frac{\partial r}{\partial r^2} \cdot \frac{\partial r^2}{\partial \varepsilon_j}$$

where the first term represents the first derivative, the second term equates to $1/2r$, and the last term is solved by squaring the equation defining the bulk strains and differentiating, giving

$$\frac{\partial r'^2}{\partial \varepsilon_j} = 2r^{\alpha} \cdot r^{\beta} + 2r^{\alpha} \cdot \boldsymbol{\varepsilon} \cdot r^{\beta}$$

Assuming the equilibrium condition, that there is no strain ($\varepsilon = 0$) then:

$$\frac{\partial r'^2}{\partial \varepsilon_j} = 2r^{\alpha} \cdot r^{\beta}$$

with

j	1	2	3	4	5	6
r^{α}	x	y	z	y	x	x
r^{β}	x	y	z	z	z	y

For a detailed discussion of the strain derivatives, see Catlow and Norgett (1976).

The next step is to calculate the constant of proportionality between the stress and the strain, the elastic compliance matrix. This is the inverse of the elastic constant matrix (the second derivative of energy with respect to strain), which is determined by again expanding the lattice energy to second order:

$$U(r') = U(r) + g \cdot \delta + \tfrac{1}{2}\delta^T \cdot \mathbf{W} \cdot \delta$$

where δ is the $3N + 6$ strain matrix containing components of internal and bulk strain:

$$\delta = \begin{bmatrix} \delta r \\ \varepsilon \end{bmatrix}$$

with ε equal to the six independent components of the symmetric strain matrix,

$\boldsymbol{\varepsilon}$, and the configuration r' defined in the normal way;

$$r' = (\mathbf{I} + \boldsymbol{\varepsilon}) \cdot r$$

The second derivative matrix, $\mathbf{W}$, now also includes mixed coordinate and strain derivatives and is of order $3N + 6$ by $3N + 6$.

$$\mathbf{W} = \begin{bmatrix} \dfrac{\partial^2 U}{\partial r^2} & \dfrac{\partial^2 U}{\partial r \partial \varepsilon} \\ \dfrac{\partial^2 U}{\partial \varepsilon \partial r} & \dfrac{\partial^2 U}{\partial \varepsilon^2} \end{bmatrix} = \begin{bmatrix} \mathbf{W}_{rr} & \mathbf{W}_{\varepsilon r} \\ \mathbf{W}_{r\varepsilon} & \mathbf{W}_{\varepsilon\varepsilon} \end{bmatrix}$$

where $\mathbf{W}_{rr}$ is the coordinate second derivative matrix ($3N$ by $3N$), $\mathbf{W}_{r\varepsilon}$ and $\mathbf{W}_{\varepsilon r}$ are the mixed coordinate and strain second derivative matrices (6 by $3N$ and $3N$ by 6), and $\mathbf{W}_{\varepsilon\varepsilon}$ is the strain second derivative matrix (6 by 6).

Applying the equilibrium condition that $g = \mathbf{0}$ (which assumes the crystal is at zero strain) and splitting δ into components of δr and ε gives

$$U(r') = U(r) + \tfrac{1}{2}\delta r \cdot \mathbf{W}_{rr} \cdot \delta r + \delta r \cdot \mathbf{W}_{r\varepsilon} \cdot \varepsilon + \tfrac{1}{2}\varepsilon \cdot \mathbf{W}_{\varepsilon\varepsilon} \cdot \varepsilon$$

Differentiating with respect to δr and applying the equilibrium condition $\partial U/\partial \delta r = 0$ gives

$$\frac{\partial U}{\partial \delta r} = 0 = \partial r \cdot \mathbf{W}_{rr} + \mathbf{W}_{r\varepsilon} \cdot \varepsilon$$

Thus

$$\partial r = -\mathbf{W}_{rr}^{-1} \cdot \mathbf{W}_{r\varepsilon} \cdot \varepsilon$$

Substituting this gives

$$U(r') = U(r) + \tfrac{1}{2}\varepsilon \cdot [\mathbf{W}_{\varepsilon\varepsilon} - (\mathbf{W}_{\varepsilon r} \cdot \mathbf{W}_{rr}^{-1} \cdot \mathbf{W}_{r\varepsilon})] \cdot \varepsilon$$

The elastic constants are defined here as the second derivative of lattice energy with respect to strain, normalized to cell volume. If the above equation is differentiated twice with respect to strain, the elastic constants are clearly given by

$$\mathbf{C} = \frac{1}{V}[\mathbf{W}_{\varepsilon\varepsilon} - (\mathbf{W}_{\varepsilon r} \cdot \mathbf{W}_{rr}^{-1} \cdot \mathbf{W}_{r\varepsilon})]$$

and the strains are given by the stress (static pressure) divided by the elastic constant matrix. Thus, assuming Hooke's law, the strains can be calculated:

$$\varepsilon = (P_{\text{static}} + P_{\text{applied}}) \cdot \mathbf{C}^{-1}$$

which can be used to evaluate new coordinates and lattice vectors. However, as the above expressions are approximate (except that relating to the calculation of the pressure) the process continues iteratively until all the strains are removed.

5 FREE ENERGY MINIMIZATION

The minimization methods so far considered have neglected temperature or can be considered to be effectively at 0 K without the zero point energy. One method of including temperature is through the use of lattice dynamics (the normal modes of vibration), a method pioneered by Born and Huang (1954). Free energy minimization, within the computer code PARAPOCS (Parker and Price, 1989) is performed with respect to the Gibbs free energy:

$$G = U_{\text{latt}} + PV + U_{\text{vib}} - TS_{\text{vib}}$$

The vibrational (or thermal) contribution ($U_{\text{vib}} - TS_{\text{vib}}$) is given by

$$U_{\text{vib}} - TS_{\text{vib}} = \frac{1}{w_t}\sum_{q=1}^{p} w_q \sum_{i=1}^{3N}\left(\frac{h\omega_{iq}}{2} - kT\ln(Q_{iq})\right)$$

where Q_{iq} is the vibrational partition function for vibration i with frequency ω at the qth wavevector in the Brillouin zone; $h\omega_{iq}/2$ is the vibrational zero point energy; N is the number of ions in the unit cell; p is the number of points in the Brillouin zone sampled; w_q is the weighting for point q and w_t is the total weighting.

$$Q_{in} = \frac{1}{1 - \exp\left(\frac{-h\omega_{iq}}{kT}\right)}$$

In principle this expression must be summed over the entire Brillouin zone, but this is impractical and thus a sampling method must be used. A number of methods have been suggested, but that of Filippini *et al.* (1976) compares favourably with other methods (Patel *et al.*, 1991). This method consists in choosing points within the Brillouin zone such that the density of points is greatest at the centre where variations in frequency are greatest, with all the points weighted (w_q) according to volume to give an even sampling scheme.

The following section describes the method used to calculate the frequencies (and eigenvectors) of all the normal modes of vibration required to calculate the free energy.

5.1 Calculation of the vibrational modes

The calculation of the frequencies for the shell model is performed using the method developed by Cochran (1977). In this approach the ions are held fixed at their equilibrium positions and therefore only the energy at the specific site is sampled.

If an atom in the crystal is displaced from its equilibrium position by an amount δr, it will experience a force, F:

$$F = \frac{-\partial U_{\text{latt}}}{\partial \delta r}$$

Thus Newton's equation of motion can be expressed as

$$\frac{\partial U_{\text{latt}}}{\partial \delta r} = m \frac{\partial^2 \delta r}{\partial t^2}$$

where m is the mass of the ion. For the shell model the mass on the shell is zero, and the assumption of the model is that the shell can instantaneously relax to its equilibrium position; thus for the shells this condition becomes

$$0 = \frac{\partial U_{\text{latt}}}{\partial w}$$

where w is the displacement of the shells and u will now represent the displacement of the cores such that

$$\delta r = \begin{bmatrix} u \\ w \end{bmatrix}$$

The differences in the treatment of the cores and shells, requires the splitting of the coordinate second derivative matrix into components of core and shells such that

$$\mathbf{W}_{rr} = \begin{bmatrix} \mathbf{W}_{uu} & \mathbf{W}_{uw} \\ \mathbf{W}_{wu} & \mathbf{W}_{ww} \end{bmatrix}$$

Before solving the equations of motion, the periodic nature of the solid must be taken into account by including the dependence of the atomic displacements on the wavevector q, and hence the displacement is modified as follows

$$\delta r = \delta r \exp\left[i(q \cdot r - \omega t)\right]$$

where r is the atom position and ω is the vibrational frequency. The second derivative matrix is also affected:

$$\mathbf{W}_{ij} = \mathbf{W}_{ij} \exp\left(iq \cdot r_{ij}\right)$$

The potential energy is now expanded to second order with respect to ion displacements (the harmonic approximation) and the equilibrium condition, $g = 0$, is applied.

$$U(r') = U(r) + \tfrac{1}{2}(u \cdot \mathbf{W}_{uu} \cdot u + 2w \cdot \mathbf{W}_{wu} \cdot u + w \cdot \mathbf{W}_{ww} \cdot w)$$

Differentiating with respect to core and shell displacements gives

$$\frac{\partial U_{\text{latt}}}{\partial u} = \mathbf{W}_{uu} \cdot u + \mathbf{W}_{uw} \cdot w$$

and

$$\frac{\partial U_{\text{latt}}}{\partial w} = \mathbf{W}_{uw} \cdot u + \mathbf{W}_{ww} \cdot w$$

with the second derivatives of the core (u) displacements with respect to t being:

$$\frac{\partial^2 u}{\partial t^2} = \omega^2 u$$

Thus the laws of motion are solved as

$$\omega^2 \mathbf{m} \cdot u = \mathbf{W}_{uu} \cdot u + \mathbf{W}_{uw} \cdot w$$

(where $\mathbf{m}$ is the diagonal matrix of core masses) for the cores and the assumption of the model is that the shells can instantaneously relax to their equilibrium positions,

$$0 = \mathbf{W}_{uw} \cdot u + \mathbf{W}_{ww} \cdot w$$

The shell displacements are removed from the equation relating to the cores by rearranging the equation relating to the shells to give the shell displacements:

$$w = - \mathbf{W}_{ww}^{-1} \cdot \mathbf{W}_{wu} \cdot u$$

Substituting this into the equations of motion for the cores gives

$$\omega^2 \mathbf{m} \cdot u = (\mathbf{W}_{uu} - \mathbf{W}_{uw} \cdot \mathbf{W}_{ww}^{-1} \cdot \mathbf{W}_{wu}) \cdot u$$

If we define the dynamical matrix as

$$\mathbf{D} = \mathbf{m}^{-1/2} \cdot (\mathbf{W}_{uu} - \mathbf{W}_{uw} \cdot \mathbf{W}_{ww}^{-1} \cdot \mathbf{W}_{wu}) \cdot \mathbf{m}^{-1/2}$$

and

$$n = \mathbf{m}^{1/2} \cdot u$$

the problem becomes

$$\omega^2 n = \mathbf{D} \cdot n$$

This is an eigenvector problem and hence the frequencies can be calculated by diagonalizing the dynamical matrix, although this procedure must be repeated for all of the wavevectors considered in the sampling method.

5.2 Minimization of the total pressure

If the thermodynamic properties are calculated within the harmonic approximation, in which the normal modes of vibration are assumed to be independent and harmonic, the cell has no thermal expansion. PARAPOCS (Parker and Price, 1989) extends this to the quasi-harmonic approximation. In this method the vibrations are assumed to be harmonic but their frequencies change with volume. This provides an approach for obtaining the extrinsic anharmonicity which leads to the ability to calculate thermal expansion.

Minimization to a specified pressure and temperature is performed by adjusting the crystal structure until the sum of the static, kinetic and applied pressure is zero. The calculation and minimization of the static pressure has already been discussed in Section 4.2. For free energy minimization, the same procedure is followed except for an extra term dependent on the thermal contribution.

The pressure is defined as the rate of change of energy with strain normalized to cell volume. For the thermal contribution this can be calculated by applying small strains to the cell and recalculating the thermal contribution to the free energy, giving rise to the thermal pressure in each strain direction:

$$P_{\mathrm{vib}} = \frac{1}{V} \frac{\partial(U_{\mathrm{vib}} - TS_{\mathrm{vib}})}{\partial \varepsilon}$$

which simplifies for the cubic case to a simple derivative with volume:

$$P_{\mathrm{vib}} = \frac{\partial(U_{\mathrm{vib}} - TS_{\mathrm{vib}})}{\partial V}$$

During the volume adjustments, constant volume minimizations are performed to ensure that the ions stay at their potential energy minima. This reduces the possibility of an atom moving to an unstable site and giving rise to imaginary frequencies (where the solution of ω^2 is negative).

Once the kinetic pressure has been calculated, the strains required to bring the cell to its minimum energy volume are given by Hooke's law:

$$\varepsilon = (P_{\mathrm{stat}} + P_{\mathrm{vib}} + P_{\mathrm{app}})\mathbf{C}^{-1}$$

where P_{stat} is the static pressure and P_{app} is the applied pressure and $\mathbf{C}$ is the elastic constant matrix.

As the calculation of the vibrational pressure is easily the most time consuming, within each interaction, Hooke's law is applied iteratively with only recalculation of the static pressure for a fixed number of iterations (usually five) or until the total pressure is zero. The resulting structure may not be fully minimized as the change in volume may have altered the vibrational pressure, and additional iterations will be required. A schematic of the iterative process of PARAPOCS is shown in Fig. 3.1.

As noted earlier, this approach assumes the quasi-harmonic approximation which includes important anharmonic effects associated with the variation of free energy with volume (extrinsic anharmonicity). It does, however, neglect intrinsic anharmonicity which becomes important at elevated temperatures. To investigate crystals at high temperatures Molecular Dynamics (MD) can be used in which intrinsic anharmonic effects are treated explicitly. This method is considered in detail in Chapter 4.

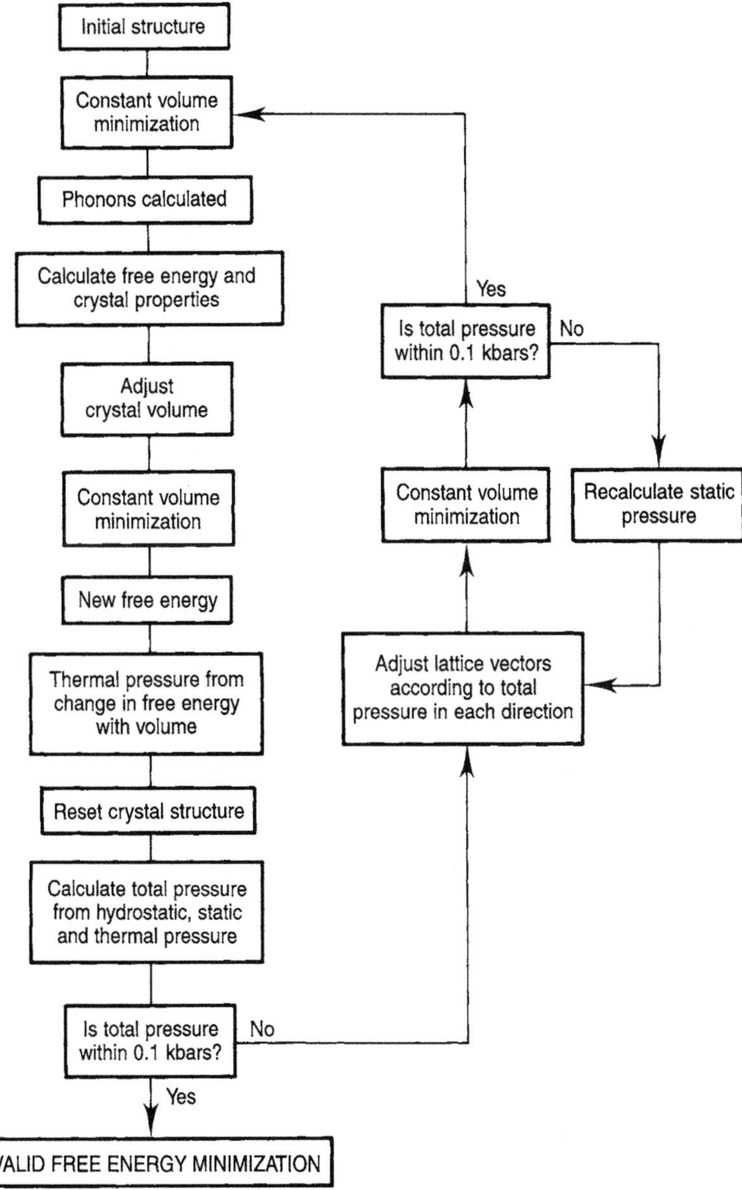

Figure 3.1 Schematic representation of the iterative process in PARAPOCS.

5.3 Calculation of thermodynamic properties

The thermodynamic properties of a crystal are obtained using a simple statistical mechanical treatment of the vibrational frequencies calculated from lattice dynamics. The thermal component of the free energy has already been defined (in Section 5) and can be written using the variable $x = h\omega_{iq}/k_B T$ as

$$U_{\text{vib}} - TS_{\text{vib}} = \frac{k_B T}{w_t} \sum_{q=1}^{p} w_q \sum_{i=1}^{3N} \frac{x}{2} - \ln\left[\frac{1}{1-\exp(-x)}\right]$$

where ω_{iq} is the vibrational frequency for vibration i at point q in the Brillouin zone, k_B is the Boltzmann constant, p is the number of points sampled in the Brillouin zone, w_q is the weighting for point q, w_t is the total weighting and N is the number of ions in the unit cell.

The components of the thermal contribution to the free energy, the vibrational energy and the vibrational entropy, can similarly be defined:

$$U_{\text{vib}} = \frac{k_B T}{w_t} \sum_{q=1}^{p} w_q \sum_{i=1}^{3N} \frac{x}{2} \left[\frac{x}{\exp(x)-1}\right]$$

$$S_{\text{vib}} = \frac{k_B}{w_t} \sum_{q=1}^{p} w_p \sum_{i=1}^{3N} \left[\frac{x}{\exp(x)-1}\right] - \ln\left[\frac{1}{1-\exp(-x)}\right]$$

Other thermodynamic properties that can be calculated directly from the frequencies include the constant volume heat capacity and the thermal Grüneisen parameter. The constant volume heat capacity is given by the derivative of vibrational energy with temperature:

$$C_v = \frac{\partial U_{\text{vib}}}{\partial T} = \frac{k_B}{w_t} \sum_{q=1}^{p} w_q \sum_{i=1}^{3N} \left[\frac{x^2 \exp(x)}{\exp(x)-1}\right]$$

The thermal Grüneisen parameter is determined from the changes in frequency with volume, by first calculating the mode Grüneisen parameters for each frequency:

$$\gamma_{iq} = \frac{-\partial \ln(\omega_{iq})}{\partial \ln(V)}$$

The thermal Grüneisen parameter is then evaluated by averaging the heat capacity weighted mode Grüneisen parameters:

$$\gamma_{\text{th}} = \frac{1}{w_t C_v} \sum_{q=1}^{p} w_q \sum_{i=1}^{3N} C_{iq}\gamma_{iq}$$

where C_{iq} is the heat capacity for each mode.

Further properties include the isothermal bulk modulus (K_T), the thermal expansion coefficient (β) and the constant pressure heat capacity (C_p). The isothermal bulk modulus is calculated first (or from corrected elastic constants as discussed in Section 6.2) and is usually defined as the Reuss bulk modulus

(Anderson, 1989). This can be subsequently used in calculating the thermal expansion coefficient and the constant temperature heat capacity,

$$K_T = K_S - \frac{C_V \gamma_{\text{th}}^2 T}{V}$$

$$\beta = \frac{\gamma_{\text{th}} C_V}{K_T V}$$

$$C_p = C_V + \beta^2 T K_T V$$

where K_s is the adiabatic bulk modulus defined as $1/\Sigma S_{nm}$ where S_{nm} is the nine elastic compliances for which $n \leqslant 3$ and $m \leqslant 3$.

6 APPLICATIONS

Three applications will be considered to illustrate the scope of the computational techniques described in this chapter. These are: (1) the calculation of the effect of temperature on framework aluminosilicate structures, showing the predictive capabilities of energy minimization; (2) simulation of elasticity at applied temperature and pressure, illustrating that one of the advantages of energy minimization techniques is that crystal properties other than structure can be calculated reliably; and (3) calculation of the phase relationship between different $MgSiO_3$ pyroxenes which demonstrates the range and scope of current techniques.

6.1 Thermal expansion of framework structures

An important consequence of using the quasi-harmonic approximation is that free energy minimization gives the temperature dependence of the cell dimensions. The reliability of this approach can be accessed by comparison of the simulated thermal expansivities with experimental data. For example, the calculated thermal expansivities for alumina, forsterite (Mg_2SiO_4) and quartz are 11, 20 and 21 per 10^6 K, respectively, whereas experimental values are 9 (Petuktov, 1973), 27 (White *et al.*, 1985) and 20 (Ackermann and Sorell, 1974) per 10^6 K. The best agreement is for alumina and quartz where the potentials were fitted to their crystal structure and elasticity data (not including thermal parameters). In contrast, the potential model for forsterite was a composite of parameters derived for magnesia and quartz. This small discrepancy probably arises from the use of the quartz potential, derived for a framework structure, in a material where the silica tetrahedra are isolated. It appears that, given good quality interatomic potentials, we may calculate accurate expansivities for silicate systems using this procedure.

This theme was further explored in calculations on a range of zeolites and aluminophosphate (ALPO) framework structured materials. As these materials are used as molecular sieves and catalysts, an understanding of the behaviour of their structure under reaction conditions is clearly important. As discussed in Chapters 1, 5 and 9, zeolites are frameworks formed from linked oxygen tetrahedra containing aluminium and silicon, while ALPOs comprise AlO_4 and PO_4 tetrahedra. The intriguing result of these simulations was that when a range of zeolites was considered many showed a contraction of heating rather than the more usual result of expansion (Tschaufeser, 1992; Tschaufeser and Parker, 1995). Most importantly, this result has since been verified by experiment (Couves *et al.*, 1993). Of the four siliceous zeolites considered, cancrinite, sodalite, zeolite L and zeolite X, the latter two gave rise to negative thermal expansion coefficients (Fig. 3.2). One of the features of sodalite and cancrinite is that they are much more dense than the others, implying that there may be a relationship between expansivity and density. This view is supported by examination of the effect of Al substitution. Most zeolites have a significant aluminium content and are charge compensated by the presence of interframework cations. Simulations on zeolite L and X containing aluminium and charge compensation cations in observed concentrations show that the magnitude of the cell contraction is much reduced in comparison with the purely siliceous zeolites (Fig. 3.2).

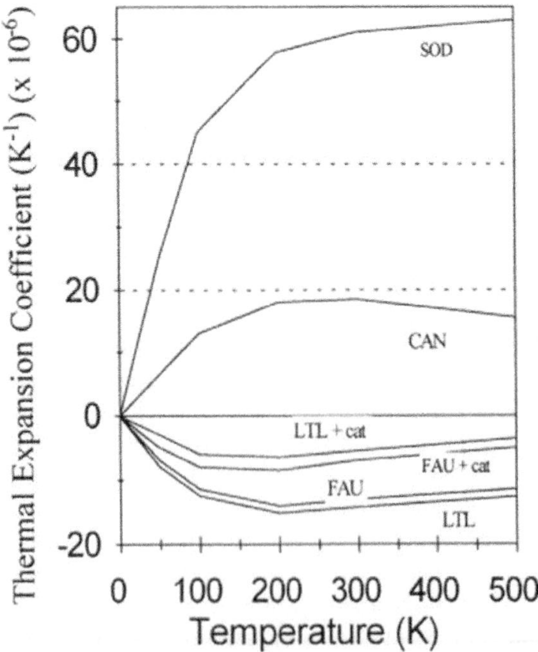

Figure 3.2 Thermal expansion coefficient of several zeolites as a function of temperature.

Table 3.1 Unit cell parameters for siliceous and non-siliceous zeolite L at various temperatures.

	Zeolite L		K-Zeolite L	
Temperature	a (Å)	c (Å)	a (Å)	c (Å)
50	18.051	7.547	18.199	7.607
100	18.048	7.545	18.200	7.607
200	18.041	7.540	18.194	7.605
300	18.034	7.534	18.190	7.604
400	18.026	7.529	18.187	7.602
500	18.020	7.524	18.184	7.601

A further feature of the nature of the contraction is exemplified by zeolite L which contracts anisotropically. Table 3.1 shows the cell parameters as a function of pressure for siliceous and non-siliceous zeolite L (the latter contains aluminium and charge compensation cations), showing the a axes to be 70% and 40% more compressible than the c axes. One might expect a similar result for the ALPOs. Indeed, some of the structures, e.g. VPI5, have a much larger channel system and hence the contraction might be even more pronounced. Those ALPOs that are isostructural with zeolite structures, e.g. ALPO 17, show similar thermal contraction, while those including VPI5 which have no aluminosilicate analogue show an expansion. This effect may again be associated with the effects of rigidity; although VPI5 has a large channel the bonding perpendicular to the channel is very dense and hence provides the necessary rigidity. Barron and co-workers (Barron *et al.*, 1980, 1982; Barron and Pasternak, 1987; Barron and Rogers, 1989) have studied thermal contraction in a number of materials and suggest that the contraction is related to a decrease in the vibrational frequency with a reduction in volume of an oxygen vibrating at right angles to two tetrahedral cations. This effect leads to a negative mode Grüneisen parameter, described in Section 5.3, which favours a unit cell contraction. However, we find that virtually all low frequency modes have negative Grüneisen parameters including those vibrations with oxygen atoms vibrating perpendicular to the two adjacent tetrahedral cations, implying complex behaviour that needs to be investigated further.

In addition to understanding crystal structure behaviour, energy minimization techniques afford a route for modelling a wide range of crystal properties including elasticity and dielectric constants. Furthermore, when attempting to select a potential for modelling crystallographic data as a function of temperature and pressure, the model that best reproduces the elastic constants will, in general, give the most reliable results. Therefore, calculation of elastic constants should be a routine undertaking when developing or using energy minimization techniques.

6.2 Calculation of elastic constants at applied pressure and temperature

The elastic constants have already been defined in Section 4.2 as the second derivative of energy at zero basis strain with respect to bulk strain normalized to the cell volume. However, when minimizing to a specified pressure or to a specific temperature using lattice dynamics, pressure and temperature corrections must be considered (Barron and Klein, 1965; Garber and Granato, 1975; Wall *et al.*, 1993).

The elastic constants are defined as the second derivative of the Gibbs free energy with respect to strain (Nye, 1985), i.e.

$$\mathbf{C}_{ij} = \frac{1}{V}\left(\frac{\partial^2 G}{\partial\varepsilon_i \partial\varepsilon_j}\right)_X$$

which can be calculated at constant temperature (isothermal, $X = T$) or constant entropy (adiabatic, $X = S$).

The traditionally used elastic constants are obtained from the second derivative of lattice energy with respect to strain.

$$\mathbf{C}_{ij} = \frac{1}{V}\left(\frac{\partial^2 U_{\text{latt}}}{\partial\varepsilon_i \partial\varepsilon_j}\right)$$

This definition neglects two terms, i.e. the contribution from the PV energy and the contribution from vibrational energy (including the zero point), giving rise to pressure and temperature corrections.

The pressure correction can be obtained from an equation derived by Barron and Klein (1965).

$$P_{\text{corr}} = \tfrac{1}{2}P(2\delta_{\alpha\beta}\delta_{\sigma\tau} - \delta_{\alpha\tau}\delta_{\beta\sigma} - \delta_{\alpha\sigma}\delta_{\beta\tau})$$

where P is the applied pressure and δ the Kronecker delta. This gives rise to pressure corrections for 6 of the 21 independent elastic constants, shown in Table 3.2.

These corrections to the elastic constants were calculated using the above equation and compared with those determined numerically. The numerical approach was to apply small stresses (s_j Voigt notation) of $+0.2$ GPa and -0.2 GPa with the elastic compliances related to the resulting strains, ε_1 to ε_6, by

$$\varepsilon_i = \mathbf{S}_{ij}\sigma_j$$

The elastic compliance tensor was then inverted to evaluate the elastic constants.

Calculations of the elastic constants were performed for fluorite (CaF_2) at 0 and 10 GPa and forsterite (Mg_2SiO_4) at 0 and 5 GPa. The calculations were made at 0 K (neglecting the zero point energy) thus avoiding the problem of

Table 3.2 Pressure corrections required for the 21 independent elastic constants.

Elastic constants	Pressure corrections
$C_{11}, C_{22}, C_{33}, C_{14}, C_{15}, C_{16}, C_{24}, C_{25}, C_{26}, C_{34}, C_{35}, C_{36}, C_{45}, C_{46}, C_{56}$	None
C_{12}, C_{13}, C_{14}	$+P$
C_{44}, C_{55}, C_{66}	$-\frac{1}{2}P$

introducing the additional temperature corrections. Table 3.3 shows the results, with U representing the uncorrected elastic constants calculated from the derivative of lattice energy, C the corrected elastic constants and N the numerical elastic constants. These clearly show that the pressure corrections of Table 3.2 are needed to reproduce the elastic constants of a crystal at non-zero pressure.

The elastic constants described above neglect the effect of temperature. The temperature component depends on whether adiabatic elastic constants (at constant entropy) or isothermal elastic constants (at constant temperature) are required.

Table 3.3 Comparison of calculated elastic constants with and without the pressure correction with numerical elastic constants (in GPa).

Crystal		P_s	C_{11}	C_{22}	C_{33}	C_{44}	C_{55}	C_{66}	C_{12}	C_{13}	C_{23}
CaF_2	U	0 GPa	237.79	–	–	24.91	–	–	38.50	–	–
	N		237.79	–	–	24.91	–	–	38.50	–	–
CaF_2	U		284.76	–	–	35.98	–	–	60.30	–	–
	C	10 GPa	284.76	–	–	30.98	–	–	70.30	–	–
	N		284.76	–	–	30.99	–	–	70.30	–	–
Mg_2SiO_4	U	0 GPa	206.66	281.21	358.78	74.56	84.35	44.34	87.73	93.92	96.23
	N		206.66	281.20	358.99	74.59	84.37	44.30	87.71	93.95	96.27
Mg_2SiO_4	U		228.08	307.41	389.96	81.25	95.51	58.86	100.51	111.81	110.05
	C	5 GPa	228.08	307.41	389.96	78.75	93.01	56.36	105.41	116.81	115.05
	N		228.05	307.34	390.15	78.73	93.02	56.31	105.44	116.79	115.15

U is uncorrected, C is corrected and N is numerical.

Table 3.4 Comparison of uncorrected and pressure and temperature corrected isothermal elastic constants of Mg_2SiO_4 at 5 GPa and 1000 K.

	$C_{\text{uncorrected}}$ (GPa)	$P_{\text{correction}}$ (GPa)	T_{Corr} (GPa)	Isothermal C (GPa)	Numerical C (GPa)
C_{11}	213.49	0	−14.65	198.84	198.85
C_{22}	288.77	0	−17.11	271.66	271.66
C_{33}	369.67	0	−21.83	347.78	347.74
C_{12}	91.29	5	−5.65	90.64	90.55
C_{13}	99.02	5	−5.87	98.15	98.10
C_{23}	99.88	5	−6.37	98.51	98.32

The full definition of the adiabatic elastic constants is given by

$$\mathbf{C}^{\mathrm{s}}_{ij} = \frac{1}{V}\left(\frac{\partial^2 U}{\partial\varepsilon_i \partial\varepsilon_j}\right) + (P_{\text{corr}})$$

$$\mathbf{C}^{\mathrm{T}}_{ij} = \frac{1}{V}\left(\frac{\partial^2 U}{\partial\varepsilon_i \partial\varepsilon_j}\right) + \frac{1}{V}\left(\frac{\partial^2(-TS_{\text{vib}})}{\partial\varepsilon_i \partial\varepsilon_j}\right) + (P_{\text{corr}})$$

where U is now the internal energy and $\mathbf{C}^{\mathrm{s}}$ and $\mathbf{C}^{\mathrm{T}}$ are the adiabatic and isothermal elastic constants respectively. The numerical approach used to check the results uses full free energy minimization to constant temperature and is therefore only applicable to calculating the isothermal elastic constants and thus only calculations of the isothermal elastic constants have been performed.

Table 3.4 shows a comparison of the uncorrected elastic constants and the pressure and temperature corrections for the isothermal elastic constants of Mg_2SiO_4 (forsterite) at 1000 K and 5 GPa with the elastic constants calculated numerically using the application of stresses. These show excellent agreement and indicate a problem in the generally used potential fitting procedures. Many potential models are fitted empirically to experimental data, often including elastic constants. Those experimental data will include contributions from the thermal component of the free energy while the calculated value will only include the lattice energy component. It is clear that future empirical fitting should include fitting to elastic constants appropriate to the temperature of the experimental data.

Once a reliable potential has been obtained which can model the elastic constants, lattice dynamical simulations can be used to predict the phase stability as a function of pressure. The next section gives an example of the calculation of phase stability for $MgSiO_3$ pyroxenes.

6.3 Phase stabilities of $MgSiO_3$ pyroxenes

The $MgSiO_3$ pyroxene structured polymorphs are not well understood even though they are important rock-forming minerals and a significant component of the Earth's upper mantle. There are at least four pyroxene polymorphs: enstatite, protoenstatite, low clinoenstatite and high clinoenstatite. Free energy minimization has been used to model the crystal structure and to investigate the stability fields of these pyroxene phases at temperatures between 0 and 1500 K and pressures between 0 and 15 GPa.

The pyroxenes are composed of single chains of corner-sharing SiO_4 tetrahedra, linked by cations in parallel edge-sharing polyhedral chains (e.g. Putnis, 1992). The polymorphs differ in the form of their SiO_4 chains and in the stacking arrangement of layers of chains parallel to (100) along the *a*-axis. Layers of chains along the *a* axis are related to the next layer by a *b* glide which can be in either a+ or a− direction. Enstatite is orthorhombic with a space group of *Pbca*. It has two types of SiO_4 chain which have the same sense of rotation, but differ in the amount by which they are twisted. The stacking sequence of chain layers along the *a* axis can be represented as $--++$. Protoenstatite is also orthorhombic with a space group of *Pbcn*. The SiO_4 chains are almost straight and the stacking sequence of layers of chains along *a* can be described as $+-+-$. High clinoenstatite and low clinoenstatite are both monoclinic and the stacking sequence of chains along their *a* axes is $----$. High clinoenstatite has a space group of *C2/c* with the SiO_4 chains all symmetry related; they thus have the same sense and the same degree of rotation. In low clinoenstatite, space group $P2_1/c$, there are two types of SiO_4 chain in which the tetrahedra are rotated in the opposite sense.

Experimentally, protoenstatite is stable at low pressure and high temperature (above 1273 K). High clinoenstatite is stable at high pressures (above 7 GPa), but is not quenchable and transforms to the low clinopyroxene structure as the pressure is reduced. The relative stability of enstatite and low clinoenstatite is uncertain. However, low clinopyroxene is known to be stabilized by shear stress and occurs naturally in highly stressed environments, including meteorites and ophiolites. It is possible that low clinopyroxene does not have a stability field in the hydrostatic pressure regime. An additional high temperature phase with the space group *Bmcm* has been proposed by Matsui and Price (1992) to occur at temperatures preceding melting. Experimental investigations to establish the relative stability of the pyroxene polymorphs are hampered by the sluggishness of some of the reactions, the difficulty in quenching some of the high pressure and high temperature phases, the occurrence of metastable structures over a wide pressure and temperature range and the sensitivity of the phase diagram to shear stress.

We used the interatomic potentials shown in Table 3.5 that were used initially to model Mg_2SiO_4 olivine (Price *et al.*, 1987). This potential has been transferred successfully by Wall and Price (1988) and Price *et al.* (1987) to model the higher pressure spinel and ilmenite polymorphs.

The pyroxene structures simulated with this potential model are shown in Table 3.6. We have compared enstatite (*Pbca*) and low clinoenstatite ($P2_1/c$) with observed structures at 300 K and 0 GPa. The protoenstatite structure is

Table 3.5 Interatomic potential used for modelling $MgSiO_3$.

Species	Charge	ij	A_{ij} (eV)	ρ_{ij} (Å)	C_{ij} (eV $Å^6$)
Mg	+2	Mg–O_{shel}	1428.500	0.29453	0.0
Si	+4	Si–O_{shel}	1283.735	0.32052	0.0
O_{shel}	−2.848	O_{shel}–O_{shel}	22764.0	0.1490	27.88
O_{core}	+0.848	O_{core}–O_{shel}	Core-shell spring constant k = 74.92 eV $Å^2$		

O_{shel}–Si–O_{shel} three-body spring constant $k = 2.09724$ eV rad.

Table 3.6 Comparison of simulated and experimental pyroxene structures.

Phase	Low clino-pyroxene		High clino-pyroxene		Enstatite		Protoenstatite	
Space group	$P2_1/c$		$C2/c$		$Pbca$		$Pbcn$	
T,P	300 K, 0 GPa		300 K, 7.9 GPa		300 K, 0 GPa		1300 K, 0 GPa	
	Expt[a]	Calc.	Expt[b]	Calc.	Expt[a]	Calc.	Expt[b]	Calc.
G (eV)		−1364.77		−1345.00		−1364.82		−1378.26
V (cm^3 mol^{-1})	31.22	31.27	28.72	28.86	31.33	31.29	33.32	32.07
a (Å)	9.61	9.54	9.2	9.34	18.23	18.06	9.31	9.15
b (Å)	8.81	8.65	8.62	8.32	8.82	8.67	8.89	8.67
c (Å)	5.17	5.32	4.91	5.11	5.18	5.31	5.37	5.35
$\beta°$	108.5	108.94	101.5	104.99	90	90	90	90
$\phi°$	137/157	174[c]	133	158	145/167	160/170	168	170
K (GPa)	–	115	–	123	108	87	114	102
α (10^{-5} K^{-1})	2.5	1.4	–	1.7	2.5	1.5	4	2

[a] Matsui and Price (1992) and reference therein.
[b] Angel *et al.* (1992).
[c] The space group of this structure has the higher symmetry of $C2/c$.

compared with the observed structure at 1300 K and 0 GPa and the high clinoenstatite structure with the observed structure at 7.9 GPa and 300 K. The experimental data is from Matsui and Price (1992) and the reference therein and Angel *et al.* (1992). The calculated volumes of enstatite, low clinoenstatite and high clinoenstatite are all within 0.5% of the experimentally measured volumes. However, the protoenstatite volume is underestimated by 3.7%. The a, b and c unit cell lengths are within 4% for all the structures. In low clinoenstatite the angle b is well reproduced, but for high clinoenstatite it is predicted to be 3.5° too large at 300 K and 7.9 GPa. However, the structure minimized at 300 K and 0 GPa from a low clinoenstatite starting structure has the higher symmetry of $C2/c$. The bulk moduli of enstatite and protoenstatite have been measured and these are also compared to our calculated values. The Hill average bulk modulus of enstatite is calculated to be 20% too low and the Hill average bulk modulus of protoenstatite is calculated to be 10% too low. The volume coefficients of thermal expansion for each of the polymorphs have also been calculated. Compared to the measured values, our calculated thermal expansion is also low, e.g. for enstatite, the most stable polymorph under ambient conditions, it is calculated to be 15 per 10^6 K compared to the experimental value of 25 per 10^6 K. As a result of these discrepancies between our predicted properties and the measured properties, we might only expect to predict qualitatively the phase relationships in the $MgSiO_3$ pyroxene system.

We have used two strategies to locate the phase boundaries. In searching for enstatite/clinoenstatite, the enstatite/protoenstatite and the protoenstatite/clinoenstatite transitions, the individual phases remain metastable outside their stability fields. We can, therefore, calculate their Gibbs free energies and search for the pressures and temperatures at which the free energies of the two phases are equal. However, the clinoenstatite polymorphs are able to transform during minimization to whichever structure is most stable and hence the same, stable structure resulted irrespective of the symmetry (*P2₁/c* or *C2/c*) of the starting coordinates.

The simulation results are summarized in the phase diagram in Fig. 3.3. Although the differences in free energy between enstatite and clinoenstatite are very small – less than 5 kJ/mol – enstatite is predicted to be the stable phase at low pressure and temperature. The low clinoenstatite structure was not predicted to occur under hydrostatic pressure conditions and hence comparison was made with the resulting high clinoenstatite structure of symmetry *C2/c*. Experimental work by, for example, Boyd and England (1965) and Sclar *et al.* (1964), also failed to find a low temperature/low pressure stability field for low

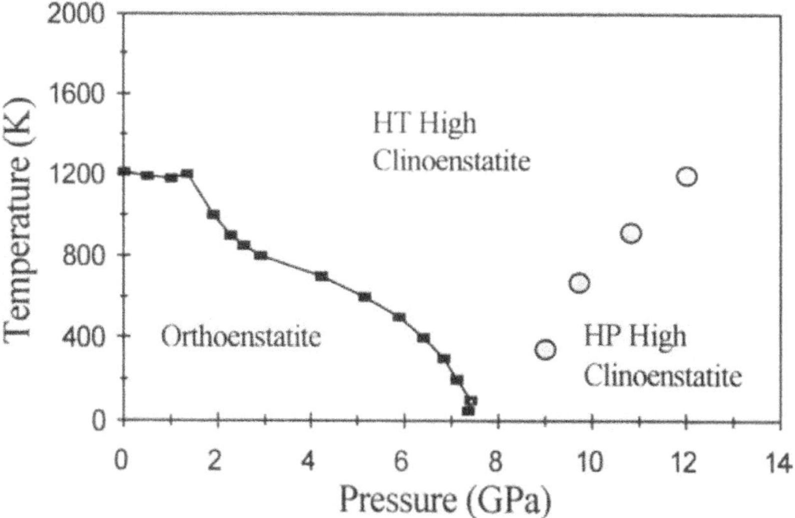

Figure 3.3 Calculated phase relationships of $MgSiO_3$-pyroxene. The solid line represents a martensitic transformation and the open circles a displasive phase transition.

clinoenstatite. The common occurrence of orthoenstatite rather than low clinoenstatite in most terrestrial rocks further supports this conclusion. However, this result contradicts the work of Boyd and England (1965), Sclar *et al.* (1964) and Grover (1972). These experiments were apparently under hydrostatic conditions and a boundary between low clinoenstatite and orthoenstatite was reported. This observation suggests that the simulation methods, or rather the reliability of the potential models, is unable to distinguish energetically between these two phases.

However, with both increasing pressure and temperature, the high clinoenstatite structure is predicted to become the most stable $MgSiO_3$ phase. Again this approach cannot precisely identify the exact pressure and temperature at which the high clinoenstatite becomes the most stable phase, but it is evident that the effect of increasing *P* and *T* stabilizes this structure. Following high pressure experimental work by Pacalo and Gasparik (1990) and Kanzaki (1991), Angel *et al.* (1992) confirmed the occurrence of the high clinoenstatite phase at pressures greater than 7.9 GPa at 300 K using *in situ* X-ray diffraction. The high pressure phase reported by Angel *et al.* (1992) may correspond to our

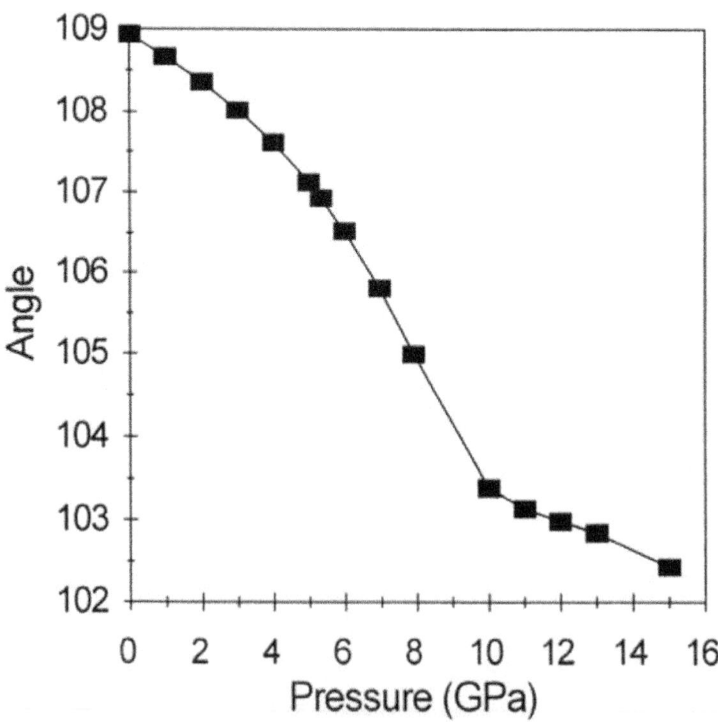

Figure 3.4 Variation in monoclinic angle β of high clinoenstatite as a function of pressure.

predicted high pressure form of high clinoenstatite. In addition, we also predict that enstatite will transform to high clinoenstatite with increasing temperature. Pacalo and Gasparik (1990) speculated on the occurrence of a high temperature polymorph with *C2/c* symmetry and concluded that it probably differs in structure from the high pressure *C2/c* phase. Our simulations do show a difference in the structure of the high pressure and high temperature high clinoenstatite phases. Even though they retain the space group of *C2/c*, the monoclinic angle β changes as shown in Fig. 3.4. This change corresponds to a change in the angle of rotation of the SiO_4 tetrahedra in the pyroxene chains. As the applied pressure is increased, the SiO_4 tetrahedra rotate, decreasing the chain length. However, at the transition there is an abrupt change in rotation angle and then the rotation does not increase further with pressure.

We predict a much smaller stability field for enstatite and a much larger stability field for the high temperature high clinoenstatite phase than reported by experimental studies. There is, however, no *in situ* observation of enstatite at high temperature and moderate pressures. Also, because the gradient of the phase transition between the high temperature and high pressure phases of high clinoenstatite is similar to that reported for the enstatite/high clinoenstatite transition by Pacalo and Gasparik (1990) and Kanzaki (1991), we speculate that the high clinopyroxene phase may transform to enstatite on cooling and decompression and hence not be observed.

In summary, we calculate that the low clinoenstatite is not stable under hydrostatic conditions and enstatite has a comparatively small stability field. The energy differences are so small that they are within the reliability of the simulations and thus the precise positions of the phase boundaries are not well located. The primary reason for this problem is the reliability of the potential models. Hence, calculating phase relationships represents the most difficult challenge for free energy minimization techniques. However, the simulations do provide valuable insights into the mechanisms of phase transitions and the effect of pressure and/or temperature on the crystal structures and the relative phase stabilities.

7 CONCLUSIONS

Free energy minimization is a powerful tool for modelling a wide range of crystal properties including structure and elasticity. When applying this approach to larger and more complex systems special attention always needs to be paid to the reliability of the potential model. However, developments in *ab initio* techniques make the prospect of more reliable potentials feasible. The other area that requires attention is the importance of intrinsic anharmonicity. One proven approach of including part of this effect is the use of a self-consistent phonon method as has been outlined for the shell model by Ball (1986). This involves the calculation of the Green's function from the dynamical matrix, the calculation of the dependence of the dynamical matrix on volume and atomic displacements and averaging over all configurations after weighting by the Green's function. This newly averaged dynamical matrix is then used to recalculate a range of properties including the free energy and a new Green's function. The process iterates until a self-consistent Green's function and a minimum free energy is obtained, and hence provides a route for incorporating part of the intrinsic anharmonicity which is absent from the quasi-harmonic simulations. By these and other methods the range of problems that can be addressed using lattice dynamics techniques will unquestionably increase.

REFERENCES

Ackermann, R.J. and Sorell, C.A. (1974) *Appl. Crystallogr.*, **7**, 461.

Allan, N.L., Braithwaite, M., Cooper, D.L., Mackrodt, W.C. and Petch, B. (1992) *Mol. Simul.*, **9**, 161.

Allan, N.L., Braithwaite, M., Cooper, D.L., Petch, B. and Mackrodt, W.C. (1993) *J. Chem. Soc. Faraday Trans.*, **89**, 4369.

Anderson, D.L. (1989) *Theory of the Earth*, Blackwell Scientific Publications, Oxford, UK.

Angel, R.J., Chopelas, A. and Ross, N.L. (1992) *Nature*, **358**, 322.

Ball, R.D. (1986) *J. Phys. C Solid State Physics*, **19**, 1293.

Baram, P.S. and Parker, S.C. (1995) *Phil. Mag.*, *B* **73**, 49.

Barron, T.H.K. and Klein M.L. (1965) *Proc. Phys. Soc.*, **85**, 523.

Barron, T.H.K. and Pasternak A.J. (1987) *J. Phys. C. Solid State Phys.*, **20**, 215.

Barron, T.H.K. and Rogers K.J. (1989) *Mol. Simul.*, **4**, 27.

Barron, T.H.K., Collins, J.G. and White G.K. (1980) *Adv. Phys.*, **29**, 609.

Barron, T.H.K., Collins, J.G., Smith T.W. and White, G.K. (1982) *J. Phys. C. Solid State Phys.*, **15**, 4311.

Born, M. and Huang K. (1954) *Dynamical Theory of Crystal Lattices*, Oxford University Press, Oxford.

Boyd, F.R. and England J.L. (1965) *Carnegie Inst. Wash. Year Book*, **64**, 117.

Bush, T.S., Gale, J.D., Catlow, C.R.A. and Battle, P.D. (1994) *J. Mater. Sci.*, **4**, 831.

Catlow, C.R.A. and Norgett, M.J. (1976) U.K.A.E.A. Report AERE-M2936, United Kingdom Atomic Energy Authority, Harwell, UK.

Catlow, C.R.A. and Stoneham, A.M. (1983) *J. Phys.*, *C* **6** 1325.

Cochran, W (1977) *Crit. Rev. Solid State Sci.*, **2**, 1.

Cormack, A.N. and Parker, S.C. (1990) *J. Am. Ceram. Soc.*, **73**, 3220.

Cormack, A.N., Lewis, G.V., Parker, S.C. and Catlow, C.R.A. (1988) *J. Phys. Chem. Solids*, **49**, 53.

Couves, J.W., Jones, R.H., Parker, S.C., Tschaufeser, P. and Catlow, C.R.A. (1993) *J. Phys. Condens. Matter*, **5**, 329.

Dick, B.G. and Overhauser, A.W. (1958) *Phys. Rev.*, **112**, 90.

Filippini, G., Gramaccioli, C.M., Simonetta, M. and Suffritti, G.B. (1976) *Acta Crystallogr.*, **A31**, 259.

Gale, J.D., Catlow, C.R.A. and Macrodt, W.C. (1992) *Modelling Simul. Mater. Sci. Engng.*, **1** 73.

Garber, J.A. and Granato, A.V. (1975) *Phys. Rev.*, **B11**, 3990.

Grover, J. (1972) *EOS Trans.* (abstract) AGU **53**, 539.

Harding, J.H. (1986) *J. Phys. C:Solid State Phys.*, **19**, L731.

Harding, J.H. (1989) *J. Chem. Soc. Faraday Trans. II*, **85**, 351.
Jackson, R.A. and Catlow, C.R.A. (1988) *Mol. Simul.*, **1**, 207.
Jackson, R.A. and Price, G.D. (1992) *Mol. Simul.*, **9**, 175.
Kanzaki, M. (1991) *Phys. Chem. Minerals*, **17**, 726.
Lewis, G.V. and Catlow, C.R.A. (1985) *J. Phys., C* **18**, 1149.
Mackrodt, W.C. and Stewart, R.F. (1979) *J. Phys., C* **12**, 5015.
Matsui, M, and Price, G.D. (1992) *Phys. and Chem. Minerals*, **18**, 365.
Norge, M.J. and Fletcher, R. (1970) *J. Phys., C* **3**, 163.
Nye, J.F. (1985) *Physical Properties of Crystals*, Oxford Scientific Publications, Oxford.
Pacalo, R.E.G. and Gasparik, T. (1990) *J. Geophys. Res.*, **95**, 15853.
Parker, S.C. (1982) PhD Thesis, University College London.
Parker, S.C. (1983) *Solid State Ionics*, **8**, 179.
Parker, S.C. and Price, G.D. (1989) *Adv. Solid State Chem.*, **1**, 295.
Parker, S.C., Titiloye, J.O. and Watson, G.W. (1993a) *Phil. Trans. R. Soc. Lond.*, **A344**, 37.
Parker, S.C., Kelsey, E.T., Oliver, P.M. and Titiloye, J.O. (1993b) *Faraday Discussions*, **95**, 75.
Patel, A., Price, G.D. and Mendelsohm, M.J. (1991) *Phys. Chem. Minerals*, **17**, 690.
Petuktov, V.A. (1973) *Teplofiz. V. Temp.*, **11**, 1083.
Price, G.D., Parker, S.C. and Leslie, M. (1987) *Phys. Chem. Minerals*, **15**, 181.
Purton, J. and Catlow, C.R.A. (1990) *Am. Mineralogist*, **75**, 1268.
Putnis, A. (1992) *Introduction to Mineral Sciences*, Cambridge University Press, Cambridge.
Sclar, C.B., Carrison, L.C. and Schwartz, C.M. (1964) *Trans. Am. Geophys. Union*, **45**, 121.
Telfer, G.B., Wilde, P.J., Jackson, R.A., Meenan, P. and Roberts, K.J. (1995) *Phil. Mag.*, in press.
Titiloye, J.O., Tschaufeser, P. and Parker, S.C. (1992) In *Spectroscopic and Computational Studies of Supramolecular Systems* (ed. J.E.D. Davies), Kluwer Academic Publishing, Dordrecht.
Tosi, M.P. (1964) *Solid State Phys.*, **16**, 1.
Tschaufeser, P. (1992) PhD Thesis, University of Bath.
Tschaufeser, P. and Parker, S.C. (1995) *J. Phys. Chem.*, in press.
Vocadlo, L., Wall, A., Parker, S.C. and Price, G.D. (1995) *Phys. Earth and Planetary Interiors*, **88**, 193.
Wall, A. and Price, G.D. (1988) *Am. Mineralogist*, **73**, 224.
Wall, A., Parker, S.C. and Watson, G.W. (1983) *Phys. Chem. Minerals*, **20**, 69.
Watson, G.W. and Parker, S.C. (1995) *Phil. Mag. Lett.*, **71**, 59.
White, G.K., Roberts, R.B. and Collins, J.G. (1985) *High Temp. High Pressure*, **17**, 61.

4

Molecular Dynamics Methods

P.W.M. Jacobs and Z.A. Rycerz

1 INTRODUCTION

Molecular Dynamics (MD) is a technique for the computer simulation of the properties of a condensed-matter system using the equations of classical (or quantum) mechanics. Since it is the aim of statistical mechanics to interpret the properties of macroscopic systems in terms of the properties of its constituents at the atomic level, molecular dynamics belongs to a class of 'computer experiments' that might be called 'numerical statistical mechanics'. Conceptually the simplest ensemble in statistical mechanics is the microcanonical ensemble with constant number of particles N, volume V and energy E. It represents a thermodynamically isolated system. Connection to a heat bath at constant temperature T replaces the (N, V, E) microcanonical ensemble by the (N, V, T) canonical ensemble; connection to a constant pressure reservoir via a piston gives the (N, P, H) constant enthalpy ensemble, or via a diathermic piston to a reservoir at constant T, P gives the (N, P, T) ensemble. All these ensembles may be realized by the MD method but by far the commonest used, and the simplest to set up, is the microcanonical ensemble. A specified number N of particles (atoms, molecules or ions) are assigned coordinates (corresponding to a lattice in simulations of crystalline systems), and the 3-D system is made pseudo-infinite by the application of periodic boundary conditions. Commonly N may be a few hundred or a few thousand or even, rarely, 10^5–10^6. The particles are initially given random velocities that are made to satisfy the requirements that the whole system has components of linear momentum that are zero and a kinetic energy that corresponds to the desired temperature of the system.

Because of the small size of the MD system (in the thermodynamic sense) and because the initial configuration will inevitably not correspond to an equilibrium state, the temperature usually fluctuates wildly at the beginning of a run. Consequently the particle velocities are rescaled (usually for a few thousand time steps) so that the temperature calculated from the kinetic energy of each species is restored to the desired preset value. With some systems exceptionally long annealing times may be required. Even with the 14-figure accuracy of

COMPUTER MODELLING IN INORGANIC CRYSTALLOGRAPHY
ISBN 0-12-164135-X

64-bit computers, the accumulation of rounding errors in very long runs can lead to small displacements of the mean positions of the particles from their perfect-lattice sites. To avoid this it is also necessary to rescale the components of linear momentum **P** for each species so as to ensure that they remain zero (by subtracting the appropriate component of the total momentum **P** of each chemical species).

The classical equations of motion are then solved in a finite difference approximation at successive time steps Δt and the velocities and positions of the particles at each time step are accumulated. This information, i.e. the position and velocity components of each particle at each time step, constitutes the primary information from the MD simulation. Usually some fundamental thermodynamic or kinetic information about the system is also calculated at run time and the velocities and positions are saved to allow for the calculation of other properties, particularly those that require a lot of computer time.

2 ALGORITHMS

The most time-consuming feature of an MD simulation is the calculation of the total force acting on each particle. This step can take up as much as 97% of the total CPU time of an MD simulation and so it must be carried out with the greatest possible efficiency. In a non-ionic system the particle interactions have only a short range amounting to a few times the lattice parameter. These short-range interactions can be limited to those between a central particle and all other particles that lie within a cut-off sphere of radius R_c. Summing over each particle in the N-particle system gives the total short-range potential energy. Often, only pair-wise interactions are considered since a full many-body treatment would add considerably to the time of the simulation. However, for some systems, such as silica and silicates, a limited three-body interaction is essential for a reasonable representation of static-lattice properties and so this should be introduced into an MD calculation as well. A simple O–Si–O bond-bending potential is adequate for the above-mentioned cases (Catlow *et al.*, 1985; Vessal *et al.*, 1989a,b; but contrast Habasaki and Okada 1992; see also Vessal in Chapter 12 of this book). Summing the pair-wise interactions between all pairs in the system would result in a simulation time of $O(N^2)$ but restricting the interactions counted to those within the cut-off sphere of radius R_c around each particle makes the calculation of $O(N)$, a very important feature for systems of large N. In practice, some additional CPU time is necessary to select from all near neighbours only those whose interactions with the central particle are to be counted. The algorithm used by the present authors is based on the construction around each particle of a local neighbourhood in the form of a *pyramid* (Rycerz and Jacobs, 1992a,b). The program based on this algorithm has been optimized for several computers; it has been described in detail (Rycerz and Jacobs, 1990) and is available from the CPC Program Library at Queen's

University of Belfast or from the Quantum Chemistry Exchange Project at Indiana University. A vectorized version of the *pyramid* algorithm and of a different, specifically vector-oriented construction called the *slab* have also been described (Rycerz and Jacobs, 1991a,b; 1992b). It is essential to restrict the pair-wise interactions only to a set of near neighbours if an efficient, workable program is to be achieved and most of the $O(N)$ algorithms that have been developed differ principally in their approach to the pre-selection problem (Allen and Tildesley, 1987). Considerable savings in computer time result from retaining, for each particle, a list of these near neighbours (called 'neighbour lists'). Because the particles generally move very little during a single time step of $O(10^{-15}$–10^{-14} s), it is not necessary to update these neighbour lists every time step but only after every NUPDA steps. In practice a value of NUPDA in the range 5–20 (commonly 10) is satisfactory.

A crucial step in the optimization of an MD code is the location of all pairs of particles within the cut-off distance R_c. An all-pairs neighbour search would be of $O(N^2)$ so some kind of cell-based search is generally employed. The computational box of side L that contains the N particles is divided by a fine 3-D grid into *subcells* of side r_{sc} and number

$$N_{sc} = (L/r_{sc})^3 \tag{1}$$

The question of the size of the subcell is an important issue but it is a question to which there is no universal answer since this depends on the algorithm. For the link-cell method (Hockney *et al.*, 1973; Fincham, 1994) the optimum size of the cell is one that contains, on average, about four particles, but for the *pyramid* algorithm the optimum number of particles per cell is two. The program searches consecutively all subcells that are within the pyramid and the occupied ones comprise the neighbour list for which the interactions with a given particle (in the central subcell at the origin of the pyramid) are calculated. Since the pyramid circumscribes a hemisphere of radius R_c about the centre of the central subcell, all the required interactions are automatically included. Why is the complication of the pyramid construction introduced? Simply because in practical cases, the pyramid contains around 230 subcells to be searched, compared to 729 in a cubic cell of side $2R_c$. An optimal choice for r_{sc} is $\sim$0.80–0.85 r_{min} where r_{min} is the minimum distance between particles, as shown, for example, by a calculation of the radial distribution function $g(r)$ (Section 4.5). For n particles per cell it is necessary to keep n sublists for each particle. In the pyramid algorithm with a maximum of two particles per cell, these two lists have very different occupancies, the first being almost filled and the second sparsely occupied. Consequently different methods are used for searching the neighbour lists with a further gain in efficiency. The optimal choice of R_c in the pyramid algorithm has been discussed in some detail (Rycerz and Jacobs, 1992a) and the same paper contains a discussion of factors affecting the efficiency of MD algorithms and performance tests for various algorithms in both scalar and vector modes. The pyramid algorithm has also proved to be very effective in the simulation of systems, like Si, SiO_2 and GaAs, that require three-body forces. A three-body code based on the pyramid algorithm which is also of $O(N)$, has been described (Rycerz, 1995) and it permits the simulation of systems containing several thousand particles even on a Pentium personal computer.

3 CALCULATION OF THE POTENTIAL ENERGY AND FORCES*

As discussed in Chapter 1, in the absence of an external field the potential energy of a system of N particles is

$$\Phi = \sum_i \sum_{j>i} \phi_2(\mathbf{r}_i, \mathbf{r}_j) + \sum_i \sum_{j>i} \sum_{k>j>i} \phi_3(\mathbf{r}_i, \mathbf{r}_j, \mathbf{r}_k) + \cdots \tag{2}$$

The second term on the right-hand side of equation (2) represents the three-body interactions and, as noted, is often omitted in simulations of systems that do not require angle-dependent potentials. In such cases ϕ_2 is usually fitted to reproduce static and dynamic properties of the material concerned and is thus an *effective* pair potential which may take some account of the non-additive many-body terms (Allen and Tildesley, 1987). The elastic, dielectric and lattice-dynamical properties of ionic crystals are commonly described in terms of the shell model in which an ion is represented by a core of charge X that contains all its mass, and a mass-less shell of charge Y, such that $X + Y = Z$, the valence of the ion. Though a convenient and effective model for static lattice calculations and for lattice dynamics, it is not often employed in MD because of the need to relax the shells at every time step. (An effective method of implementing shell-model MD has been described by Lindan and Gillan (1992)). So henceforth we take the potential energy Φ as a sum of rigid-ion pair potentials $\phi_2(r_{ij})$,

$$\Phi = \sum_{j>i} \phi_2(r_{ij}), \quad r_{ij} = |\mathbf{r}_j - \mathbf{r}_i| \tag{3}$$

$\phi_2(r_{ij})$ includes both electrostatic and short-range contributions. Integration of the classical equations of motion requires a knowledge of the total force

$$\mathbf{F}_i = -\nabla_{\mathbf{r}_i} \Phi \tag{4}$$

acting on each particle i, at each time step. It is usual to use the minimum image convention in evaluating r_{ij}, which amounts to choosing one of the periodic images of j, j', whenever $r_{ij'} < r_{ij}$. A particularly clear description is given by Allen and Tildesley (1987). If the minimum image convention is used then $R_c \leqslant L/2$.

*Note that to avoid inconsistencies with other symbols used elsewhere in this chapter, a slightly different symbolism is employed for interatomic potentials from that employed in Chapter 1.

There may, therefore, be some advantages in using non-cubical simulation boxes (Adams, 1979).

For charged particles the Coulomb interactions are of long range so that the Coulomb contribution to the potential energy and to the force $\mathbf{F}_i$ must be summed over all particles $j \neq i$ in the simulation box and also the periodic images of these particles, including those of i. The Coulomb contribution Φ^c to the potential energy is

$$\Phi^c = \frac{1}{2}\left(\frac{1}{4\pi\varepsilon_0}\right)\sum_{i,j}^{N}{}' \sum_{\mathbf{n}} \frac{Q_i Q_j}{\mathbf{r}_{i,j\mathbf{n}}} \tag{5}$$

where Q_i, Q_j are the charges on ions i and j, $\mathbf{n}$ is the ordered triple of integers which define the periodic images of the simulation box (which is therefore located at $\mathbf{n} = (0,0,0)$) and $r_{i,j\mathbf{n}} = |\mathbf{r}_j - \mathbf{r}_i + L\mathbf{n}|$, where L is the length of one side of the cubical simulation box containing N particles. The prime on the first summation indicates that the term with $j = i$ is to be omitted when $\mathbf{n} = \mathbf{0}$. In reality, macroscopic systems contain a very large number of particles but they are not infinite in extent: real crystals have surfaces. So one must imagine a very large (macroscopic) sphere that contains the simulation box at $\mathbf{n} = \mathbf{0}$ and its periodic images. The box, and its images, each contain N particles with N perhaps a few hundred, a few thousand or even a few hundred thousand but always $\ll N_A$, the Avogadro constant. The Coulomb energy of this system depends on how it is terminated, i.e. on the relative permittivity, ε', of the medium surrounding the sphere. Unless $\varepsilon' = \infty$ (as for a metallic conductor) there is an additional term Φ^s in Φ, where

$$\Phi^s = \frac{2\pi}{(2\varepsilon' + 1)L^3}\left|\sum_i Q_i \mathbf{r}_i\right|^2 \tag{6}$$

(de Leeuw *et al.*, 1980; de Leeuw and Perram, 1981). However, de Leeuw *et al.* (1980) have shown that the thermodynamic properties of the system are independent of ε', though difficulties may arise in the calculation of the relative permittivity of a system of dipoles. The current–current correlation function and the conductivity are affected only at wavelengths comparable to the size of the system (de Leeuw and Perram, 1981). So for MD calculations of the structure of, and diffusion in, crystalline ionic materials Φ^s need not be a concern and the usual Ewald method of summing Φ^c should be adequate; this standard procedure referred to in Chapters 1 and 3 and which is employed in static as well as dynamical simulations will now be discussed in some detail.

In the Ewald method the lattice of point charges is replaced by one in which each point charge is associated with a spherically symmetric charge distribution of opposite sign, which is usually (but not necessarily) taken to be a Gaussian distribution (Heyes, 1981). These extra charges screen the point charges so that their interactions are now short ranged and may be summed directly in $\mathbf{r}$-space.

To restore the original actual situation of an array of point charges, we must also add a second set of identical charge distributions of the same sign as the original point charges. The energy Φ^q of this set of charge distributions is evaluated by summing their Fourier transforms in reciprocal **q**-space and transforming the result back into **r**-space. This result includes the unwanted self-term from the interaction of the cancelling distribution at $\mathbf{r}_i$ with itself, and so a correction

$$\Phi^0 = -\frac{\alpha}{(4\pi\varepsilon_0)\sqrt{\pi}} \sum_{i=1}^{N} Q_i^2 \tag{7}$$

for this (constant) self-term must be included in Φ^q; α is the Ewald parameter which defines the width of the Gaussian charge distributions and its value has a strong influence on the rate of convergence of both the **r**-space and **q**-space summations, allowing the possibility of choosing α so as to optimize the speed of the MD simulation. Thus in the Ewald method of summing the long-range Coulomb interactions (with the surface term Φ^s omitted for reasons explained above)

$$\Phi^c = \Phi^r + \Phi^q \tag{8}$$

The real space sum is

$$\Phi^r = \frac{1}{2}\left(\frac{1}{4\pi\varepsilon_0}\right) \sum_{i,j}^{N}{}' \sum_{\mathbf{n}} Q_i Q_j \,\mathrm{erfc}\,(\alpha r_{i,j\mathbf{n}})/r_{i,j\mathbf{n}} \tag{9}$$

where the prime denotes the omission of the term with $j=i$ when $\mathbf{n}=\mathbf{0}$; erfc(x) is the complementary error function

$$\mathrm{erfc}\,(x) = 1 - (2/\sqrt{\pi}) \int_0^x du \exp(-u^2) \tag{10}$$

and, because erfc $(x) \to 0$ with increasing x, it is the reason for the rapid convergence in real space. The reciprocal-space term is

$$\Phi^q = -\frac{\alpha}{(4\pi\varepsilon_0)\sqrt{\pi}} \sum_{i=1}^{N} Q_i^2 + \frac{2\pi}{(4\pi\varepsilon_0)L^3} \sum_{i,j}^{N} Q_i Q_j \sum_{\mathbf{q}}{}' (1/q^2) \exp(-q^2/4\alpha^2) \cos(\mathbf{q}\cdot\mathbf{r}_{i,j}) \tag{11}$$

The last sum is over reciprocal-lattice vectors $\mathbf{q} = 2\pi\mathbf{n}/L$ and the prime indicates that the term with $\mathbf{n} = \mathbf{0}$ is to be omitted. The double sum over i,j in equation (11) may be converted to a single sum over i (Allen and Tildesley, 1987; Rycerz, 1992; Fincham, 1994). For small MD systems of a few hundred particles, $\bar{\alpha} = \alpha L$ is set so that only the term with $\mathbf{n} = \mathbf{0}$ need be included in the real-space sum, in which case $\bar{n}$, the maximum value of $|\mathbf{n}| = |L\mathbf{q}/2\pi|$ would be about 5–10. The advantage of this procedure is that the real space contribution to the potential energy and the forces may be included along with the short-range contributions in look-up tables that are calculated only once at the beginning of the simulation.

The accuracy of the Ewald prescription in simulations with cubic periodic boundary conditions has been considered by Kolafa and Perram (1992) and by Fincham (1994), who has discussed important questions relating to the execution time. In the minimum image convention $\bar{R} = R_c/L$ cannot be greater than 0.5 so this is the value that must be used for 'small' MD systems with N a few hundred particles. Requiring comparable accuracy in the two sums (9) and (11) then gives $\bar{n} = \bar{\alpha} = 2\pi$, which are similar to values commonly employed in MD simulations (Sangster and Dixon, 1976). The time taken to sum the Coulomb forces then scales with N as $N^{3/2}$ (Perram *et al.*, 1988; Fincham, 1994). Because of the $N^{3/2}$ dependence, other methods that are of $O(N)$ have been proposed (Eastwood *et al.*, 1980; Greengard and Rokhlin, 1987; Rycerz, 1992; Ding *et al.*, 1992; see also Lekner, 1991 and Sperb, 1994 for an alternative to the Ewald transformation). The long-range real-space contribution to the forces is calculated from equation (4), either numerically or using a standard formula obtained by differentiating equation (9), once only at the beginning of the simulation and stored in look-up tables. The major requirement in the reciprocal-space sum is the evaluation of either the complex exponential $\exp(i\mathbf{q}\cdot\mathbf{r})$ or cosine and sine functions of $\mathbf{q}\cdot\mathbf{r}$. Again this step is conveniently precomputed at suitable intervals of r and the results stored in look-up tables.

The accuracy required of the Ewald sum in MD is not an easy problem. Ultimately, one is interested in the accuracy of the total resultant force (including short-range forces) acting on a particular ion. The choice of $\bar{\alpha}$, $\bar{n}$, made above may in fact be too pessimistic. Forester and Smith (1994) have made proposals for speeding up the Ewald procedure in MD and Fincham (1994) has pointed out that significant reductions in the ranges of **r**-space and **q**-space sums may be possible without a noticeable deterioration of the MD results, but with significant reductions in the simulation time. The effects on correlation functions are more complex. For example, we have shown that approximations that hardly affect the potential energy may have significant effects on the velocity autocorrelation function (Rycerz and Jacobs, 1995). The best practical solution seems to be to perform several (relatively short) MD runs with different values of N, $\bar{\alpha}$, $\bar{R}$, $\bar{n}$ and to observe the behaviour of calculated quantities. In general $\bar{\alpha}$ should lie in the range that gives a constant Coulomb energy (Rycerz, 1992) with $\bar{R}$, $\bar{n}$ as small as possible (for an efficient simulation) but large enough to reproduce all calculated quantities with acceptable accuracy.

4 INTEGRATION OF THE EQUATIONS OF MOTION

When the potential energy is independent of the velocities of the particles and the time t, Hamilton's equations of motion in Cartesian coordinates reduce to

$$\dot{\mathbf{p}} = -\nabla_i\Phi = \mathbf{F}_i \tag{12}$$

$$\mathbf{v}_i = \dot{\mathbf{r}}_i = \mathbf{p}_i/m \tag{13}$$

where $\mathbf{v}_i$, $\mathbf{r}_i$, $\mathbf{p}_i$ are the velocity, position and linear momentum of particle i and $\mathbf{F}_i$ is the net force acting on particle i. There is clearly one such pair of equations for each of the N particles ($i = 1, \ldots, N$). For a sufficiently short time interval Δt (called in MD the *time step*) the integration of equations (12) and (13) may be approximated by

$$\mathbf{v}_i^{n+1/2} = \mathbf{v}_i^{n-1/2} + \Delta t \mathbf{F}_i^n / m \tag{14}$$

$$\mathbf{r}_i^{n+1} = \mathbf{r}_i^n + \Delta t \mathbf{v}_i^{n+1/2} / m \tag{15}$$

For $n = 1$, the $\{\mathbf{v}_i^{1/2}\}$ are the initial velocities given to the particles and the $\{\mathbf{F}_i^1\}$ are the forces calculated from the initial positions (lattice sites). So the mid-step velocities are calculated from equation (14) and then the particle positions from equation (15). The velocities at each time step are, because of the linearization employed,

$$\mathbf{v}_i^n = [\mathbf{v}_i^{n+1/2} + \mathbf{v}_i^{n-1/2}]/2 \tag{16}$$

This approach to solving the equations of motion is called the *leap-frog algorithm* because, by calculating the velocities $\mathbf{v}_i^{n+1/2}$ first, and then the $\mathbf{r}_i^{n+1}$, the velocities are 'leaping over' the known positions at $\mathbf{r}_i^n$. A large number of alternatives have been proposed, but the present authors have found the leap-frog algorithm very effective, being both reliable and stable with good energy conservation. To be sufficiently accurate it requires a short Δt and many of the algorithms that have been proposed aim at a more accurate solution to the equations of motion and therefore at permitting a larger time step, Δt (Verlet, 1967; Beeman, 1976; Hockney and Eastwood, 1981; Allen and Tildesley, 1987).

The other main class of algorithms are the 'predictor–corrector' algorithms (Gear, 1971). New positions, velocities, accelerations and higher time derivatives of $\mathbf{r}$ at $(n+1)\Delta t$ are *predicted* using Taylor expansions and the current values at $n\Delta t$. But these are not correct, and will eventually fail, because the forces have not been updated. So the accelerations at $(n+1)\Delta t$ are now calculated using the predicted positions, and hence the forces at $(n+1)\Delta t$, and these *corrected* accelerations are used to correct the positions, velocities, ... These correction terms involve numerical coefficients (Gear, 1971; see also Allen and Tildesley, 1987) chosen to give optimum stability and accuracy. Ideally, the corrector step would be repeated to improve the accuracy of the estimates at $(n+1)\Delta t$ but each correction involves a new evaluation of the forces, which is the most time-consuming part of an MD simulation. So in practice just one or two corrector steps are carried out. Other forms of predictor–corrector algorithm exist. A discussion of various MD algorithms has been given by Berendsen and van Gunsteren (1986).

By far the largest part of the total time required is in evaluating the net force on each of the N particles and this must be done at least once at each time step. Any algorithm will do that provides the necessary stability and accuracy; and the constancy of the total energy, E, and kinetic energy, K, and their standard deviations must always be tested. If E or K drifts, then almost certainly Δt is too long and must be shortened. A useful initial guess for Δt is about 10 fs, though adjustments may have to be made to satisfy the constancy of E and K. For crystals with light particles Δt will need to be shorter, for example, for Li_2O at room temperature, 2 fs is satisfactory but at high temperatures in the superionic range $\Delta t = 0.5$ fs will be more appropriate, necessitating longer runs to evaluate the physical properties of the system. The ingenuity of the programmer is more usefully devoted to shortening the calculation of the forces, e.g. by the use of dimensionless internal units, avoidance of square roots and special functions, use of look-up tables and vectorization or parallelization of the code, if appropriate (Rycerz and Jacobs, 1992c). Useful discussions of many points are given by Allen and Tildesley (1987).

5 EVALUATION OF PROPERTIES AT RUN TIME

We consider it advisable to evaluate only certain basic quantities at run time. These include the potential, kinetic and total energies, the temperature:

$$T = 2\,K/3\,Nk_B \tag{17}$$

the virial:

$$\Psi = \sum_{j>i} \mathbf{r}_{ij} \nabla \Phi(r_{ij}) \tag{18}$$

and the pressure

$$P = (2\,K - \Psi)/3V \tag{19}$$

We also calculate routinely histograms of particle displacements for each species, α, β..., which are plotted as part of the output. Also plotted are the projections of the positions of the particles in the simulation box on XY, YZ and ZX planes. This information is contained in the position output file and so may be transferred to a microcomputer where the plots may be viewed or printed.

The van Hove self-correlation function for ions of type α is

$$G_\alpha^s(\mathbf{r},\, t) = \frac{1}{N_\alpha} \left\langle \sum_{j=1}^{N_\alpha} \delta[\mathbf{r} + \mathbf{r}_j(0) - \mathbf{r}_j(t)] \right\rangle \tag{20}$$

where $\mathbf{r}_j(t) - \mathbf{r}_j(0)$ is the displacement of a particular ion of type α in time t and the ensemble average implies an average over all initial times (ergodic hypothesis). $G_\alpha^s(\mathbf{r},t)$ is the probability that in the time interval t, the displacement of a particular ion of type α is within the volume element $d\mathbf{r}$ centred on $\mathbf{r}$. The even moments of $G_\alpha^s(\mathbf{r},t)$ are

$$\langle \Delta r_\alpha(t)^{2n} \rangle = \int \mathrm{d}\mathbf{r} (\Delta r)^{2n} G^{\mathrm{s}}_\alpha(\mathbf{r}, t) \tag{21}$$

For $n = 1$

$$\langle \Delta r_\alpha(t)^2 \rangle = \int \mathrm{d}\mathbf{r} (\Delta r)^2 G^{\mathrm{s}}_\alpha(\mathbf{r}, t) \tag{22}$$

which is the mean square displacement of ions of type α, often abbreviated to $\mathrm{MSD}_\alpha(t)$. For a set of ions that simply vibrate about the lattice sites on which they were placed originally the mean square displacement is constant (after some short initial time t_0). The increase in this constant value with T causes the intensity of the coherent diffraction (elastic Bragg scattering) to decrease exponentially with increasing temperature. This exponential factor is the Debye–Waller factor written exp $(-2W)$ with W defined by

$$W = \tfrac{1}{6}(\mathrm{MSD}(\Delta\mathbf{k})^2) \tag{23}$$

where $\Delta\mathbf{k}$ is the change in wave vector of the scattered radiation which must satisfy the diffraction condition that $\Delta\mathbf{k}$ equals a reciprocal lattice vector. Thus, mean square displacements from MD simulations can be used to predict Debye–Waller factors for particular reflections. One should add a cautionary note: the results from any MD simulation can only be accurate if the pair potential used gives a satisfactory description of the forces between the particles. Thus before any simulation, some attention must be given to the potential to be used by calculating the cohesive energy and as many of the elastic, dielectric and lattice-dynamical properties of the crystal as possible. The major disadvantage of the rigid-ion model is that the high frequency permittivity is 1.0 and consequently the high frequency longitudinal optic modes will generally be poorly represented, especially if the static permittivity, ε_s, is made to agree with the experimental value. This is generally accomplished by suitably altering the hardness parameter, ρ, in the nearest-neighbour Born–Mayer repulsive potential:

$$\phi^{\mathrm{R}}(r) = A \exp(-r/\rho) \tag{24}$$

from its calculated shell-model value, to fit ε_s. If ions of a particular species hop between lattice sites (by exchanging positions with vacancies) or move into interstitial positions, then the MSD of that species will not oscillate about a constant value but will increase linearly with time according to the Einstein equation:

$$\mathrm{MSD}_\alpha(t) = \mathrm{MSD}_\alpha(0) + 6\, D_\alpha\, t \tag{25}$$

where D_α is the diffusion coefficient of α. A finite diffusion rate will be immediately apparent from the line pointer plots (or video realization) of $\mathrm{MSD}_\alpha(t)$. Mean square displacements are conveniently fitted to the Einstein equation (25) and diffusion coefficients D_α and intercepts $\mathrm{MSD}_\alpha(0)$ may then be output from the program. The plots of $MSD_\alpha(t)$ versus t are not always perfectly linear and if the other conditions concerning the stability of the simulation have been met satisfactorily (short enough Δt; constant T; sufficiently long run) the most likely reason is that the ensemble average has not sampled a sufficient number of initial times. For a sufficiently long run, initial times starting at every 10th time step, with t extending over half the run, may be sufficient. The ratio of fourth moment of $G^s_\alpha(\mathbf{r},t)$ to the square of the second moment yields a very useful quantity:

$$P_\alpha(t) = \frac{3\langle \Delta r_\alpha(t)^4 \rangle}{5\langle \Delta r_\alpha(t)^2 \rangle^2} \tag{26}$$

which may be routinely calculated and plotted. If the van Hove self-correlation function obeys a Gaussian distribution, then $P_\alpha(t)$ is unity and for ions that are simply vibrating about lattice sites P_α is indeed usually close to 1. Hopping motion results in P_α showing strong departures from unity so that it is an excellent criterion of whether or not diffusive motion is occurring.

The final property that is clearly well worth calculating at run time is the radial distribution function. Let $n^{(2)}_{\alpha\beta}(\mathbf{r}_1, \mathbf{r}_2)\,d\mathbf{r}_1\,d\mathbf{r}_2$ be the probability that there is an ion of type α in $d\mathbf{r}_1$ at $\mathbf{r}_1$ and an ion of type β in $d\mathbf{r}_2$ at $\mathbf{r}_2$. The pair correlation function is

$$g^{(2)}_{\alpha\beta}(\mathbf{r}_1, \mathbf{r}_2) = n^{(2)}_{\alpha\beta}(\mathbf{r}_1, \mathbf{r}_2)/\rho_\alpha\,\rho_\beta = g_{\alpha\beta}(\mathbf{r}) \tag{27}$$

where ρ_α, ρ_β are the bulk densities of species α, β. The pair correlation function tells us how the local density of β, $\rho_\beta g_{\alpha\beta}(r)$ around an ion of type α, differs from the bulk density ρ_β, and similarly for species α; $g_{\alpha\beta}(r)$ is generally known as the radial distribution function. Even though its calculation is of $O(N^2)$ it is still a very useful quantity to compute for all pairs α, β because of the information it gives about the structure of the material. Fortunately $g_{\alpha\beta}(r)$ is not a particularly sensitive property and so it need be calculated only every NG time steps. NG and the corresponding NP, which determines the number of initial times for the MSD calculation, are two of several parameters that can be set to suit individual simulations. For a crystalline material at low T, $g_{\alpha\beta}(r)$ displays a succession of sharp peaks that correspond to the various coordination shells around a particular ion and the integral of $g(r)$ in successive shells gives the coordination number. At higher temperatures the peaks become smaller in height and correspondingly broader as the amplitude of the vibration of the ions about lattice sites increases; $g_{\alpha\beta}(r)$ is also a good qualitative indicator of atomic motion, for if rapid diffusion is occurring on one sublattice (e.g. Li^+ in Li_2O or Cl in $SrCl_2$) then the self-correlation function $g_{\alpha\alpha}(r)$ shows this by the presence of small, broad overlapping peaks.

6 FURTHER ELABORATION OF THE PRIMARY DATA

The solution of the equations of motion during an MD simulation produces a record of the positions and velocities of the particles at each time step. These are used to evaluate some thermodynamic and kinetic properties of the system at run time, as described in the previous Section 5, and are stored in files for further elaboration after the run. This huge amount of information can be halved by preserving only the particle positions and regenerating the velocities when required. We now describe some useful basic structural and dynamical information that may be generated from the particle positions and velocities.

6.1 The velocity autocorrelation function

Integration of the equation of motion (12) gives for the displacement of a particle in time t:

$$\mathbf{r}(t) - \mathbf{r}(0) = \int_0^t \mathrm{d}t' \mathbf{v}(t') \tag{28}$$

Therefore the ensemble average of the square of the displacements is given by

$$\langle [\mathbf{r}(t) - \mathbf{r}(0)]^2 \rangle = \int_0^t \mathrm{d}t' \int_0^t \mathrm{d}t'' \langle \mathbf{v}(t') \cdot \mathbf{v}(t'') \rangle \tag{29}$$

Because the equilibrium ensemble average does not change with time and because of time-reversal symmetry, this may be written as

$$\langle [\mathbf{r}(t) - \mathbf{r}(0)]^2 \rangle = \int_0^t \mathrm{d}t' \int_0^t \mathrm{d}t'' \langle \mathbf{v}(t'' - t') \cdot \mathbf{v}(0) \rangle \tag{30}$$

Integrating the right-hand side of the equation once, gives

$$\mathrm{MSD} = 2t \int_0^t \mathrm{d}\tau \left(1 - \frac{\tau}{t}\right) \langle \mathbf{v}(0) \cdot \mathbf{v}(\tau) \rangle \tag{31}$$

For times t much greater than the time t_{m} in which the velocity autocorrelation function (VAF) $\langle \mathbf{v}(0) \cdot \mathbf{v}(t) \rangle$ decays to zero, equation (31) becomes:

$$D = (1/3) \int_0^t \mathrm{d}\tau \langle \mathbf{v}(0) \cdot \mathbf{v}(\tau) \rangle = \frac{k_{\mathrm{B}} T}{m} \int_0^\infty \mathrm{d}t\, Z(t) \tag{32}$$

where we use the Einstein equation MSD $= 6\,Dt$, and where the upper limit $t \to \infty$ simply means $t \gg t_{\mathrm{m}}$; $Z(t)$ is the normalized VAF:

$$Z(t) = \frac{\langle \mathbf{v}(0) \cdot \mathbf{v}(t) \rangle}{\langle \mathbf{v}(0) \cdot \mathbf{v}(0) \rangle} \tag{33}$$

The Fourier transform (FT) of the normalized VAF (also known as the spectral density) is given by

$$Z(\omega) = \frac{1}{2\pi} \int_{-\infty}^{\infty} dt \exp(-i\omega t) Z(t) \tag{34}$$

and since $Z(t)$ is a symmetric function of t, the zero-frequency limit of the Fourier transform $Z(\omega)$ of $Z(t)$ is

$$Z(0) = \frac{1}{\pi} \int_{0}^{\infty} dt\, Z(t) = mD/\pi k_B T \tag{35}$$

on using equation (32). Thus the diffusion coefficient D may also be evaluated from the zero-frequency limit of the FT of the normalized VAF. For a harmonic solid the spectral density $Z(\omega)$ gives the frequency distribution of phonon states (density of states) $g(\omega)$. However, $Z(\omega)$ may predict too rich a distribution at high frequencies if a rigid-ion potential has been used in the MD simulation.

6.2 Incoherent inelastic neutron scattering

The Fourier transform of $G^s_\alpha(\mathbf{r}, t)$ is given by

$$F^s_\alpha(\mathbf{q}, t) = \int_{-\infty}^{\infty} d\mathbf{r} \exp(i\mathbf{q}\cdot\mathbf{r}) G^s_\alpha(\mathbf{r}, t) \tag{36}$$

Using the definition of the van Hove self-correlation function (equation (20)) in equation (36) gives:

$$F^s_\alpha(\mathbf{q}, t) = \frac{1}{N_\alpha} \langle \exp(-i\mathbf{q}\cdot\mathbf{r}_j(0)) \exp(i\mathbf{q}\cdot\mathbf{r}_j(t)) \rangle \tag{37}$$

which has the appropriate form for a time-correlation function (Jacobs *et al.*, 1991). By considering successive motions of a particular particle which are slow enough for correlations with the previous jump to have died away it may be shown that $G^s_\alpha(\mathbf{r},t)$ obeys the diffusion equation

$$D_\alpha \nabla^2 G^s_\alpha(\mathbf{r}, t) = \frac{\partial G^s_\alpha(\mathbf{r}, t)}{\partial t} \tag{38}$$

which holds also in the macroscopic limit when $G^s_\alpha(\mathbf{r},t)$ becomes the concentration of the diffusing species. The inverse transform is given by

$$G^s_\alpha(\mathbf{r}, t) = \frac{1}{(2\pi)^3} \int_{-\infty}^{\infty} d\mathbf{q} \exp(-i\mathbf{q}\cdot\mathbf{r}) F^s_\alpha(\mathbf{q}, t) \tag{39}$$

and so

$$\nabla^2 G^s_\alpha(\mathbf{r}, t) = \frac{1}{(2\pi)^3} \int_{-\infty}^{\infty} d\mathbf{q} \exp(-i\mathbf{q}\cdot\mathbf{r})(-q^2) F^s_\alpha(\mathbf{q}, t) \tag{40}$$

and consequently

$$\frac{\partial F_\alpha^s(\mathbf{q}, t)}{\partial t} = \mathrm{FT}[D_\alpha \nabla^2 G_\alpha^s(\mathbf{r}, t)] = -q^2 D_\alpha F_\alpha^s(\mathbf{q}, t) \tag{41}$$

The solution to this differential equation is

$$F_\alpha^s(\mathbf{q}, t) = \exp(-q^2 D_\alpha t) \tag{42}$$

on using the initial condition

$$G_\alpha^s(\mathbf{r}, 0) = \delta(\mathbf{r}) \tag{43}$$

as follows from the definition of the self-correlation function, equation (20). On substituting from equation (42) in equation (39) and integrating, one finds the *Gaussian approximation* for the self-correlation function:

$$G_\alpha^s(\mathbf{r}, t) = \frac{1}{(4\pi D_\alpha t)^{3/2}} \exp(-r^2/4Dt) \tag{44}$$

which is based on the assumption that the time required for a particular jump is sufficiently long for all memory of the previous jump to be lost—and on the approximation that the Taylor expansion of $G_\alpha^s(\mathbf{r}, t)$ may be cut off at the term of second order in r. (This is a step in the derivation of equation (38).) Several results follow from the Gaussian approximation (equation (44)):

1. The moments ratio, P_α (equation (26)) is unity, which is obeyed reasonably well for non-diffusing species in ionic materials and for ionic melts.
2. The mean square displacement (equation (22)) obeys the Einstein equation, which is true for times significantly longer than t_m, the time required for the VAF to decay to zero (see equation (31) and note the first remark under equation (44)).
3. From equation (42)

$$(1/q^2)\ln F_\alpha^s(\mathbf{q}, t) = -D_\alpha t \tag{45}$$

and so the plot of the left side (LS) versus t, for a particular wave vector $\mathbf{q}$, is linear with slope $-D_\alpha(q)$. The result is valid provided the conditions on the differential equation (38) are satisfied, namely for times $t > t_m$ (the time required for loss of memory) and neglecting higher order terms in the Taylor expansion of $G_\alpha^s(\mathbf{r}, t)$. $D_\alpha(\mathbf{q})$ from equation (45) is one of the Fourier components of D_α, which in the hydrodynamic limit of small wave vectors ($q \to 0$) and long times, become equal to the macroscopic diffusion coefficient D_α.

The *incoherent* inelastic neutron scattering arises from the interference of waves scattered from the *same ion* at different times. The differential scattering cross-section per unit solid angle and per unit interval of outgoing energy is

$$\frac{\partial^2 \sigma^{\mathrm{inc}}}{\partial \Omega \partial E} = \frac{q_f}{\hbar q_i} \sum_\alpha N_\alpha [\langle b_\alpha^2 \rangle - \langle b_\alpha \rangle^2] S_\alpha^s(\mathbf{q}, \omega) \tag{46}$$

where $S_\alpha^s(\mathbf{q}, \omega)$ is the incoherent scattering function, b_α is the scattering length (a property of the nucleus) $\mathbf{q}_i$, $\mathbf{q}_f$ are the wave vectors of the incident and scattered neutrons. $S_\alpha^s(\mathbf{q}, \omega)$ is the FT of $F_\alpha^s(\mathbf{q}, t)$:

$$S_\alpha^s(\mathbf{q}, \omega) = \frac{1}{2\pi}\int_{-\infty}^{\infty} dt \exp(-i\omega t) F_\alpha^s(\mathbf{q}, t) \tag{47}$$

and so $F_\alpha^s(\mathbf{q}, t)$ is called the *intermediate scattering function*. $S_\alpha^s(\mathbf{q}, \omega)$ is thus the double Fourier transform ($r \to q$, $t \to \omega$) of $G_\alpha^s(\mathbf{r}, t)$. On using the Gaussian approximation (42) for $F_\alpha^s(\mathbf{q}, t)$ and integrating,

$$S_\alpha^s(\mathbf{q}, \omega) = \frac{q^2 D_\alpha/\pi}{(q^2 D_\alpha)^2 + \omega^2} \tag{48}$$

which shows that the line shape of $S_\alpha^s(\mathbf{q}, \omega)$, for a particular $\mathbf{q}$, is Lorentzian with half-width at half-height of $q^2 D_\alpha$. The D_α in equation (48) is one of the Fourier components of D_α, $D_\alpha(\mathbf{q})$, and in the limit of $\mathbf{q} \to 0$, $\omega \to 0$ (in a manner such that q^2/ω remains constant) it becomes equal to the macroscopic D_α. Thus MD provides several different methods of finding diffusion coefficients: from MSD, from the time integral of the VAF, from the zero-frequency limit of the FT of the VAF, from the intermediate scattering function F_α^s and from the half-width at half-height of the incoherent inelastic scattering function S_α^s. The last two yield $D_\alpha(\mathbf{q})$ rather than the macroscopic diffusion coefficient D_α, which, however, becomes equal to D_α in the hydrodynamic limit of small $\mathbf{q}$ and long t.

From its nature (same particle, different times) we see that the *incoherent* scattering will provide information about the motion of particles. In contrast, information about *structure* is contained in the *coherent* part of the total cross-section and this will be described in Section 6.3 below.

6.3 Coherent inelastic neutron scattering

The van Hove correlation function, defined by

$$G_{\alpha\beta}(\mathbf{r}, \mathbf{r}', t) = \left\langle \sum_{i=1}^{N_\alpha} \delta[\mathbf{r} - \mathbf{r}_{\alpha i}(t)] \sum_{j=1}^{N_\beta} \delta[\mathbf{r}' - \mathbf{r}_{\beta j}(0)] \right\rangle \tag{49}$$

is the joint probability density of an ion of type β being at $\mathbf{r}'$ at time $t = 0$ and an ion of type α being at $\mathbf{r}$ at time t. The FT of $G_{\alpha\beta}(\mathbf{r}, \mathbf{r}', t)$ is the intermediate scattering function:

$$F_{\alpha\beta}(\mathbf{q}, t) = (N_\alpha N_\beta)^{-1/2} \int_{-\infty}^{\infty} d\mathbf{r}\, d\mathbf{r}' \exp[i\mathbf{q}\cdot(\mathbf{r} - \mathbf{r}')] G_{\alpha\beta}(\mathbf{r}, \mathbf{r}', t) \tag{50}$$

$$= (N_\alpha N_\beta)^{-1/2} \sum_{i=1}^{N_\alpha} \sum_{j=1}^{N_\beta} \exp[i\mathbf{q}\cdot(\mathbf{r}_{\alpha i}(t) - \mathbf{r}_{\beta j}(0))] \tag{51}$$

Equation (51) shows that $F_{\alpha\beta}(\mathbf{q}, t)$ *is* a time correlation function. The dynamical structure factor $S_{\alpha\beta}(\mathbf{q}, \omega)$ is the FT of $F_{\alpha\beta}(\mathbf{q}, t)$:

$$S_{\alpha\beta}(\mathbf{q}, \omega) = \frac{1}{2\pi}\int_{-\infty}^{\infty} \mathrm{d}t \exp(-\mathrm{i}\,\omega t) F_{\alpha\beta}(\mathbf{q}, t) \tag{52}$$

and the coherent inelastic neutron scattering cross-section is given by

$$\frac{\partial^2\sigma}{\partial\Omega\partial E} = \frac{q_\mathrm{f}}{\hbar q_\mathrm{i}} \sum_{\alpha,\beta} (N_\alpha N_\beta)^{1/2} \langle b_\alpha\rangle \langle b_\beta\rangle S_{\alpha\beta}(\mathbf{q}, \omega) \tag{53}$$

The coherent inelastic scattering arises from the interference of waves scattered from two different ions, one (of type α) at $\mathbf{r}$ at time t and one (of type β) at $\mathbf{r}'$ at time $t = 0$. The density of ions of type α is

$$\rho_\alpha = \sum_{i=1}^{N_\alpha} \langle \delta(\mathbf{r} - \mathbf{r}_{\alpha i})\rangle \tag{54}$$

and its FT is

$$\begin{aligned} \rho_\alpha(\mathbf{q}) &= (N_\alpha)^{-1/2} \sum_{i=1}^{N_\alpha} \int_{-\infty}^{\infty} \mathrm{d}\mathbf{r} \exp(\mathrm{i}\,\mathbf{q}\cdot\mathbf{r}) \delta(\mathbf{r} - \mathbf{r}_{\alpha i}) \\ &= (N_\alpha)^{-1/2} \sum_{i=1}^{N_\alpha} \exp(\mathrm{i}\,\mathbf{q}\cdot\mathbf{r}_{\alpha i}) \end{aligned} \tag{55}$$

Using equation (55) in equation (51) shows that the intermediate scattering function

$$F_{\alpha\beta}(\mathbf{q}, t) = \langle \rho_\alpha(\mathbf{q}, t)\, \rho_\beta(\mathbf{q}, 0)^* \rangle \tag{56}$$

is the density–density correlation function (at wave vector $\mathbf{q}$). As $t \to 0$, $F_{\alpha\beta}(\mathbf{q}, t)$ becomes

$$F_{\alpha\beta}(\mathbf{q}, 0) = \langle \rho_\alpha(\mathbf{q}, 0)\, \rho_\beta(\mathbf{q}, 0)^* \rangle \equiv S(\mathbf{q}) \tag{57}$$

$S(\mathbf{q})$ is called the *structure factor* and it describes the structure in $\mathbf{q}$-space just as its FT, the radial distribution function $g(r)$, describes the structure in real space. Two methods of calculating the structure factor are available: from the FT of $g(r)$ and from the density–density correlation function $F_{\alpha\beta}(\mathbf{q}, 0)$ which is the zero-time limit of the intermediate scattering function $F_{\alpha\beta}(\mathbf{q}, t)$. The problem with the FT of $g(r)$ is that accuracy in $S(\mathbf{q})$ at small values of $\mathbf{q}$ requires high accuracy in $g(r)$ at large r. But $g(r)-1$ is very small at large r and so high accuracy is hard to attain even with a large system that extends sufficiently far in r. Thus various integral equation methods have been devised for extending $g(r)$ to large r, but the method we recommend is to use the FT of $g(r)$ to give $S(\mathbf{q})$ at intermediate and large q, and to calculate the small-q values of $S(\mathbf{q})$ from $F_{\alpha\beta}(\mathbf{q}, 0)$. The procedure still requires a large system in order to have sufficient resolution at small $\mathbf{q}$. The major numerical problem is that of averaging over a sufficient number of directions in $\mathbf{q}$-space. From equations (56) and (52):

$$S_{\alpha\beta}(\mathbf{q},\,\omega) = \frac{1}{2\pi}\int_{-\infty}^{\infty} \mathrm{d}t\, \mathrm{e}^{-i\omega t}\, \langle \rho_\alpha(\mathbf{q},\,t)\,\rho_\beta(\mathbf{q},\,0)^*\rangle \tag{58}$$

So $S_{\alpha\beta}(\mathbf{q},\,\omega)$ describes the spectrum of density fluctuations at wave vector $\mathbf{q}$. At low temperatures the crystal dynamics consist of phonon vibrations and $F_{\alpha\beta}(\mathbf{q},\,t)$ is a superposition of harmonic oscillations; so $S_{\alpha\beta}(\mathbf{q},\,\omega)$ should consist of a set of δ-functions at the phonon frequencies corresponding to wave vector $\mathbf{q}$. At higher temperatures translational motion occurs and the associated correlations should simply decay in time, giving rise to a peak in $S_{\alpha\beta}(\mathbf{q},\,\omega)$ that is centred on $\omega = 0$, and therefore called the quasi-elastic peak, with subsidiary phonon peaks at the appropriate values of ω. A nice example is shown in a paper by Gillan (1986).

7 OTHER ENSEMBLES

In Section 4 we focused on the implementation of the MD method for the microcanonical ($N,\ V,\ E$) ensemble. Other ensembles are distinguished by the use of different independent variables which function as the control parameters during the simulation. In the canonical ensemble the independent variables are ($N,\ V,\ T$). The calculated value of the energy E is the same for both ensembles in the thermodynamic limit $N \to \infty$ at constant N/V. However, different formulae may apply for thermodynamic properties that are derivatives of a thermodynamic potential (Allen and Tildesley, 1987).

Thus it is necessary to use the appropriate ensemble-dependent expressions in calculating the heat capacity, compressibility or thermal expansion.

7.1 Constant temperature molecular dynamics

In the canonical ensemble, the temperature of the system is controlled by its contact with a constant T reservoir of large heat capacity, generally called a 'heat bath'. One way of including the thermal contact with the heat bath is to introduce an additional degree of freedom that represents the reservoir: the simulation is then carried out for the *extended system* that includes the thermodynamic system of interest plus the reservoir. In the original implementation of this method (Nosé, 1984a,b) thermal contact was achieved by scaling the particle velocities by a heat-bath variable s, so that

$$\mathbf{v}_i = s\dot{\mathbf{r}}_i \tag{59}$$

An equivalent implementation due to Hoover (1985) introduces a constraint variable ζ into the equations of motion, which become

$$\mathbf{v}_i = \dot{\mathbf{r}}_i = \mathbf{p}_i/m_i \tag{60}$$

$$\dot{\mathbf{p}}_i = -\nabla_i\Phi - \zeta\mathbf{p}_i \tag{61}$$

Compared to equation (12), equation (61) contains the additional force $-\zeta\mathbf{p}_i$, which is a feedback term analogous to a frictional force. The time dependence of ζ is controlled by

$$(Q/2)\,\mathrm{d}\zeta/\mathrm{d}t = \sum_i p_i^2/2m_i - g/2k_\mathrm{B}T \tag{62}$$

where g is the number of degrees of freedom of the system ($= 3N - 3$ if the total momentum is fixed). The right-hand side of equation (62) shows that changes in ζ arise from the imbalance between the kinetic energy of the system and its average value. Q is a heat-bath parameter that controls the speed of this response to kinetic energy fluctuations in the system. So the dynamics of the system are controlled by equations (60)–(62). Extended systems have been discussed by Evans and Holian (1985), by Bulgac and Kusnezov (1990), and by Nosé (1991), who describes factors affecting the choice of the parameter Q.

Alternatively, the temperature of the simulation system may be held constant by adding to the equation of motion (12), a constraint force which is proportional to the velocity, thereby giving

$$\dot{\mathbf{p}}_i = -\nabla_i\Phi - \lambda\mathbf{p}_i \tag{63}$$

(Hoover *et al.*, 1982; Evans, 1983). Equations (61) and (63) may seem to have a formal resemblance, but their origins are quite different. In the extended system, ζ supplies (positive or negative) feedback to the system to maintain the kinetic energy of the system close to its average value. In equation (63) λ is a parameter that is going to hold the kinetic energy constant; λ plays the role of a Lagrange undetermined multiplier chosen to satisfy Gauss' principle of least constraint: this yields (Nosé, 1991)

$$\lambda = -\sum_i \mathbf{v}_i\cdot\mathbf{F}_i/gkT = -\frac{\mathrm{d}\Phi}{\mathrm{d}t}/gkT \tag{64}$$

In an atomic system $g = 3N - 4$ since the three components of total momentum and the kinetic energy are fixed. In the initial step the velocities are adjusted so that the total kinetic energy is $gkT/2$ and thereafter it is maintained constant by equations (60) and (63) with equation (64). Earlier, a simpler algorithm for maintaining the kinetic energy constant was proposed by Woodcock (1971). In this method the mid-step velocities of the leap-frog algorithm are scaled by a factor s so that they satisfy the constraint of constant kinetic energy. It turns out that this procedure is approximately equivalent to that which satisfies Gauss' principle (equations (60), (63) and (64)) but that it introduces an error of order Δt, which is one order less than the accuracy of the ordinary leap-frog (equivalent to Verlet) algorithm in equations (14) and (15). The problem is that in Newtonian dynamics energy and momentum are conserved and velocity rescaling has to be but an approximate method of solving equations (60) and (63), which differ from the Newtonian form (12) and (13). A different method of velocity rescaling has been proposed by Berendsen *et al.* (1984). It does not generate canonical ensemble states but is very useful for effecting a change in state and equilibrating the system at the new temperature. Stochastic methods have been described by Andersen (1980) and several other methods of undertaking constant temperature molecular dynamics are discussed in the book by Allen and Tildesley (1987).

7.2 Constant pressure molecular dynamics

Constant pressure MD can be implemented either by extended system methods that couple the dynamical system to an external variable V which is the volume of the simulation box (Andersen, 1980) or by constraint methods that use a Lagrange multiplier determined from Gauss' principle (Evans and Morriss, 1983, 1984). The equations of motion are:

$$\dot{\mathbf{r}}_i = \mathbf{p}_i/m + \chi\mathbf{r}_i \tag{65}$$

$$\dot{\mathbf{p}}_i = \mathbf{F}_i - \chi\mathbf{p}_i \tag{66}$$

$$\dot{V} = 3V\chi \tag{67}$$

On differentiating the virial equation for the pressure,

$$3PV = 2K + \sum_i \mathbf{r}_i \cdot \mathbf{F}_i \tag{68}$$

$$3\dot{P}V + 3P\dot{V} = \sum_i (2/m)\mathbf{p}_i \cdot \dot{\mathbf{p}}_i + \dot{\mathbf{r}}_i \cdot \mathbf{F}_i + \mathbf{r}_i \cdot \dot{\mathbf{F}}_i \tag{69}$$

But since P is constant, $\dot{P} = 0$ and substituting for $\dot{V}$ from equation (67) in equation (69) yields the Lagrange multiplier χ. The equations of motion (65)–(67) can then be solved using predictor–corrector methods. The presence of the constraint term in equation (65) means that derivatives of position change discontinuously when a particle crosses a boundary.

The ensemble that corresponds to these equations is (N, P, H) and it seems not to be commonly used. Of much more interest in materials science is the (N, P, T) ensemble. MD at constant pressure and temperature may be achieved by the addition of a second Lagrange multiplier (Evans and Morriss, 1983):

$$\dot{\mathbf{p}}_i = \mathbf{F}_i - \chi\mathbf{p}_i - \xi\mathbf{p}_i \tag{70}$$

where $\chi + \xi$ is given by:

$$\chi + \xi = \sum_i \mathbf{p}_i \cdot \mathbf{F}_i / \sum_i p_i^2 \tag{71}$$

and

$$\chi = \frac{\sum_i 1/m_i \sum_{j>i} (\mathbf{r}_{ij} \cdot \mathbf{p}_{ij}) f(r_{ij}) / r_{ij}^2}{\sum_i \sum_{j>i} f(r_{ij}) + 9PV} \tag{72}$$

where

$$f(r) = r \frac{\mathrm{d}w(r)}{\mathrm{d}r} \tag{73}$$

and

$$w(r) = r \frac{\mathrm{d}\phi(r)}{\mathrm{d}r} \tag{74}$$

is the intermolecular virial. The equations of motion are solved by the predictor–corrector method. Care must be taken to include the long-range corrections in evaluating $f(r_{ij})$ and in the PV term when evaluating χ.

7.3 Molecular dynamics at constant stress

The MD simulation box is a parallopiped of shape and size determined by the three vectors $\mathbf{b}_1$, $\mathbf{b}_2$, $\mathbf{b}_3$ which need not be equal in magnitude nor mutually orthogonal. The volume of the box is $V = |\mathbf{B}|$ where $\mathbf{B}$ is the matrix $\{\mathbf{b}_1, \mathbf{b}_2, \mathbf{b}_3\}$. Since there are no symmetry constraints on $\mathbf{B}$, a system of N particles has $3N + 9$ degrees of freedom. Following Tallon (1988), the positions of the particles are described by position vectors

$$\mathbf{r}_i = \mathbf{B}\mathbf{s}_i, \quad -\tfrac{1}{2} \leq \mathbf{s}_{i1}, \mathbf{s}_{i2}, \mathbf{s}_{i3}, \leq \tfrac{1}{2}, \quad i = 1, N \tag{75}$$

and the relative separation of two particles is therefore given by

$$r_{ij}^2 = (\mathbf{s}_i - \mathbf{s}_j)^{\mathrm{T}} \mathbf{M} (\mathbf{s}_i - \mathbf{s}_j) \tag{76}$$

where the metric $\mathbf{M}$ is $\mathbf{B}^{\mathrm{T}}\mathbf{B}$. The Parrinello and Rahman (1980, 1981, 1982) Lagrangian is

$$L = \frac{1}{2} \sum_i m_i \dot{\mathbf{s}}_i^{\mathrm{T}} \mathbf{M} \dot{\mathbf{s}}_i - \sum_i \sum_{j>i} \phi(r_{ij}) + (\tfrac{1}{2}) \mu \,\mathrm{Tr}\, (\dot{\mathbf{B}}^{\mathrm{T}} \dot{\mathbf{B}}) - PV \tag{77}$$

The parameter μ, which has the dimensions of mass, determines the time scale of the fluctuations in box shape and size described by $\dot{\mathbf{B}}$. Particle momenta are given by $m_i \dot{\mathbf{r}}_i = m_i \mathbf{B} \dot{\mathbf{s}}_i$. The equations of motion in Lagrangian form are

$$m_i \ddot{\mathbf{s}}_i = \sum_{j \neq i} \chi(r_{ij})(\mathbf{s}_i - \mathbf{s}_j) - m_i \mathbf{M}^{-1} \dot{\mathbf{M}} \dot{\mathbf{s}}_i, \quad i, j = 1, N \tag{78}$$

and

$$\ddot{\mathbf{B}} = \mu^{-1} (\mathbf{\Pi} - P\mathbf{I}) \boldsymbol{\sigma} \tag{79}$$

where $\chi(r) = -(1/r)\mathrm{d}\phi/\mathrm{d}r$, $\mathbf{I}$ is the 3×3 unit matrix, $\boldsymbol{\sigma} = \{\mathbf{b}_2 \times \mathbf{b}_3,\ \mathbf{b}_3 \times \mathbf{b}_1,\ \mathbf{b}_1 \times \mathbf{b}_2\}$ and $\boldsymbol{\Pi}$ is the internal stress tensor.

$$V\boldsymbol{\Pi} = \sum_i m_i \mathbf{v}_i \cdot \mathbf{v}_i + \sum_i \sum_{j>i} \chi(r_{ij})(\mathbf{r}_i - \mathbf{r}_j)\cdot(\mathbf{r}_i - \mathbf{r}_j) \tag{80}$$

which can be regarded as a generalization of the virial equation for the pressure (equation (68)). Tallon (1988) chooses μ by equating the transit time of a longitudinal wave with the period of isotropic oscillation of the box, which yields

$$\mu = (3/4\pi^2) \sum_i m_i \tag{81}$$

The conserved thermodynamic function for this ensemble is the generalized enthalpy

$$\mathcal{H} = H + (\tfrac{1}{2})\mu \,\mathrm{Tr}\,(\dot{\mathbf{B}}^{\mathrm{T}}\dot{\mathbf{B}}) \tag{82}$$

where the enthalpy is given by

$$H = K + \Phi + PV \tag{83}$$

The last term in equation (82) is the box kinetic energy which at equilibrium is $9k_{\mathrm{B}}T/2$. The right-hand side of equation (78) is the force acting on particle i in position $\mathbf{s}_i$, so the Verlet algorithm for this system becomes

$$\begin{aligned}\{\mathbf{I} + \tfrac{1}{2}\Delta t\mathbf{M}^{-1}\dot{\mathbf{M}}\}\mathbf{s}_i(t+\Delta t) &= 2\mathbf{s}_i(t) - \{\mathbf{I} - \tfrac{1}{2}\Delta t\mathbf{M}^{-1}\dot{\mathbf{M}}\}\mathbf{s}_i(t-\Delta t)\\ &+((\Delta t)^2/m_i)\sum_{j>i}\chi(r_{ij})[\mathbf{s}_i(t) - \mathbf{s}_j(t)]\end{aligned} \tag{84}$$

Again the implementation of the nearest-image convention needs some care: at each iteration *both* $\mathbf{s}_i(t + \Delta t)$ and $\mathbf{s}_i(t)$ must be replaced by $\mathbf{s}_i(t + \Delta t) - U$ and $\mathbf{s}_i(t) - U$ where U is the integer part of $2\,\mathbf{s}_i(t + \Delta t)$. Constant stress MD has been used by Tallon (1988) in a detailed study of the phase diagram of silver iodide.

8 SOME EXAMPLES OF APPLICATIONS

There will be space only to describe or reference but a few of the numerous papers that have appeared on the use of MD in the study of materials (Catlow *et al.*, 1990; Ronchetti and Jacucci, 1990; Meyer and Pontikis, 1991). Most authors have been interested in elucidating structural problems (Gillan, 1986; Dalba *et al.*, 1994; Sonwalkar *et al.*, 1994) including phase stability (Tallon, 1988; Somayazulu *et al.*, 1994) and melting (Vessal *et al.*, 1994); in studying defects (Mościński and Jacobs, 1985; Ngoepe and Catlow, 1991); in the mechanisms of atom transport (Wolf *et al.*, 1984; Gillan *et al.*, 1987; Sindzingre and Gillan, 1988; Smith and Gillan, 1992); in the study of transport coefficients (Sindzingre and Gillan, 1990; Lindan and Gillan, 1992); and in surface phenomena (Beaudet *et al.*, 1993). For convenience, the illustrative examples are taken from our own work. As discussed in Chapter 1, MD techniques are of greatest utility in modelling materials showing the phenomenon of 'superionic' or 'fast-ion' transport, i.e. high ionic mobilities within the solid state. Two topical and important examples will be discussed, the first an oxygen and the second a silver conducting solid.

8.1 $\delta-Bi_2O_3$

The δ-phase of bismuth oxide Bi_2O_3 is stable between 1002 and 1097 K. This has a fluorite structure so the oxide ions occupy a simple cubic array of sites with only 3/4 occupancy. Alternate cubes are occupied by Bi^{3+} ions. Generally in the fluorite structure, the vacant cubes favour the formation of anion interstitials, as in CaF_2 and $SrCl_2$, but in the δ-phase there are already 25% vacant anion sites, which accounts for its extremely high oxide ion conductivity. MD simulations have been undertaken by Jacobs and MacDónaill (1987) and Jacobs *et al.* (1990). Radial distribution functions for Bi–Bi and O–O obtained from MD simulations of the material are shown in Fig. 4.1. The Bi peaks are sharp and the first four coordination shells are clearly resolved. In contrast, the O peaks are broader with smaller maxima, and the second shell is barely resolved. This is indicative of greater disorder on the oxide sublattice. The FT of $g(r)$ yields the structure factor, but unfortunately experimental data are not available for comparison. (However, for liquid lead there are very detailed and accurate measurements of $S(q)$ at small values of q and a comparison of these measurements to $S(q)$ calculated from an MD simulation at 621 K with N = 21 952 is

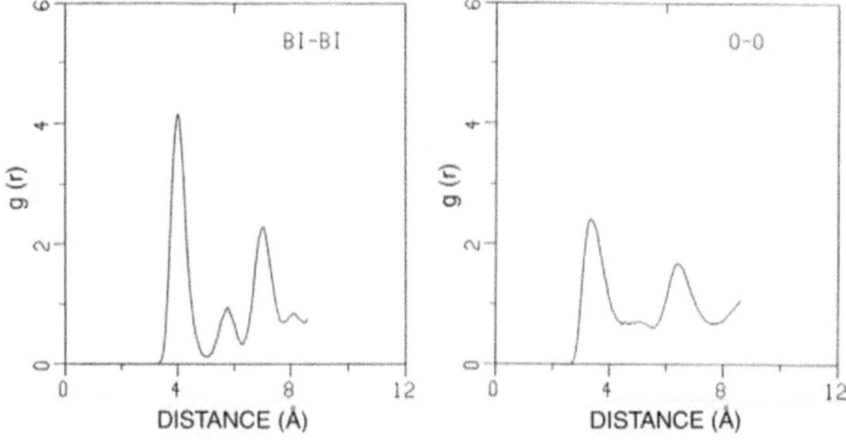

Figure 4.1 Radial distribution function $g(r)$ for Bi–Bi and O–O in δ-Bi_2O_3.

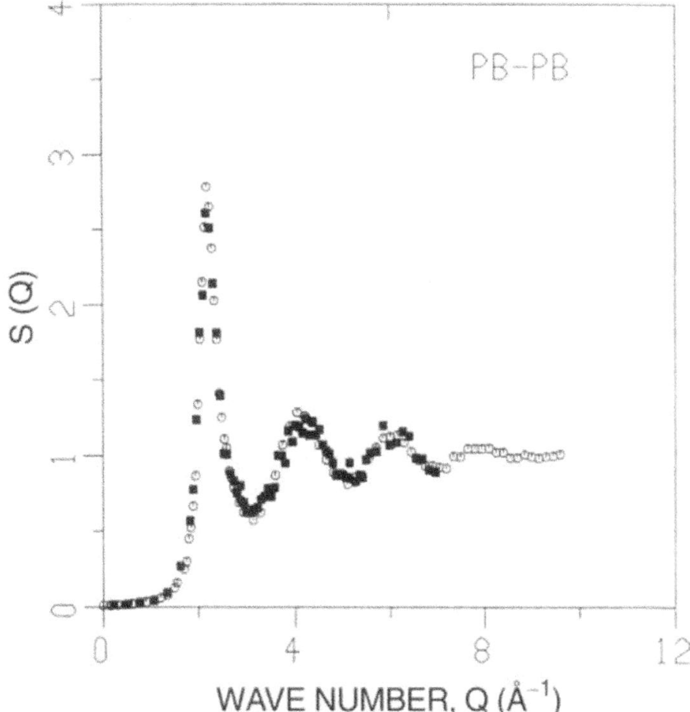

Figure 4.2 Structure factor $S(q)$ for liquid lead at 621 K. Filled squares from MD simulation; small circles from neutron scattering measurements of Dahlborg and Ollson (1980).

shown in Fig. 4.2. The agreement is excellent, even on a much larger scale that shows that $S(q)$ tends to the proper thermodynamic limit as $q \to 0$.)

The VAF $Z(t)$ for O^{2-} in δ-Bi_2O_3 at 1094 K is shown in Fig. 4.3, together with its time-integral $\xi(t)$ and FT $Z(f)$. Note the use of frequency units of THz for $f = \omega/2\pi$. The VAF decays to zero very rapidly in not much more than 0.2 ps. Its decay is clearly modulated at a frequency of about 10 THz which corresponds to one of the peaks in $Z(f)$ (Fig. 4.3) and which is probably the transverse optic mode although a little higher than the only known optic mode measurements: those of Pettsol'd (1984) for the α (distorted fluorite) phase. $\xi(t)$ tends to a finite limit for $t > 0.5$ ps and this yields a value for D of 1.8×10^{-5} cm^2/s in agreement (as it must be) with the zero frequency limit of $Z(f)$. The high frequency limit of $Z(f)$ is probably too high because of the (dielectrically adjusted) rigid-ion potential used. The density of oxide ions along the $\langle 100 \rangle$, $\langle 110 \rangle$ and $\langle 111 \rangle$ directions is shown in Fig. 4.4. The last shows the absence of cube-centre interstitials, as expected, while the first two show that the predominant mechanism of oxide ion motion is along $\langle 100 \rangle$, with just a small fraction of

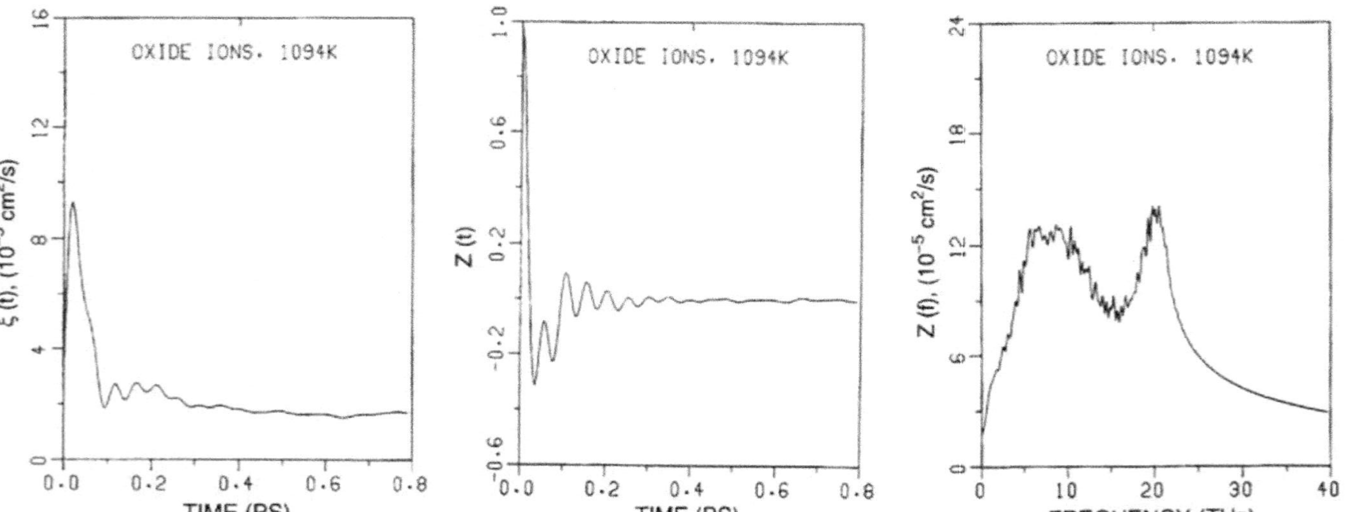

Figure 4.3 Velocity autocorrelation function $Z(t)$, its time-integral $\xi(t)$ and its Fourier transform $Z(f)$ for oxide ions in δ-Bi_2O_3 at 1094 K.

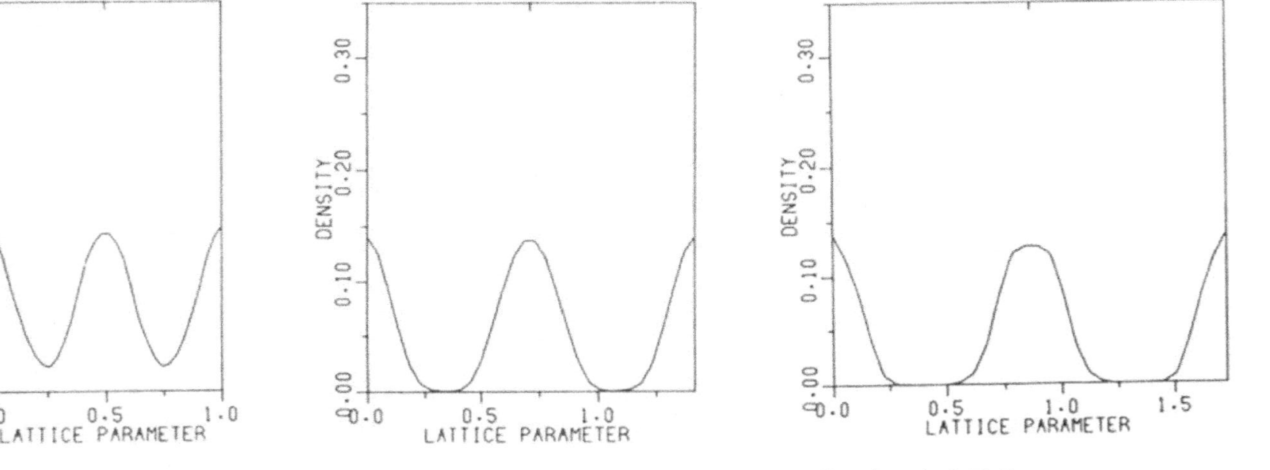

Figure 4.4 Density of oxide ions along ⟨100⟩, ⟨110⟩ and ⟨111⟩ directions in δ-Bi_2O_3.

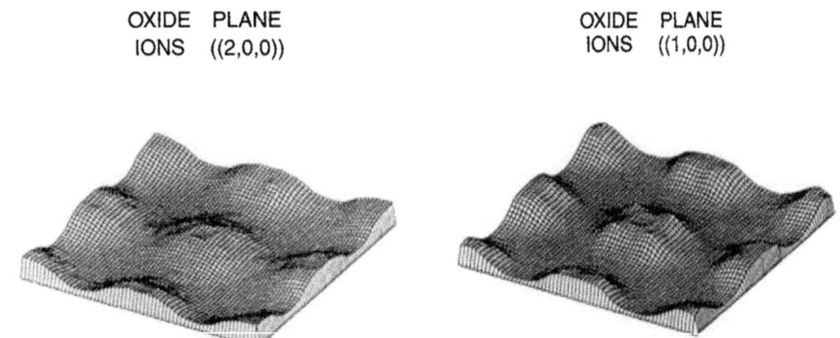

Figure 4.5 Density of oxide ions in (200) and (110) planes of δ-Bi_2O_3.

⟨110⟩ jumps. Some jumps occur directly between lattice sites, but others occur via interstitial sites halfway along a cube edge. There is a significant oxide ion density in the (400) Bi^{3+} plane with maxima at cube-edge interstitial sites (MacDońaill *et al.*, 1990) and the interstitials are also seen clearly in individual ion trajectories. Some of these suggest a ⟨110⟩ displacement of the interstitial towards an empty cube centre, as predicted by static lattice simulations (Jacobs and Mac Dońaill, 1987) which were in close agreement with results of neutron diffraction measurements (Battle *et al.*, 1983, 1986). The ⟨111⟩ density, in particular, hints at disorder of the type originally proposed by Willis (1965) in which anions in fluorite structures undergo ⟨111⟩ tetrahedral displacements towards the centres of the four nearest neighbour (nn) vacant cubes. These displacements are very clearly seen in 3-D plots of oxide ion density in (200) and (110) planes (Fig. 4.5).

8.2 α-AgI

The structure of the superionic α-phase of AgI (Fig. 4.6) consists of a body-centred arrangement of I^- ions with the two Ag^+ ions per unit cell distributed over 42 possible sites comprising 6 octahedral b-sites, 12 tetrahedral d-sites and 24 trigonal h-sites. Earlier MD studies of this system were reported by Vashishta and Rahman (1978). Figure 4.7 shows histograms of the displacements from lattice sites of the I^- and Ag^+ ions obtained in our own recent work. The larger MSDs of Ag^+ are immediately apparent; in addition several Ag^+ ions are undergoing much larger displacements, consistent with rapid Ag^+ diffusion. Figure 4.8 shows plots of the coherent (S_T) and incoherent (S_s) scattering functions against frequency $f = \omega/2\pi$. S_s has the expected Lorentzian form and the Ag^+ diffusion coefficient $D(q)$ could be evaluated from the HWHM (equation (48)). S_T shows the quasi-elastic peak due to translational motion of Ag^+ as well as several phonon peaks. There is a strong qualitative resemblance

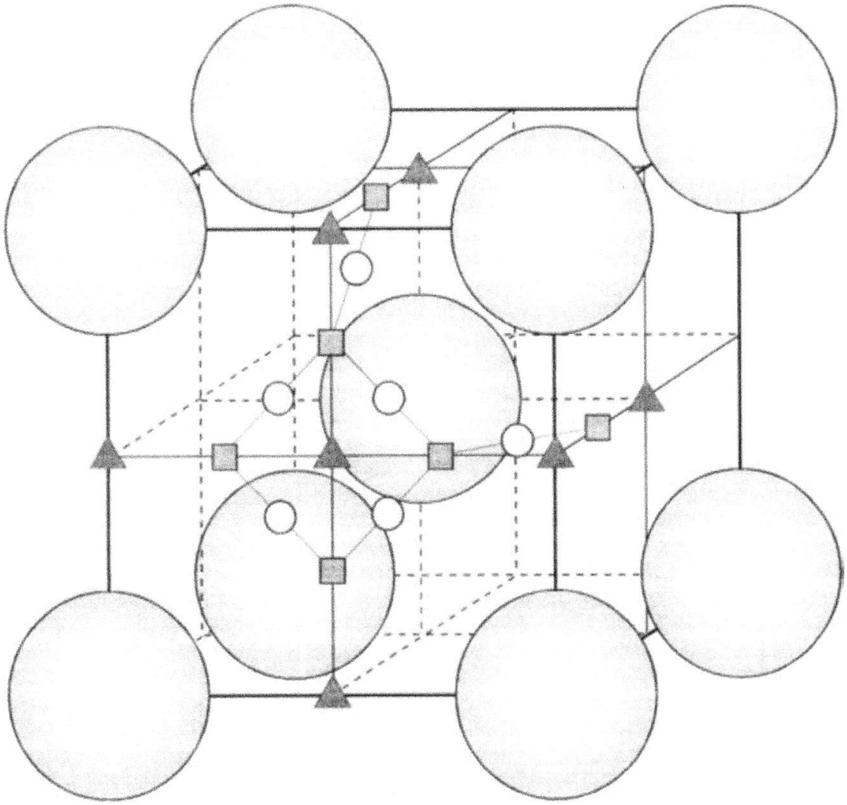

Figure 4.6 Unit cell of α-AgI showing octahedral b-sites, tetrahedral d-sites and trigonal h-sites. ▲ 6(b), ■ 12(d), ● 24(h).

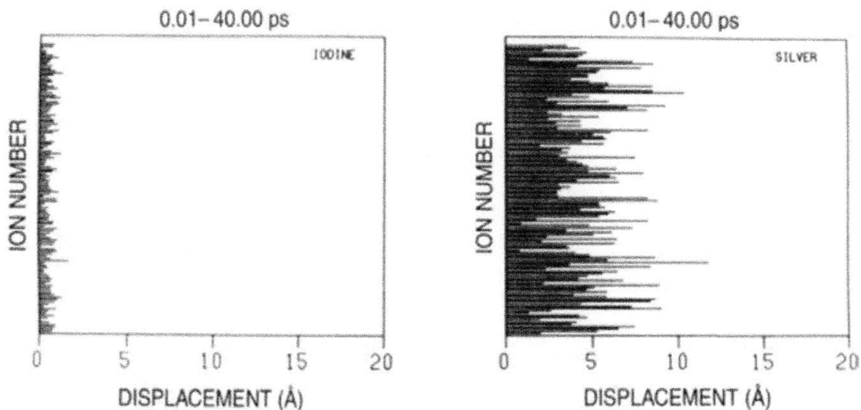

Figure 4.7 Histograms of ionic displacements of I^- and Ag^+ ions in the superionic α-phase AgI. The largest of the Ag^+ displacements exceed twice the cube edge lattice parameter, 5.07 Å.

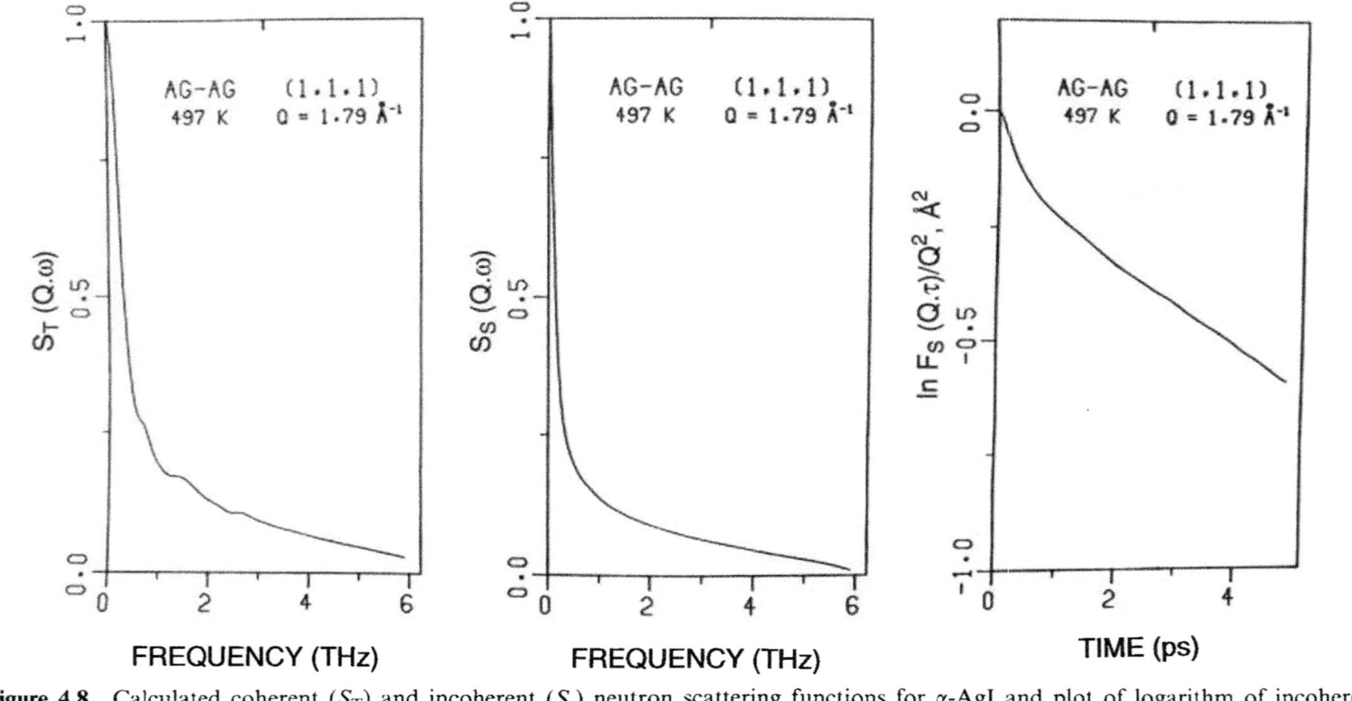

Figure 4.8 Calculated coherent (S_T) and incoherent (S_s) neutron scattering functions for α-AgI and plot of logarithm of incoherent intermediate scattering function, F_s, divided by q^2, against t.

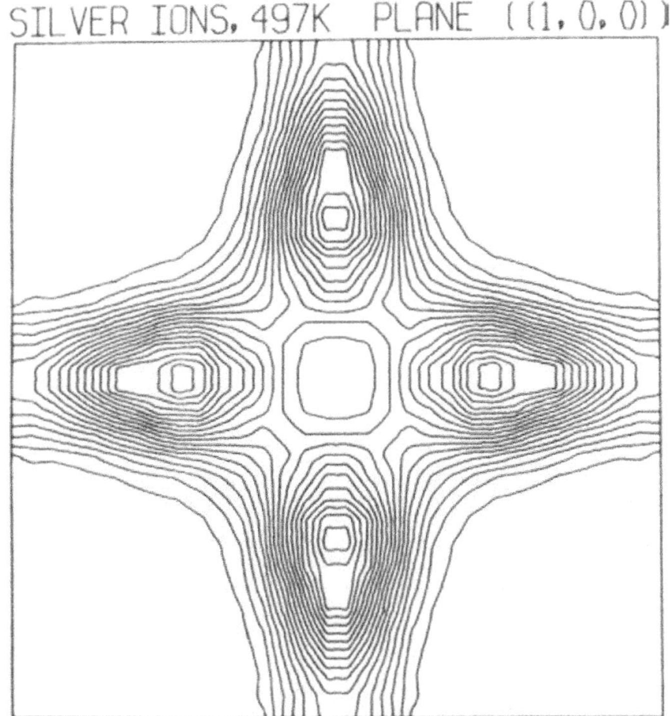

Figure 4.9 Contour plot of Ag^+ ion density in (100) planes of α-AgI.

to the experimental results of Eckold *et al.* (1976a,b) who have formulated a model involving local motion, plus translational diffusion of the Ag^+ cations. The occurrence of two different linear sections in the plots of $\ln F_s(q, t)/q^2$ versus t is consistent with a two-step process in the diffusion of Ag^+ in α-AgI. There are differences in interpretation of the neutron diffraction (Cava *et al.*, 1977) and neutron scattering (Eckold *et al.*, 1976a,b) data. Both envisage a two-step process: Cava *et al.* suppose a vibration of Ag^+ in tetrahedral d-sites followed by hopping to an adjacent d-site in about 1/3 of the total time to make the jump, estimated from D as 2.5 ps. Eckold *et al.* regard the local motion as a rapid oscillation between adjacent d-sites, followed by the migration step between second-neighbour octahedral sites, as though a face-centred set of one b- and four d-sites constituted a local cage in which an Ag^+ would be restrained (temporarily) by the local Coulomb field.

MD provides information about the local ion densities and therefore gives useful insights into both structure and diffusion. Figure 4.9 shows a contour plot of Ag^+ ion density in (100) planes of α-AgI. This clearly shows the preferred occupancy of the tetrahedral d-sites and bears a striking resemblance to the similar plot, from experimental neutron diffraction data, in a paper by

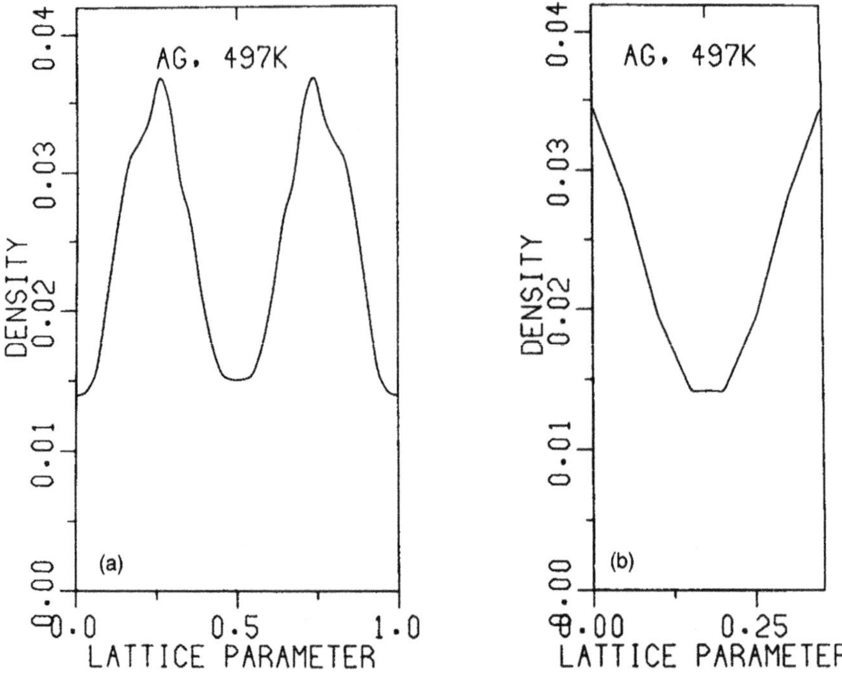

Figure 4.10 Density of Ag^+ ions (a) along ⟨100⟩ in a cube face between two cube-edge octahedral b-sites and (b) along ⟨110⟩ between two d-sites via a trigonal h-site (cf. Fig. 4.6). The density at the two tetrahedral d-sites is about $2\frac{1}{2}$ times that at the three b-sites, which is comparable with that at the h-sites.

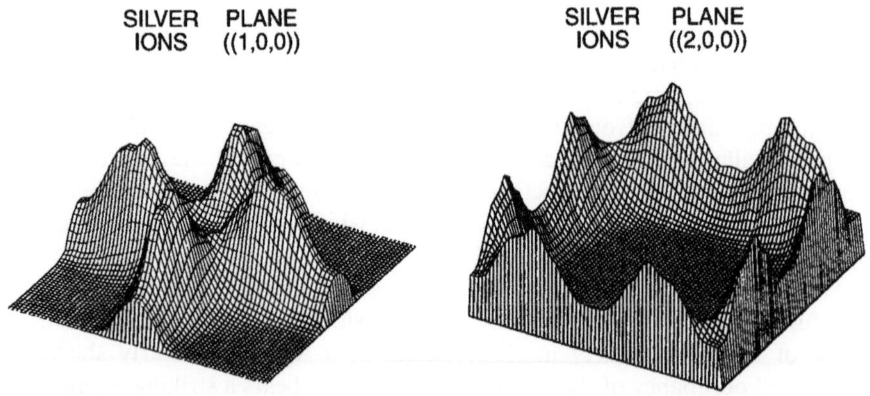

Figure 4.11 3-D plots of Ag^+ ion density in (100) and (200) planes of α-AgI.

Cava *et al.* (1977). Figure 4.10 shows the Ag^+ ion density along ⟨100⟩ between cube-edge b-sites and along ⟨110⟩ between two d-sites. These show the preferred occupancy of tetrahedral d-sites, though not to the exclusion of b- and h-sites. Translational motion between d-sites must proceed through h- and/or b-sites. Finally, Fig. 4.11 shows Ag^+ ion density in (100) and (200) planes. The highest peaks correspond to the tetrahedral sites, ridges between these to the trigonal sites and the slightly lower saddle points to the octahedral sites. These figures suggest a rapid motion between nn d-sites via h-sites and a slower diffusion via cube-edge b-sites to the d, h 'cage' in the next face. Although diffusion via d- and h-sites is possible in this structure, these density plots certainly suggest the involvement of the b-sites, at least as part of the diffusion path.

ACKNOWLEDGEMENT

We are grateful to the Natural Sciences and Engineering Research Council of Canada for financial support.

REFERENCES

Adams, D.J. (1979) *Chem. Phys. Lett.*, **62**, 329.

Allen, M.P. and Tildesley, D.J. (1987) *Computer Simulation of Liquids*, Clarendon Press, Oxford.

Andersen, H.C. (1980) *J. Chem. Phys.*, **72**, 2384.

Battle, P.D., Catlow, C.R.A., Derennan, J. and Murray, A.D. (1983) *J. Phys. C*, **16**, L561.

Battle, P.D., Heap, J.W. and Moroney, L.M. (1986) *J. Solid State Chem.*, **63**, 8.

Beaudet, Y., Lewis, L.J. and Persson, M. (1993) *Phys. Rev.* B, **47**, 4127.

Beeman, D. (1976) *J. Comput. Phys.*, **20**, 130.

Berendsen, H.J.C. and van Gunsteren, W.F. (1986) In *Molecular Dynamics Simulation of Statistical Mechanical Systems*, Proc. Enrico Fermi Summer School, Varenna, Soc. Italiana di Fisica, Bologna, p. 43.

Berendsen, H.J.C., Postma, J.P.M., van Gunsteren, W.F., Di Nola, A. and Haak, J.R. (1984) *J. Chem. Phys.*, **81**, 3684.

Bulgac, A., and Kusnezov, D. (1990) *Phys. Rev. A*, **42**, 5045.

Catlow, C.R.A., Freeman, C.M. and Royle, R.L. (1985) *Physica B+C*, **131**, 1.

Catlow, C.R.A., Parker, S.C. and Allen, M.P. (1990) *Computer Modelling of Fluids, Polymers and Solids*, Kluwer, Dordrect.

Cava, R.J., Reidinger, F. and Wuensch, B.J. (1977) *Solid State Commun.*, **24**, 411.

Dahlborg, V. and Ollson, L.G. (1980) *J. Physique*, **41**, C8.

Dalba, G., Fornasini, P., Gotter, R., Cozzini, S., Ronchetti, M. and Rocca, F. (1994) *Solid State Ionics*, **69**, 13.

de Leeuw, S.W. and Perram, J.W. (1981) *Physica A*, **107**, 179.

de Leeuw, S.W., Perram, J.W. and Smith, E.R. (1980) *Proc. R. Soc. Lond. A*, **373**, 27.

Ding, H-Q, Karasawa, N. and Goddard, W.A. (1992) *J. Chem. Phys.*, **97**, 4309.

Eastwood, J.W., Hockney, R.W. and Lawrence, D. (1980) *Comput. Phys. Commun.*, **19**, 215.
Eckold, G., Funke, K., Kalus, J. and Lechner, R. (1976a) *Phys, Lett. A*, **55**, 125.
Eckold, G., Funke, K., Kalus, J. and Lechner, R. (1976b) *J. Phys. Chem. Solids*, **37**, 1097.
Evans, D.J. (1983) *J. Chem. Phys.*, **78**, 3297.
Evans, D.J. and Holian, B.L. (1985) *J. Chem. Phys.*, **83**, 4069.
Evans, D.J. and Morriss, G.P. (1983) *Chem. Phys.*, **77**, 63.
Evans, D.J. and Morriss, G.P. (1984) *Comput. Phys. Reports*, **1**, 297.
Fincham, D. (1994) *Mol. Simul.*, **13**, 1.
Forester, T. and Smith, W. (1994) *Mol. Simul.*, **13**, 195.
Gear, C.W. (1971) *Numerical Initial Value Problems in Ordinary Differential Equations*, Prentice-Hill, Englewood Cliffs, N.J.
Gillan, M.J. (1986) *J. Phys. C*, **19**, 3391.
Gillan, M.J., Harding, J.H. and Tarento, R-J. (1987) *J. Phys. C.*, **20**, 2331.
Greengard, L. and Rokhlin, V. (1987) *J. Comput. Phys.*, **73**, 325.
Habaski, J. and Okada, I, (1992) *Mol. Simul.*, **9**, 319 (and ref. therein).
Heyes, D.M. (1981) *J. Chem. Phys.*, **74**, 1924.
Hockney, R.W. and Eastwood, J.W. (1981) *Computer Simulation Using Particles*, McGraw Hill, New York.
Hockney, R.W., Goel, S.P. and Eastwood, J.W. (1973) *Chem. Phys. Lett.*, **21**, 589.
Hoover, W.G. (1985) *Phys. Rev. A*, **31**, 1695.
Hoover, W.G., Ladd, A.J.C. and Moran, B. (1982) *Phys. Rev. Lett.*, **48**, 1818.
Jacobs, P.W.M and MacDónaill, D.A. (1987) *Solid State Ionics*, **23**, 279, 295, 307.
Jacobs, P.W.M, Rycerz, Z.A. and Mościński, J. (1991) In *Advances in Solid State Chemistry*, Vol. 2 (ed. C.R.A. Catlow) JAI Press, London, p. 113.
Kolafa, J. and Perram, J.W. (1992) *Mol. Simul.*, **9**, 351.
Lekner, J. (1991) *Physica A*, **176**, 485.
Lindan, P.J.D. and Gillan, M.J. (1992) *J. Phys. Cond. Matter*, **5**, 1019.
MacDónaill, D.A., Jacobs, P.W.M. and Rycerz, Z.A. (1990) *Mol. Simul.*, **5**, 193.
Meyer, M. and Pontikis, V. (1991) *Computer Simulation in Materials Science*, Kluwer, Dordrecht.
Mościński, J. and Jacobs, P.W.M. (1985) *Proc. R. Soc. Lond. A*, **398**, 141, 173.
Ngoepe, P.E. and Catlow, C.R.A. (1991) *Rad. Eff. Def. Solids*, **119–121**, 399.
Nosé, S. (1984a) *Mol. Phys.*, **52**, 255.
Nosé, S. (1984b) *J. Chem. Phys.*, **81**, 511.
Nosé, S. (1991) In *Computer Simulation in Materials Science* (eds M. Meyer and V. Pontikis) Kluwer, Dordrecht, p. 21.
Parrinello, M. and Rahman, A. (1980) *Phys. Rev. Lett*, **45**, 1196.
Parrinello, M. and Rahman, A. (1981) *J. Appl. Phys.*, **52**, 7182.
Parrinello, M. and Rahman, A. (1982) *J. Appl. Phys.*, **76**, 2662.
Perram, J.W., Petersen, H.G. and de Leeuw, S.W. (1988) *Mol. Phys.*, **65**, 875.
Pettsol'd, E.G. (1984) Deposited document 811–84, VINITI in Russian. Cf. *Chem. Abstracts*, **102**, 119887f.
Ronchetti, M., and Jacucci, G. (1990) *Simulation Approach to Solids*, Kluwer, Dordrecht.
Rycerz, Z.A. (1992) *Mol. Simul.*, **9**, 327.
Rycerz, Z.A. (1995) *Mol. Simul.*, **15**, 381.
Rycerz, Z.A. and Jacobs, P.W.M. (1990) *Comp. Phys. Commun.*, **60**, 53.
Rycerz, Z.A. and Jacobs, P.W.M. (1991a) *Comp. Phys. Commun.*, **62**, 125.
Rycerz, Z.A. and Jacobs, P.W.M. (1991b) *Comp. Phys. Commun.*, **62**, 145.
Rycerz, Z.A. and Jacobs, P.W.M. (1992a) In *Grand Challenges in Supercomputing, Proc. Supercomputing Symposium '92*, Atmospheric Environment Science, Environment Canada, Montréal, p. 241.

Rycerz, Z.A. and Jacobs, P.W.M. (1992b) In *Grand Challenges in Supercomputing, Proc. Supercomputing Symposium '92*, Atmospheric Environment Science, Environment Canada, Montréal, pp. 277.
Rycerz, Z.A. and Jacobs, P.W.M. (1992c) *Mol. Simul.*, **8**, 249.
Rycerz, Z.A. and Jacobs, P.W.M. (1995) unpublished work.
Sangster, M.J.L. and Dixon, M. (1976) *Adv. Phys.*, **25**, 247.
Sindzingre, P. and Gillan, M.J. (1988) *J. Phys. C*, **21**, 4017.
Sindzingre, P. and Gillan, M.J. (1990) *J. Phys. Condens. Matter*, **2**, 7033.
Smith, W. and Gillan, M.J. (1992) *J. Phys. Cond. Matter*, **4**, 3215.
Somayazulu, M.S., Sharma, S.M. and Sikka, S.K. (1994) *Phys. Rev. Lett.*, **73**, 98.
Sonwalkar, N., Yip, S. and Sunder, S.S. (1994) *J. Chem. Phys.*, **101**, 3216.
Sperb, R. (1994) *Mol. Simul.*, **13**, 189.
Tallon, J.L. (1988) *Phys. Rev.* B, **38**, 9096.
Vashishta P. and Rahman A. (1978) *Phys. Rev. Lett.*, 40, 1337.
Verlet, L. (1967) *Phys. Rev.*, **159**, 98.
Vessal, B., Amini, M., Fincham., D. and Catlow, C.R.A. (1989a) *Phil. Mag. B*, **60**, 753.
Vessal, B., Leslie, M. and Catlow, C.R.A. (1989b) *Mol. Simul.*, **3**, 123.
Vessal, B., Amini, M. and Akbarzadeh, H. (1994) *J. Chem. Phys.*, **101**, 7823.
Willis, B.T.M. (1965) *Acta Crystallogr.*, **18**, 75.
Wolf, M.L., Walker, J.R. and Catlow, C.R.A. (1984) *J. Phys. C*, **17**, 6623.
Woodcock, L.V. (1971) *Chem. Phys. Lett.*, **10**, 257.

5

Simulated Annealing and Structure Solution

C.M. Freeman, A.M. Gorman and J.M. Newsam

1 INTRODUCTION

The use of computational techniques in modelling inorganic materials is the central theme of this book and indeed structure is frequently an essential starting point for computer modelling studies. Simulated annealing has emerged as a methodology which, when combined with energy functions of the types described in Chapter 1, is capable of yielding good models of condensed inorganic systems. The present chapter focuses on annealing based optimization procedures, highlighting their role as a complement to experimental approaches in the determination of structure in the solid state.

Of course for many systems, in particular those for which good quality crystals can be obtained, structural information can be derived with comparative ease from single crystal diffraction studies. Indeed, there are many thousands of structure determinations made each year, often as a routine component of the characterization of new materials. However, many technologically important materials cannot be, or are not in typical production conditions, prepared as single crystal samples suitable for routine crystal studies. For example, the pharmaceutical industry generally requires polycrystalline products of well-defined crystal polymorphs and most heterogeneous catalysts are obtained and used as polycrystalline materials. In these instances, powder diffraction methods, for which polycrystalline samples are a necessity, are routinely used as a means of confirming the presence of a given phase or crystal polymorph. The development of full diffraction profile Rietveld (Rietveld, 1969; Cheetham and Taylor, 1977; Young, 1993) refinement techniques has made powder diffraction a viable route to structure determination in many instances (Cheetham and Taylor, 1977) with a growing number of powder based structural determinations published annually. The use of the technique in structure determination demands an initial model, or solution, which is then refined against the experimental pattern. This initial model can be obtained from Direct Methods (Gilmore *et al.*, 1991; Morris *et al.*, 1992) using an appropriate decomposition of the pattern's data (David, 1987; Estermann and Gramlich, 1991) into a synthetic single crystal data set. Alternatively, structural

COMPUTER MODELLING IN INORGANIC CRYSTALLOGRAPHY
ISBN 0-12-164135-X

analogy with known compounds or model building methodologies can provide the required starting point.

Computational modelling techniques, such as interactive model building, analysis, energy minimization, molecular dynamics and simulated annealing can also provide the necessary starting points for structural refinement and model completion. Figure 5.1 provides a representation of the model building steps required to generate a sequence of plausible models based upon a common structural theme, illustrated in this case by the construction of a number of structures from ribbons of edge- and face-sharing octahedrally coordinated cations, the basic building blocks of the Hollandite and related structures (DeGuzman *et al.*, 1994). The computational techniques, upon which such analyses are based, mirror closely the methodologies employed in physical model building. However, interactive molecular, or solid state modelling environments, permit many important extensions, such as the rapid calculation of properties ranging from densities, steric accessibility and symmetry to the prediction of diffraction, infrared and NMR characteristics, which are seldom available in physical model building (Newsam and Li, 1995).

As discussed in several chapters of this book, complementing model construction and analytical techniques are specialized simulation methods which employ energy based methodologies to determine properties (as described in Chapter 3 of this book). These methods take either quantum or classical energy descriptions of the physical system, functions of system structure, and probe system energetics, stabilities, vibrational and dynamical properties depending on the type of simulation performed. The reliability and the computational expense of such simulations is determined both by the quality of the potential energy function employed and by the simulation methodology used. In many instances it is necessary to compromise between the most accurate means of determining the energy of the system and the most accurate and assumption free simulation protocol.

Differing simulation methodologies make differing demands of the potential energy function. Clearly, simulation of structure, which can be readily compared with available experimental evidence, may only need relatively simple energy functions which describe the location of minima with reasonable accuracy, as the essential focus of this form of simulation lies in the determination of the minimum position of the energy function: for structural properties, calculation of relative energetics, forces and second derivatives (leading to vibrational properties) are unimportant.

If the chemical connectivity of a given system can be anticipated, as for example in the case of framework structured solids, reasonable bounds for first and second neighbour bond distances can be readily defined. Distance targets can, therefore, be set up for such a system and the resulting distance least squares procedure (Meier and Villiger, 1969) (DLS) takes as input a trial or random set of coordinates, the crystal symmetry and possibly site symmetry constraints, which, when combined with defined connectivity and target bond

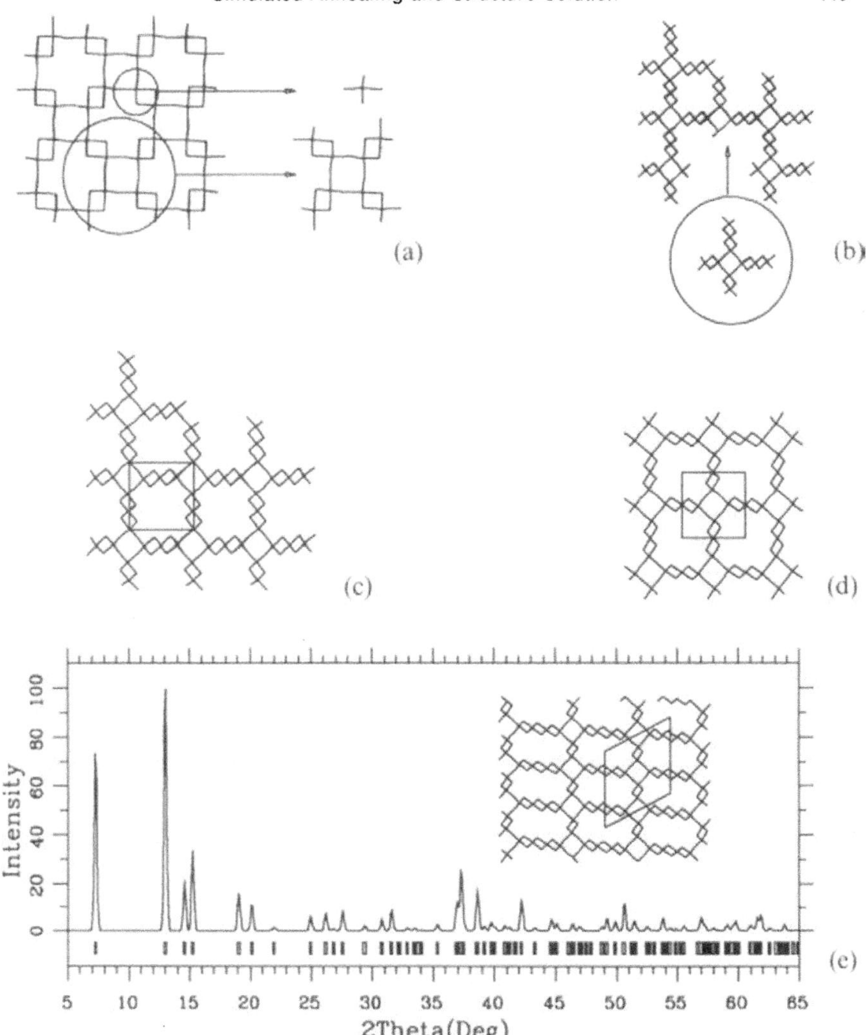

Figure 5.1 Computational model building. In this example structural elements from the Hollandite structure (highlighted in (a)) are combined (b) to produce analogue structures. Once the structural elements have been reassembled in the appropriate form, a new unit cell can be defined (either automatically or by selecting translationally related atoms) (shown in (c)) and the symmetry of the new structure, constructed from regular and symmetrical elements, automatically determined (d) (in this case the new, enlarged, model is $P4_2/m$; Hollandite itself is $I4/m$). Once a new model is constructed its properties can be readily characterized as in (e) where the calculated X-ray powder diffraction pattern of one of many alternative Hollandite derivative structures is shown.

distances, provides a route to reliable structural models. The method has become a standard technique in zeolite structural analysis and has been successfully employed in studies of various inorganic crystal materials (Tillmanns *et al.*, 1973; Khan, 1976; Catlow and Cormack, 1987). The function to be minimized is a sum of squared deviations from prescribed target bond distance values, making specialized least squares minimization methods applicable. The DLS procedure provides an efficient refinement technique once a structure is sufficiently well understood for its connectivity to be input to the program. In DLS the focus is on the location of a structure which maximizes the agreement between calculated and target bond lengths and little emphasis is placed on the ranking of competing structural models. More sophisticated procedures for building in chemical knowledge relating to connectivity are discussed by Brown in Chapter 2.

2 MULTIDIMENSIONAL ENERGY SURFACES

Despite the fact that structural simulation places limited demands upon the description of the potential energy of the system, there are complexities in the exploration of this surface. A simple energy surface for an inorganic system is illustrated in Fig. 5.2. Here the energy has been calculated, using an empirically derived generalized force field (the Extensible Systematic Force Field, ESFF (Discover, 1995; Barlow *et al.*, 1996a,b), as a function of the value of two variable side chain torsional angles. The resulting two-dimensional energy surface possesses a large number of minima and maxima and illustrates the difficulty of analysing energy surfaces for problems in high dimensions. Even when simplifying assumptions are made, such as the use of rigid

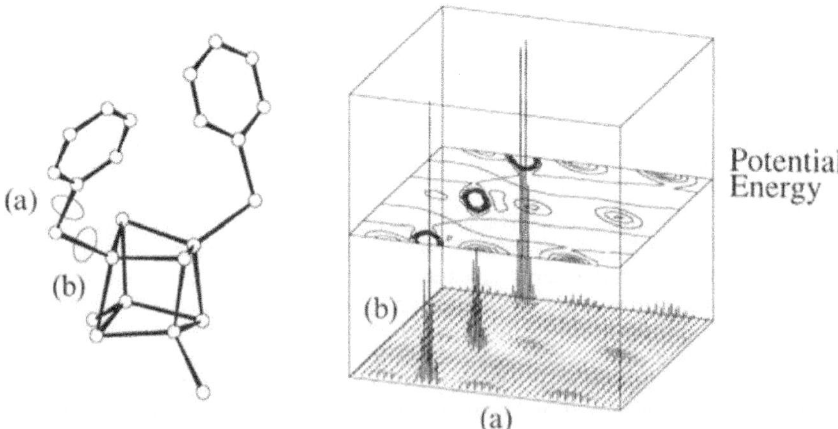

Figure 5.2 Energy surface associated with two variable torsional angles in one of the side chains of an iron sulphur complex (structure taken from Kanatzidis *et al.*, 1984). The resulting energy map, a function of the two variable side chain angles highlighted, a and b, is shown on the right-hand side of the figure, as both a contour map and a three-dimensional histogram. A variety of maxima and minima are readily apparent in the figure.

rotations about dihedral angles as in Fig. 5.2, for most systems of chemical interest the energy surface will be highly complex. In the case of determining a minimum energy configuration of a rigid molecule within a triclinic cell the problem will be nine-dimensional (three rotational degrees of freedom for the molecule and six variable degrees of freedom to describe the lattice). It may be possible to perform a brute force determination of the location of minima for a two-dimensional problem, with a suitably economical energy function and perhaps 100 energy evaluations per variable—a total of 10 000 energy evaluations (100^2). However, the corresponding approach is not effective in higher dimensions. For example 1 000 000 000 000 000 000 (100^9) energy evaluations are required for the rigid molecule within a triclinic cell possessing just nine possible variables, which is beyond the scope of present and prospective computer systems.

A second two-dimensional functional surface is illustrated in Fig. 5.3. This surface, which resembles the torsional map of Fig. 5.2, is a slice through the R value surface, a measure of the match of a calculated and experimental powder X-ray diffraction pattern, as a function of a single atom's position in two dimensions. R is calculated according to the following expression

$$R = \frac{\sum ||Y_{\mathrm{c}}| - |Y_0||}{\sum |Y_0|} \tag{1}$$

where Y_{c} and Y_0 are calculated and observed intensities, respectively, and the sum extends over all profile points. The figure shows R as a function of the movement of a single atom in an inorganic crystal structure on a 10 Å square. Again the surface has a multitude of minima and maxima and is considerably more complicated when the full dimensionality of the system is taken into account.

Although direct scans of the energy surface can be effective for low dimensional systems, such as a torsional energy map, or probing the passage of a particle from one site to another, a more efficient mechanism for exploring an energy surface stems from the development of automatic gradient based minimization methods (Fletcher and Powell, 1963; Gill *et al.*, 1981; Powell, 1982; Scales, 1985; Sedgewick, 1988; Press *et al.*, 1989). Here a given starting point is selected and is used as input to the optimization procedure. On the basis of gradient and energy information, by analogy to quadratic and simple functions, the initial structure is then iteratively improved. Unconstrained gradient based optimization procedures, are, however, generally confounded by functions that possess many local minima, a characteristic, as noted above, of the types of function that are encountered in structural simulations. This problem is highlighted schematically in Fig. 5.4(a) for a one-dimensional function where the path of a gradient based optimizer from a starting point, S, to a minimum, M, is highlighted. As Fig. 5.4(b) illustrates, the complete function in fact possesses several minima including a lower minimum, M,

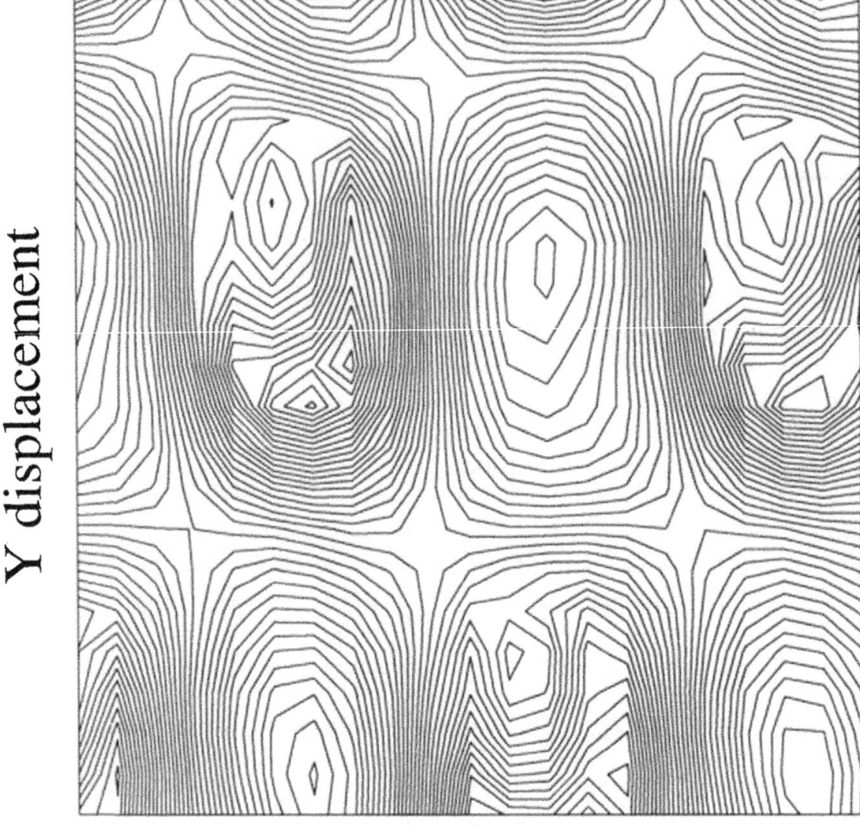

Figure 5.3 Residual, *R*, value for the match of calculated and observed powder X-ray diffraction patterns as a function of displacement for a single atom over a 10 Å square grid in an arbitrary structural model. The large number of minima on this two-dimensional surface are combined with similar complexity in all the degrees of freedom for the system.

which cannot be located by a gradient based optimizer from the supplied starting point, S.

The problem of local minima in function minimization or optimization problems has given rise to the development of a variety of algorithms which are able to seek global minima. Traditional gradient based optimizers proceed by the selection of fruitful search directions and subsequent numerical one-dimensional optimization along these paths. Such methods are therefore inherently prone to the discovery of local minima in the vicinity of their starting point, as illustrated in Fig. 5.4(b). This property is in fact desirable in

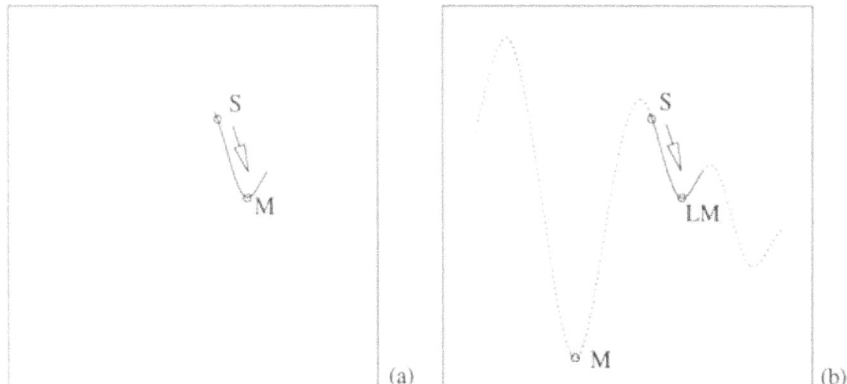

Figure 5.4 (a) Gradient based minimization methods proceed 'downhill' from a starting point, S, towards the nearest minimum, M. In 4(b) this behaviour can only find the nearer local minimum, LM, and not the deeper minimum, M.

the case of the refinement of structures from reasonable starting points, as a prelude to vibrational analysis or as a means of analysing potential relaxation to lower symmetry (Shannon *et al.*, 1991; Catlow *et al.*, 1993). However, global optimization remains a taxing problem and protocols for efficient analysis remain a subject of considerable research effort (Berg, 1993). For example numerical procedures based upon 'gentlest ascent' (as opposed to the steepest descent minimization procedure) are able to walk between minima (Baker, 1986) and a variety of methods have been described which apply modified and adaptive direct search methodologies with encouraging results (Cvijovic and Klinowski, 1995). One strategy which has succeeded in finding many minima for a given system involves repeated gradient optimizations from a large number of starting points (Freeman *et al.*, 1991). More sophisticated global optimization methods that use knowledge of the types of function to be minimized have also been reported. For example, as noted in Chapter 1, model calculations by Scheraga and coworkers (Piela *et al.*, 1989) have used the sequential deformation of the potential energy surface, exploiting the readily calculable second derivative of some functions, to generate surfaces that possess a single minimum and upon which that minimum can be readily located using standard techniques. The energy surface is then sequentially transformed back to its original form, with the previous minimum providing the starting point for successive minimizations. The deformation of the surface is illustrated schematically in Fig. 5.5, where the application of second derivative smoothing transforms a complete surface with many minima into a simple form upon which the minimum can be readily located.

An alternative class of scheme which has been shown to be effective in chemical, structural and docking simulations (Hartke, 1993; Brodmeier and

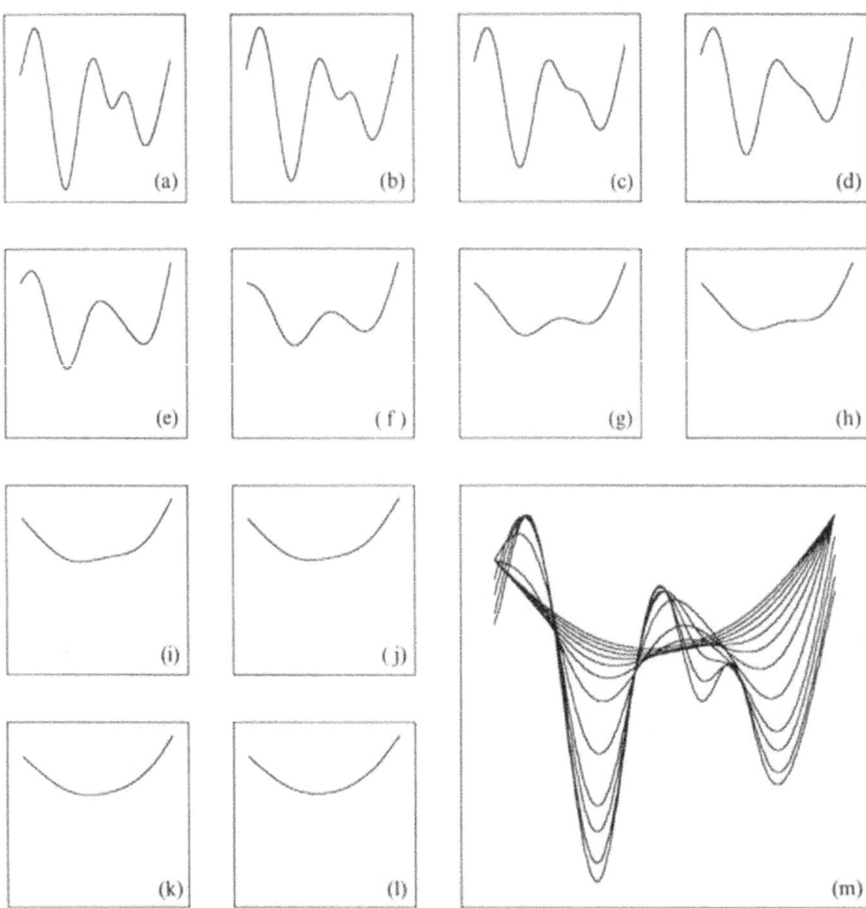

Figure 5.5 Deformation of an energy surface (represented here as a single variable function) using second derivatives. The original function, a, is successively transformed into a function, l, (in images a–l) which possesses a single minimum; $f^{\text{new}}(x)$ is given by $f(x) + \beta f''(x)$, where β is a constant. Points of inflection (where $f''(x)$ is zero) are unaffected by this transformation, maxima are decreased and minima raised. Eventually a function with a single minimum is obtained. This minimum can be readily located and used as a starting point in the minimization of *k*. The effect of the second derivative deformation on the original function is shown in m, where images a–l are superimposed.

Pretsch, 1994; Xiao and Williams, 1994) uses genetic algorithms. These methodologies maintain a set of configurational states which are combined and retained or discarded on the basis of a variety of selection criteria. Genetic algorithm based optimization combined with an appropriate energy function has recently been shown to provide reasonable structural models and, indeed, permitted the structure solution from X-ray powder diffraction data of a novel lithium ruthenate compound (Bush *et al.*, 1995), as described in Chapter 1.

3 SIMULATED ANNEALING

Many global minimization methods avoid the local bias of gradient methods by permitting function-increasing moves (as shown schematically in Fig. 5.6). This aspect of global minimization methodologies leads to an appealing analogy with the process of annealing a material in the formation of its crystalline state. In such a physical system the equilibrium distribution of states accessible at a given temperature constitutes a well-defined phase for the system. Each phase is made up of a collection of states and their associated probabilities. For a physical system the relative probability of two states is simply related to the temperature via Boltzmann statistics. At high temperature (when the system is molten) there will be little difference in the probability of any two states, while at low temperature, the state with the lowest potential energy will dominate. The temperature of a system may thus be used as a convenient control of the extent to which states higher in energy than the global minimum can be accessed. The simulation is commenced at high temperature and is gradually cooled. When the temperature of the system reaches zero, only energy-reducing moves are accepted.

As for all optimization procedures, the function minimized need not be directly related to the potential energy of the physical system. For example, the residual R of equation (1), a measure of the fit of measured and observed diffraction patterns, can form the object function. Similarly the temperature of the system simply determines the relative probability of the simulation permitting two successive states, and provides a convenient means of controlling the degree of searching permitted during the course of the simulation. As discussed in Chapter 1, it should not be equated in this context with the thermodynamic temperature.

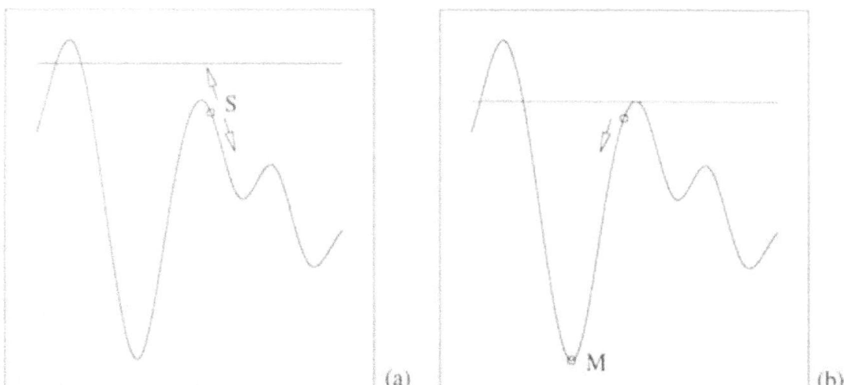

Figure 5.6 (a) In simulated annealing simulations, there is a finite probability of accepting energy-increasing moves. In these images a particular probability 'temperature' is shown schematically, as a horizontal line. In (b) the temperature has been lowered substantially, and now only energy-reducing moves are effectively allowed.

Locating the minimum requires the definition of a path over the parameter space surface: given an initial starting point, one must be able to generate a new configuration from it. For molecular systems a convenient approach employs the molecular dynamics method (reviewed in this volume by Jacobs and Rycerz in Chapter 4), in which positions and velocities for the particles of the system are allowed to evolve by solving Newton's equations of motion. The simulation, which is initiated at high temperature, is then subjected to an annealing schedule, specifying the number of steps spent in the molten state and the subsequent rate of cooling, allowing the system to settle into an energy minimum configuration. Molecular dynamics based simulated annealing has found greatest use for macromolecular systems where the choice of mass for variable (structural) parameters is unambiguous, such as in protein structure solution and refinement from experimental diffraction information (Brunger *et al.*, 1991; Brunger and Rice, 1995) and in the extraction of tertiary protein structure from NMR derived distance constraints (Crippen and Havel, 1988). However, it is worth noting that judicious choice of masses for non-structural parameters and temperature control protocols permit the use of the molecular dynamics method in optimization of both structural and electronic parameters (Car and Parinello, 1985).

An alternative to molecular dynamics based simulated annealing is provided by Metropolis importance sampling Monte Carlo (Metropolis *et al.*, 1953) which has been widely exploited in the evaluation of configurational integrals (Ciccotti *et al.*, 1987) and in simulations of the physical properties of liquids and solids (Allen and Tildesley, 1987). Here, as outlined in Chapters 1 and 2, a particle or variable is selected at random and displaced; both the direction and magnitude of the applied displacement within standard bounds are randomly selected. The energy of this new state, E_{new}, is evaluated and the state accepted if it satisfies either of the following criteria:

$$E_{\text{new}} < E_{\text{initial}} \tag{2}$$

or if

$$\exp\left\{\frac{E_{\text{initial}} - E_{\text{new}}}{kT}\right\} > \text{RND} \tag{3}$$

where E_{initial} is the total energy of the system before a parameter adjustment or step, E_{new} is the energy of the system after this modification, T is the temperature, k a constant term and RND represents a random number between 0 and 1. Failure to satisfy either of the criteria results in the rejection of the move (Fig. 5.7 provides a flow diagram of the Metropolis Monte Carlo algorithm). Metropolis and coworkers demonstrated that this prescription leads to the generation of a chain of configurations representative of a Boltzmann distribution for the system. At low temperatures the probability of accepting energy increasing movements is low; at high temperatures, however, this

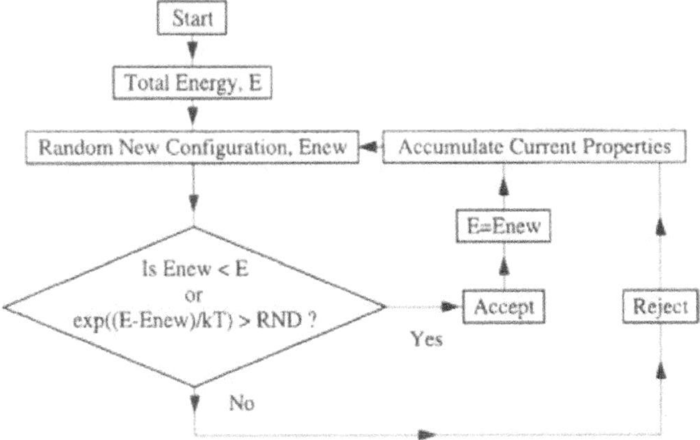

Figure 5.7 Flow diagram of the Metropolis Monte Carlo scheme. RND represents a random number between 0 and 1.

probability becomes dominant, permitting substantial sampling of the parameter space for the system.

In simulated annealing using the Metropolis Monte Carlo technique, the temperature is raised from an arbitrary point in increments or powers until the system becomes molten. This melting transition can be identified by recognizing a maximum in the heat capacity of the system as a function of temperature, the heat capacity, C_v, being a direct function of the degree of energy fluctuation present in the system at a given temperature:

$$C_v = \frac{\langle E_T^2 \rangle - \langle E_T \rangle^2}{kT^2} \tag{4}$$

where E_T is the system energy at a given temperature. Alternatively the melting transition can be detected as a point at which 50%, for example, of attempted moves are accepted. From the molten state, the temperature of the system is reduced slowly according to a defined schedule, the details of which can be customized for particular systems (Press *et al.*, 1989). Often, for example, the rate of cooling is reduced in the vicinity of a change in the system's heat capacity, indicating the onset of the sought-after freezing transition. The annealing process corresponds to the physical process of solidification. In the limit of low temperature, only energy-reducing movements are accepted and gradient based optimization procedures will be significantly more efficient.

Although the terminology of simulated annealing is derived from physical simulations, the method can be applied to general optimization and minimization problems (Kirkpatrick *et al.*, 1983). For example the method has found application to the travelling salesman problem (Press *et al.*, 1989) where the optimal path between a number of locations, each of which must be visited, is sought, and analogously in the design of integrated circuits (Kirkpatrick *et al.*,

1983). The method has also been employed extensively in the simulation of lattice models (Binder, 1987) and ordering in aluminosilicate networks (Soukoulis, 1984; Newsam, 1992).

4 SIMULATION OF STRUCTURE

Structural simulations, based upon a suitable energy function, begin with a postulated trial model. This model is then iteratively adjusted using one of the gradient or global optimization procedures described in the preceding section, or alternatively using the molecular dynamics procedure, to yield a refined structural model. Where there is no long-range order, and clusters of atoms may be considered in isolation from their environment, for example in the case of the conformation properties of an organic molecule, energy minimization using empirical energy functions is frequently referred to as Molecular Mechanics (Burkert and Allinger, 1982), and has become a standard technique. Molecular Mechanics' strength lies in the transferability of potential functions for organic molecules. Considerable effort has been expended in developing accurate descriptions of inter- and intra-molecular interactions for the study of such materials (Dinur and Hagler, 1991; Hwang *et al.*, 1994; Maple *et al.*, 1994). The power of Molecular Mechanics based methods in probing the structure and stability of molecular crystals has also received considerable attention (Williams, 1967; Pertsin and Kitaigorodskii, 1987) and has achieved notable successes (Ramdas and Thomas, 1978; Gavezzotti, 1991; Leusen *et al.*, 1995).

Organometallic systems necessarily exhibit greater diversity in the forms of bonding that must be accounted for by empirical potential functions than do their organic counterparts; and Molecular Mechanics calculations have accordingly played a smaller role for these materials. However, a number of studies have now been reported which extend the isolated description (Allured *et al.*, 1991; Castonguay and Rappe, 1992; Rappe *et al.*, 1992) of such compounds to the solid state (Talamo *et al.*, 1995).

In the study of inorganic systems as discussed in several chapters in this book, classical simulations techniques have been highly successful in describing structural and energetic properties (Catlow and Mackrodt, 1982; Catlow and Cormack, 1987). It is notable that the success of such methods in the calculation of bulk properties has encouraged their application in defect (Catlow *et al.*, 1982), transport (Catlow, 1986), and surface studies (Tasker, 1990; Gay and Rohl, 1995). Structural applications have included the study of symmetry lowering temperature controlled distortions of high symmetry materials, such as the orthorhombic form of silicalite, the siliceous analogue of the zeolite ZSM-5 (Bell *et al.*, 1990). Similar applications have been described resulting in structure determination for a number of complex materials (Shannon *et al.*, 1991; Catlow *et al.*, 1993). Simulations of structure can also be made where the concentration of a particular component is low, such as in the analysis of framework iron substitution in ZSM-5 (Lewis *et al.*, 1995) and the examination of atomic cluster (Jentys *et al.*, 1993) properties in microporous materials.

5 NON-FRAMEWORK CATION LOCATION IN ALUMINOSILICATE ZEOLITES

An illustration of the ability of modelling methods to provide information on the structural properties for a complex system is provided by an investigation of non-framework cation positions in aluminosilicate materials (Newsam *et al.*, 1996). The position of charge balancing, extra-framework cations within zeolite structures frequently governs sorptive and catalytic properties. Direct measurement of such positions often requires protracted analysis of good quality diffraction data. However, a simple potential model (detailed in Table 5.1) provides a means of simulating cation locations and relative populations of different sites in accord with available experimental knowledge.

Results for two systems (Li-A(BW) and zeolite -4A), chosen for the reliability of their structural data and their precisely known Si–Al distributions, are shown in Tables 5.2 and 5.3 and Figs 5.8 and 5.9. The procedure used in this study is an extension of a method originally developed to probe the preferred binding of molecular sorbates within host structures (Freeman *et al.*, 1991) and takes as input a suitable framework model and the number of guest items to be incorporated (Kerr, 1974; Pluth and Smith, 1980; Newsam, 1986). For zeolite -4A, the reported crystallographic coordinates were used, but the cell dimensions reduced to an a = 12.305Å, triclinic, $P1$, subcell. The resulting subcell possesses 'anti-Lowensteinian', Al–O–Al linkages across unit cell faces, but nevertheless proves suitable for probing approximate non-framework cation arrangements. It is worth emphasizing that, as a result of their electrostatic influence, aluminium sites have a pronounced influence on

Table 5.1 Potential parameters used in placement of Li^+ and Na^+ extra-framework cations in zeolites Li-A(BW) and 4A.

Ion	$q(e)$	A(kcal mol^{-1} Å^{-12})	C(kcal mol^{-1} Å^{-6})
Si	2.4	103.80	0.000 69
Al	1.4	278.39	1 690.149 59
O	−1.2	388 611.37	0.189 28
Na^+	1.0	67 423.64	0.000 46
Li^+	1.0	1 252.15	0.034 37

Potential energy, V, is calculated according to the following expression:

$$V = \sum_{ij} \frac{q_i q_j}{r_{ij}} + \frac{A_{ij}}{r_{ij}^{12}} - \frac{C_{ij}}{r_{ij}^{6}}$$

where r_{ij} is the distance between two particles, $A_{ij} = (A_i A_j)^{1/2}$ and $C_{ij} = (C_i C_j)^{1/2}$.

Table 5.2 Results for automatic placement of four lithium cations in Li-A(BW) aluminosilicate framework structure.

Run	Relative energy kcal mol^{-1}	Space group[a]
1	−7946.54	33
14	−7946.54	33
15	−7946.54	33
17	−7946.54	33
21	−7946.54	33
22	−7946.54	33
23	−7946.54	33
24	−7946.54	33
25	−7946.54	33
26	−7946.54	33
28	−7946.54	33
30	−7946.54	33
6	−7946.54	33
7	−7946.54	33
8	−7946.54	33
12	−7936.29	14
16	−7925.89	7
29	−7925.89	7
2	−7920.58	7
13	−7919.06	1
20	−7918.67	1
10	−7899.43	1
27	−7893.01	33
9	−7891.61	1
18	−7891.23	1
11	−7890.13	1
4	−7887.90	1
19	−7887.09	1
5	−7884.56	1
3	−7881.09	1

[a]Automatically determined from *P1* optimized atomic coordinates using an individual atom catchment radius of 0.2 Å.

extra-framework cation locations (den Ouden *et al.*, 1990; Brennan *et al.*, 1994).

Having established a definition of the framework model, a charge balancing complement of Li^+, and in the zeolite -4A case, Na^+, cations was introduced using a direct random insertion procedure (Freeman *et al.*, 1991) providing trial models without excessive steric hindrance between cations and framework. For each zeolite structure 30 trials were generated and used as a starting point for full lattice energy minimization (the calculations were performed with the Discover 3.2 package (Discover, 1995)). For the zeolite Li-A(BW) calculations the coordinates of all atoms within the unit cell were allowed to vary

Table 5.3 Results for automatic placement of 12 sodium cations in a subcell of the zeolite-4A aluminosilicate framework structure showing binding site populations.

Run	Correct[a]	6	8	4	D4R	Other	Energy kcal mol $^{-1}$
2	•	8	3	1	0	0	-10737.92
29	•	8	3	1	0	0	-10737.53
30	•	8	3	1	0	0	-10737.37
10	•	8	3	1	0	0	-10737.16
28	•	8	3	1	0	0	-10737.11
4	•	8	3	1	0	0	-10733.01
20	•	8	3	1	0	0	-10733.01
9	•	8	3	1	0	0	-10731.94
5	•	8	3	1	0	0	-10731.25
27	•	8	3	1	0	0	-10731.24
3	•	8	3	1	0	0	-10729.37
15	–	8	3	0	1	0	-10688.17
26	–	8	3	0	1	0	-10685.40
14	–	7	1	0	0	4	-10658.03
25	–	7	3	1	1	0	-10648.22
11	–	7	3	1	1	0	-10638.41
23	–	8	2	0	1	1	-10633.06
1	–	6	2	4	0	0	-10625.05
21	–	7	2	2	1	0	-10623.37
19	–	6	3	2	1	0	-10613.70
17	–	8	2	0	2	0	-10610.35
8	–	6	3	2	1	0	-10607.16
12	–	7	3	0	2	0	-10598.51
7	–	7	3	0	2	0	-10597.92
16	–	7	3	1	1	0	-10596.71
18	–	7	3	0	2	0	-10589.08
22	–	5	2	3	1	1	-10553.46
24	–	5	2	2	1	2	-10550.70
13	–	5	2	2	2	1	-10331.13
6	–	6	1	3	2	0	-10325.74

[a]• signifies correct structures, – signifies incorrect structures.

independently during the minimization (a total of 28 moving species or 84 variables). Analysis of the structures obtained (Table 5.2), employing automatic symmetry detection, revealed that half of the 30 trial calculations converged to an identical minimum energy structure with space group $Pna2_1$, the experimentally observed arrangement for this material. For zeolite -4A, given the structural complexity of the system, framework atom positions were fixed in the structural optimization, requiring the treatment of 12 moving species or 36 independent variables. The zeolite -4A structures were analysed, by inspection, in terms of the site populations of each non-framework cation site. Of the 30 trials the 11 lowest energy structures have the 'correct' configuration of 8

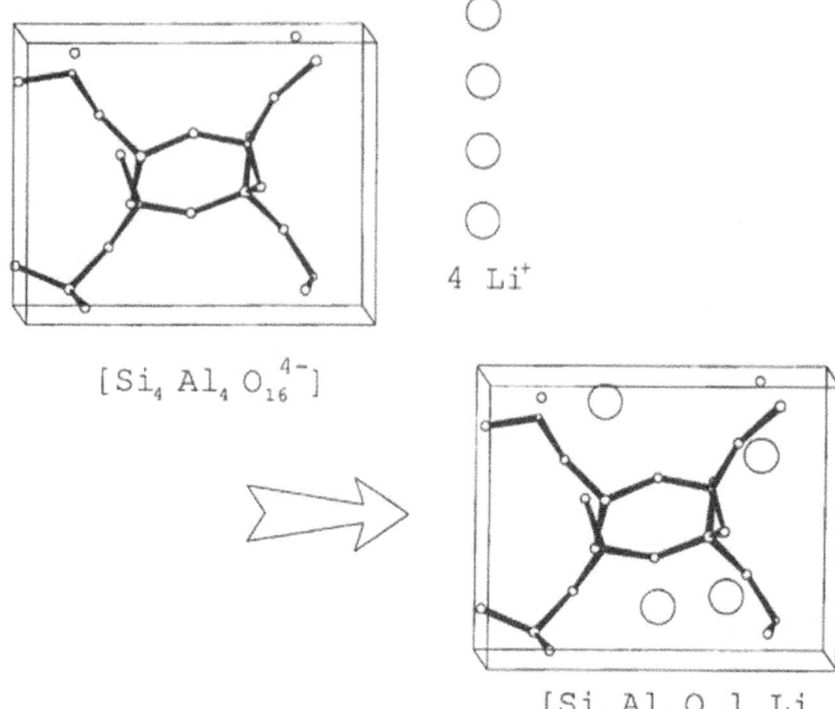

Figure 5.8 Lithium ion binding sites identified in the zeolite Li-A(BW) using energy minimization; the initial step illustrated here introduces Li^+ cations at non-clashing positions.

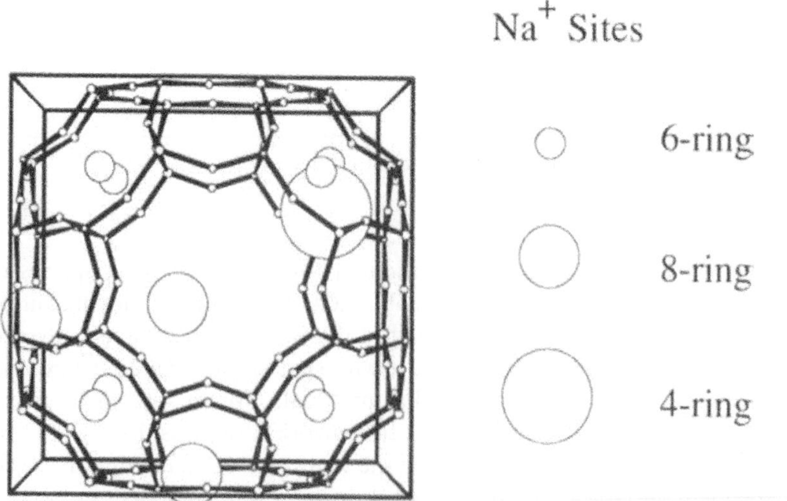

Figure 5.9 Typical model obtained using packing and minimization methods for the distribution of sodium cations in one zeolite -4A framework.

cations on 6-rings, 3 cations on 8-rings and 1 cation adjacent to a 4-ring. A representative structure obtained by the procedure is shown in Fig. 5.9.

The structures obtained in these cases are of sufficient accuracy to form the foundation for model refinement against powder diffraction data using the Rietveld method or further development using more computationally intensive or locally approximate potentials. These calculations required only a matter of minutes on a graphical workstation and employed standard (Vessal, 1995) and transferable potentials. In the case of zeolite -4A, where the detailed local structure provided by the simulations and the sample-averaged diffraction results necessarily differ, the modelling results provide the opportunity to explore the effects of site disorder. Both simulations demonstrate the potential value of modelling procedures in maximizing the information obtainable from experimental investigations. We note that there are now many such examples of the complementarity of simulation and experimental investigations (Bull *et al.*, 1993; Laszkur *et al.*, 1993; Santilli *et al.*, 1993).

'Real space' explorations of energetic and steric possibilities for chemical systems have long been used as a means of maximizing and supplementing the information recoverable from experiments (Rabinovick and Schmidt, 1966; Kitaigorodskii, 1973; Attfield, 1988; Catlow *et al.*, 1993). In the following section we focus on the application of simulated annealing procedures to problems in structural chemistry.

6 SIMULATED ANNEALING APPLICATIONS

Simulated annealing provides a particularly effective method for the development of trial structural models. Its ability to explore energy hypersurfaces, to cross barriers, and to search for regions with low energy structures permits a high degree of latitude in the development of initial starting points. An illustrative application in inorganic chemistry focused on the polymorphs of TiO_2 (Freeman *et al.*, 1992) and showed that simulated annealing combined with a potential model effective in the simulation of the rutile phase could reproduce known TiO_2 polymorphs with few initial assumptions. The course of a calculation for the Brookite polymorph is illustrated in Fig. 5.10. The sequence of images shows the emergence of the regular Brookite structure and the superimposition of the resulting asymmetric unit upon the extended structure viewed in orthogonal directions. The application of the method to systems with a large number of structural degrees of freedom (Brookite has 24 variable atom positions, 72 variables to be optimized) generally demands that several trial calculations be conducted (Freeman and Catlow, 1991). Where powder diffraction data are available, they may be readily used to discriminate between competing low energy models. Thus although the method may require several repetitions, it is in practice straightforward to select reasonable models and, indeed, the software module written to interface to this procedure provides

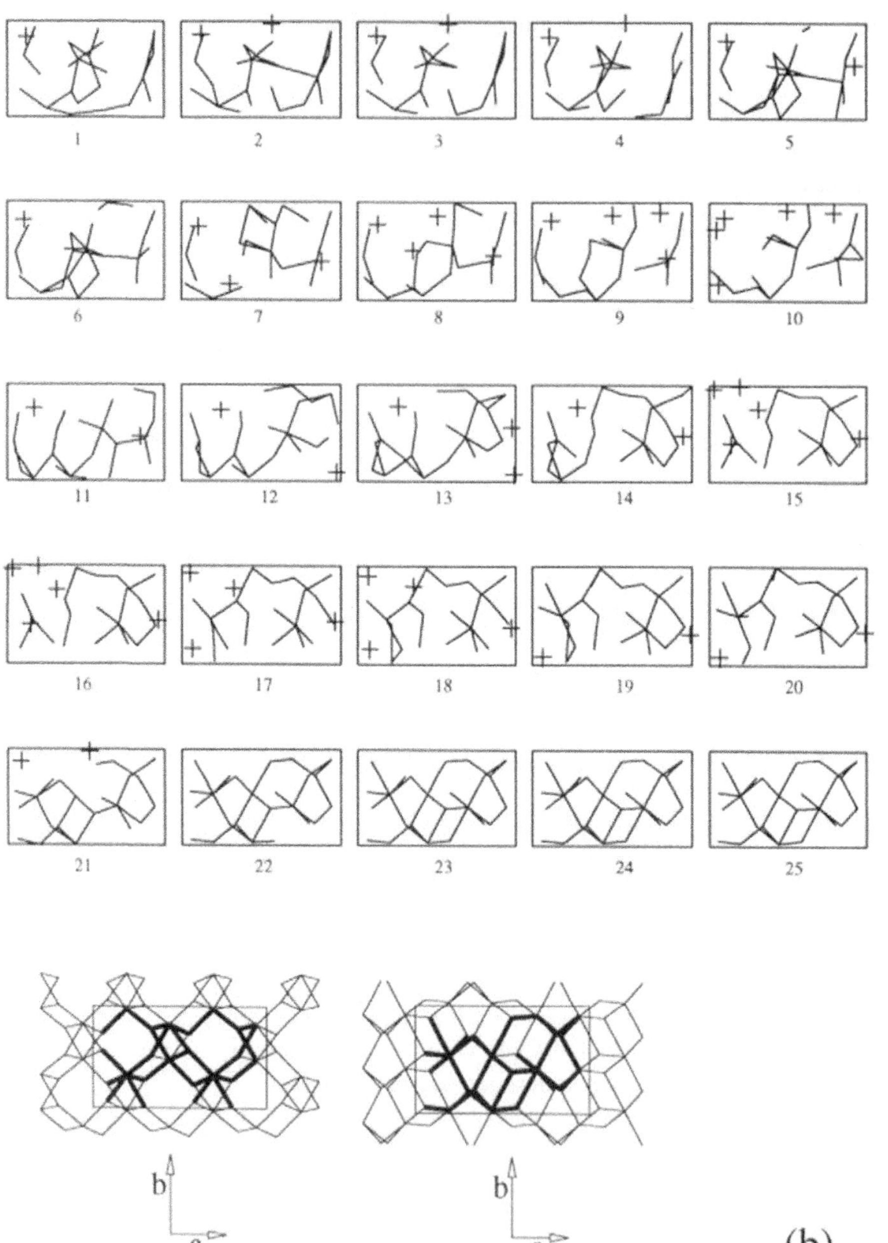

Figure 5.10 (a) Images 1–25 show stages in annealing and energy minimization of the brookite structure. TI^{4+} and O^{2-} ions are introduced randomly into a cell corresponding to the observed unit cell and allowed to relax into minimum energy positions corresponding to the brookite structures. Symmetry is not imposed in this calculation. (b) Orthogonal views of the resulting low energy structure.

integrated and automatic analysis capabilities. Figure 5.11 illustrates the structures obtained from 100 trials using the reported stoichiometry and cell parameters for the lithium ruthenate, Li_3RuO_4. The calculation, which required a total of 6 h on a graphical workstation, yields a large number of chain-like structures with one structure in particular (highlighted in Figs 5.11 and 5.12) in excellent agreement with the experimental structure which was determined, *ab initio*, from powder diffraction data employing a novel genetic algorithm optimization procedure (Bush *et al.*, 1995), as discussed in Chapter 1. Improvements in the quality of empirical potentials will improve the reliability

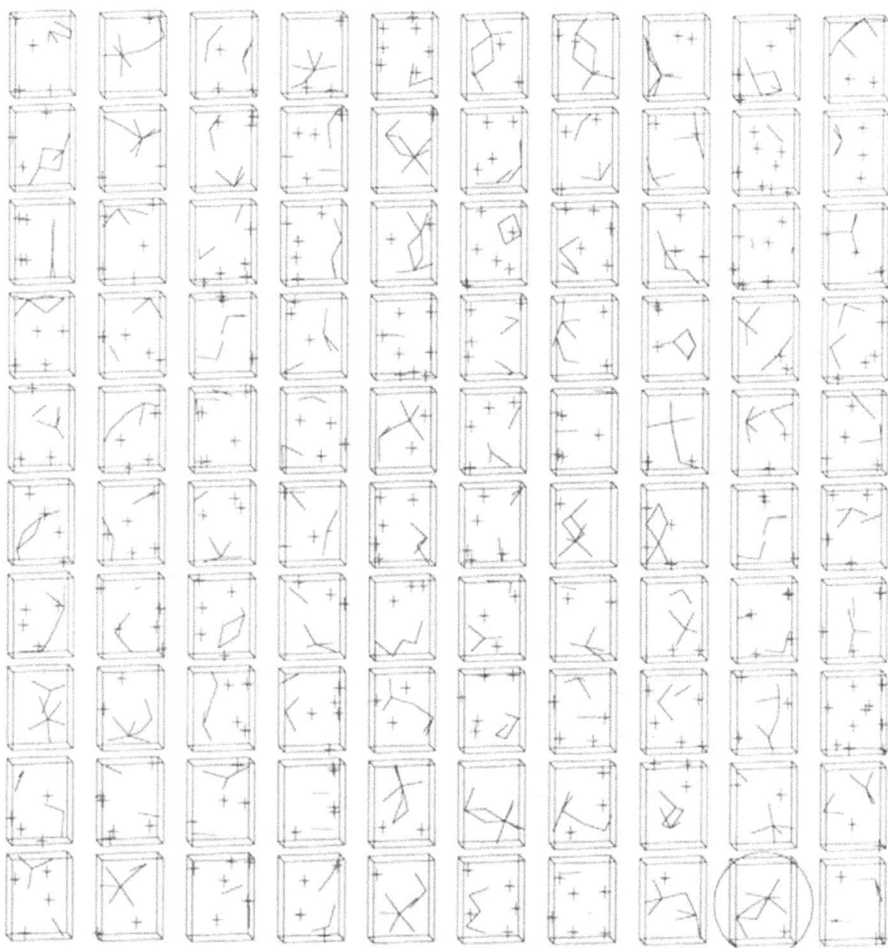

Figure 5.11 Results of 100 combined simulated annealing and energy minimization calculations for Li_3RuO_4 (Bush *et al.*, 1995). Several of the low energy structures have chains of edge- and face-shearing RuO_6 octahedra. The highlighted structure is in good agreement with the experimental structure (Fig. 5.12).

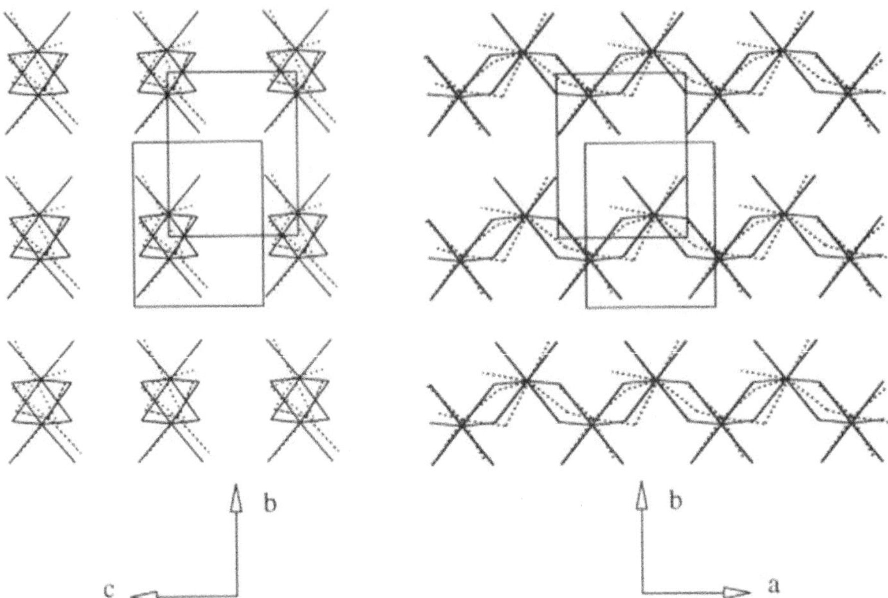

Figure 5.12 Comparison of experimental (Bush *et al.*, 1995) (solid line) and calculated (dashed line) Li_3RuO_4 structures.

of such calculations. However, for structure prediction it is important that improvements in the accuracy of interatomic potentials are not won at the expense of their transferability. Thus simple potential models, which are derived using generalized combining rules, will be the most effective in structural calculations where chemical connectivity, for example, may not initially be known. Where chemical connectivity can be presumed, this information can be exploited in the potential function or calculation procedure. Examples of such uses are afforded first by calculations on zeolite frameworks (discussed below) where the nature of the framework bonding is well defined and second by the use of predefined molecular configurations in rigid body molecular packing calculations.

As discussed in Chapters 1 and 9, zeolite structural chemistry is based upon tetrahedrally coordinated framework cations which are linked, through bridging anions, to form three-dimensional lattices (see Fig. 5.13). The resulting structures can be viewed as three-dimensional, four connected networks with nodes centred at framework cation positions (Smith, 1988). Deem and Newsam (1989) have demonstrated that inversion of the geometric expectations for zeolitic materials, combined with simulated annealing, provides an efficient means for the generation of model networks which conform to known zeolite characteristics. The procedure can be used predictively in solving structures (Nenoff *et al.*, 1993) on the basis of experimentally defined constraints. A related procedure has also been described by Price and coworkers (Price *et al.*, 1992)

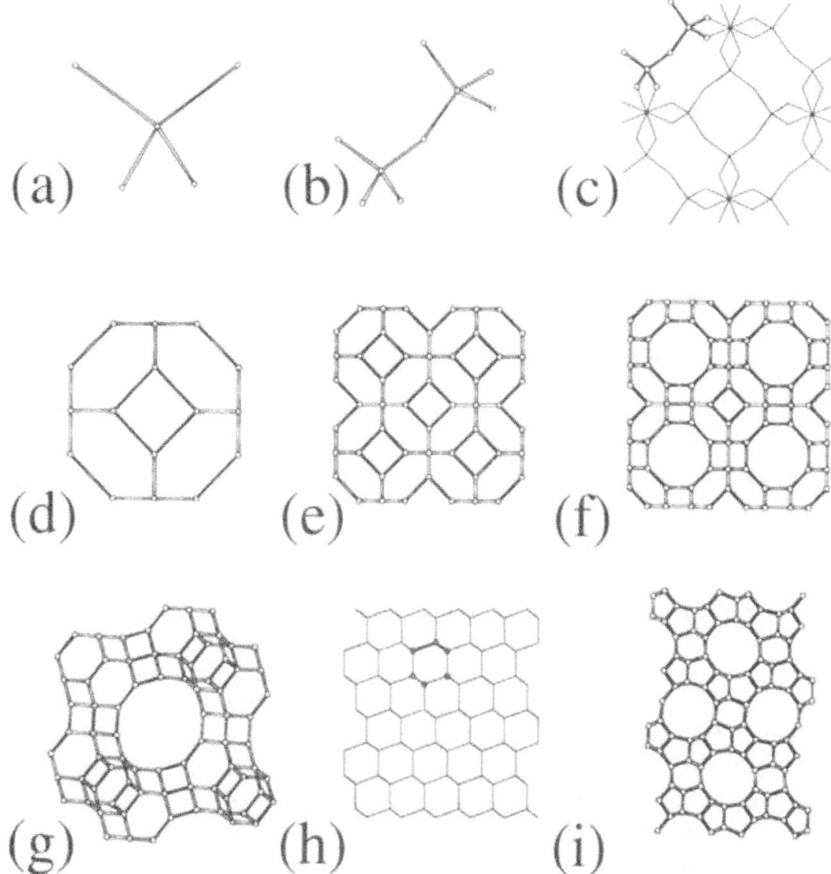

Figure 5.13 Some elements of zeolite structural chemistry (not drawn to scale). TO_4 tetrahedra (a) linked through shared oxygen atoms (b) to form cages, chains and sheets. Illustrated here is the sodalite cage (c) showing framework cation positions only (d). Also shown are the framework structures of sodalite (e), zeolite A (f), faujasite (g), the layered connections of the faujasite framework (h) and the framework of silicalite (i).

which permits the generation of two-dimensional, three connected networks, from which the majority of zeolite frameworks can be assembled by simple stacking operations (Akporiaye and Price, 1989a,b). Such methods succeed by minimizing a function which is a sum of terms representing the expected values of a range of geometric properties. In the case of three-dimensional crystal structure simulations, this function is derived by direct inspection of geometry (bond–length, bond–angle, and a variety of related quantities) distributions. A logical extension of this approach is to include the match between calculated and observed diffraction information, a path successfully employed in the development of reasonable models of amorphous systems by McGreevy and coworkers (McGreevy and Pusztai, 1988, 1990; Keen and McGreevy, 1990) as discussed in Chapter 6. In the case of zeolite structural investigations, the resulting procedure is significantly improved by the addition of diffraction information (Deem and Newsam, 1992). We note that studies of low energy configurations for inorganic materials using simulated annealing have been reported for silicate systems by Gibbs and coworkers (Boisen *et al.*, 1994) and for general inorganic systems such as NbF_4 by Pannetier and coworkers (Lacorre and Pannetier, 1987; Pannetier *et al.*, 1990).

Where there are reasonable assumptions concerning the structural chemistry of the system, approximations may be made which, combined with the match between experimental and calculated diffraction patterns, may be sufficient to determine the structure (Newsam *et al.*, 1992). For example in the case of brookite, 72 structural variables can be reduced to 6, if we assume that titanium is octahedrally coordinated and if we take into account the space group symmetry of the system. The course of a simulated annealing optimization calculation of such a system is illustrated in Fig. 5.14, where comparison is also made between the 'molecular' system and the true experimental structure. Clearly such approaches represent an effective means of accessing the structural information content of the powder diffraction pattern. The results of the application of the method to the compound $BiMo_2O_7OH$ are shown in Fig. 5.15. Here the structure was originally solved from powder diffraction using a combination of direct methods and Rietveld refinement by Hriljac and coworkers (Hriljac and Torardi, 1993). With an assumed space group symmetry, and with the expectation that molybdenum is octahedrally coordinated, simulated annealing driven matching of the experimental pattern provides an approximate model which agrees well with the reported structure (Hriljac and Torardi, 1993), as shown in Fig. 5.15.

7 MOLECULAR CRYSTALS

As discussed in Chapter 11, the simulation of structure has focused for a number of years on the development of models for the crystalline phases of molecular systems (Desiraju, 1989; Holden *et al.*, 1993; Perlstein, 1994; Gibson and Scheraga, 1995). Simulated annealing procedures have been employed in many of these investigations (Gdanitz, 1992; Karfunkel and Gdanitz, 1992; Karfunkel *et al.*, 1993a,b). It is important to recognize that such calculations are founded on the assumption that the observed structure will correspond to the internal energy minimum for a given system (Gavezzotti, 1994) and that seeding (Dunitz and Bernstein, 1995) and the kinetic nature of crystal growth can be effectively ignored. Encouraging results indicate that for many systems these assumptions may be reasonable (Harris *et al.*, 1994; Ramprasad *et al.*, 1995). Where systems possess internal degrees of freedom (as in the case of aspartame (Hatada *et al.*, 1985) shown in Fig. 5.16) these calculations are especially

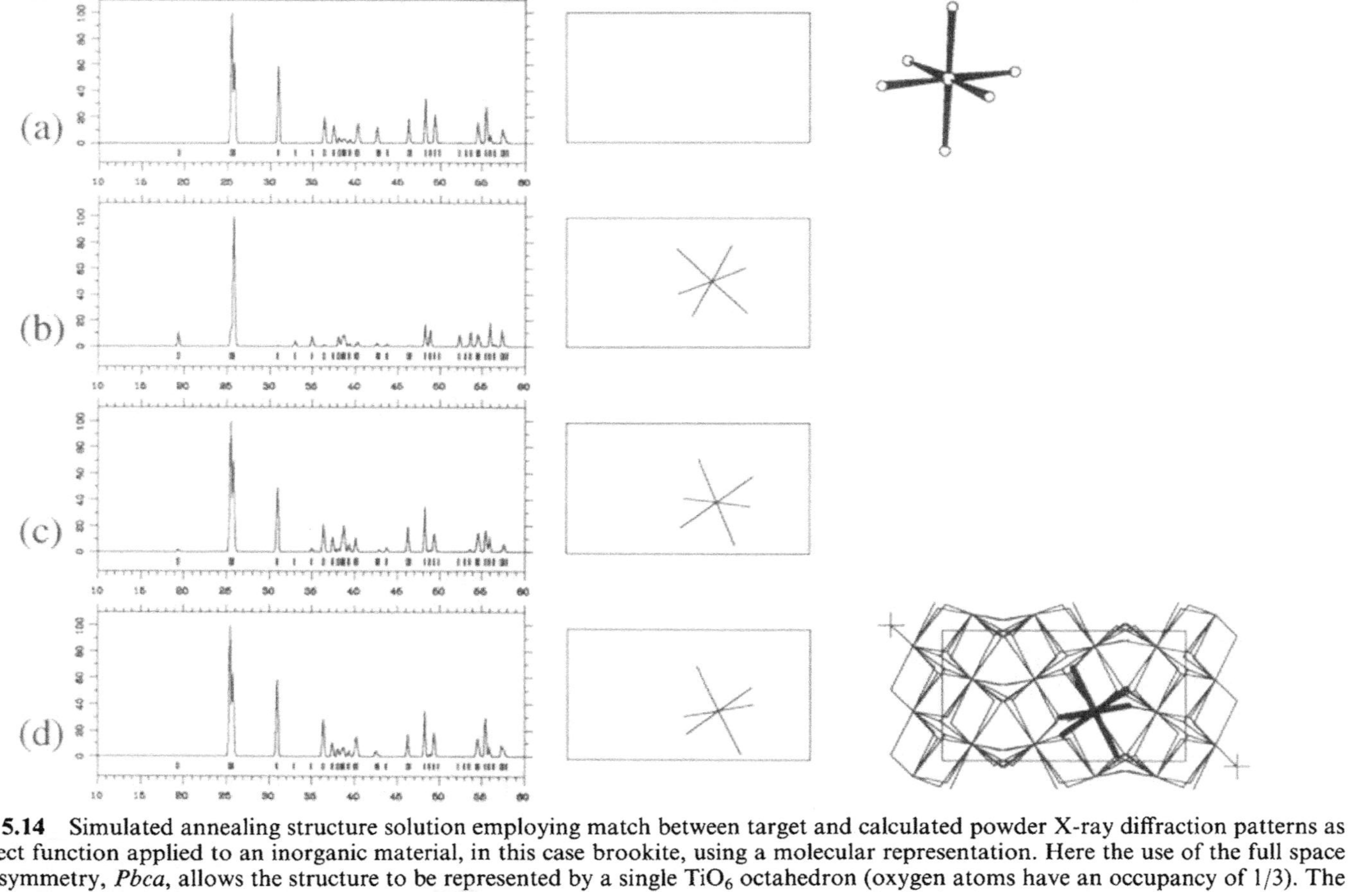

Figure 5.14 Simulated annealing structure solution employing match between target and calculated powder X-ray diffraction patterns as an object function applied to an inorganic material, in this case brookite, using a molecular representation. Here the use of the full space group symmetry, *Pbca*, allows the structure to be represented by a single TiO_6 octahedron (oxygen atoms have an occupancy of 1/3). The target pattern is shown in (a), initial starting point in (b), an intermediate step in (c) and the final calculated pattern in (d); with the agreement between calculated and experimental structures shown on the right.

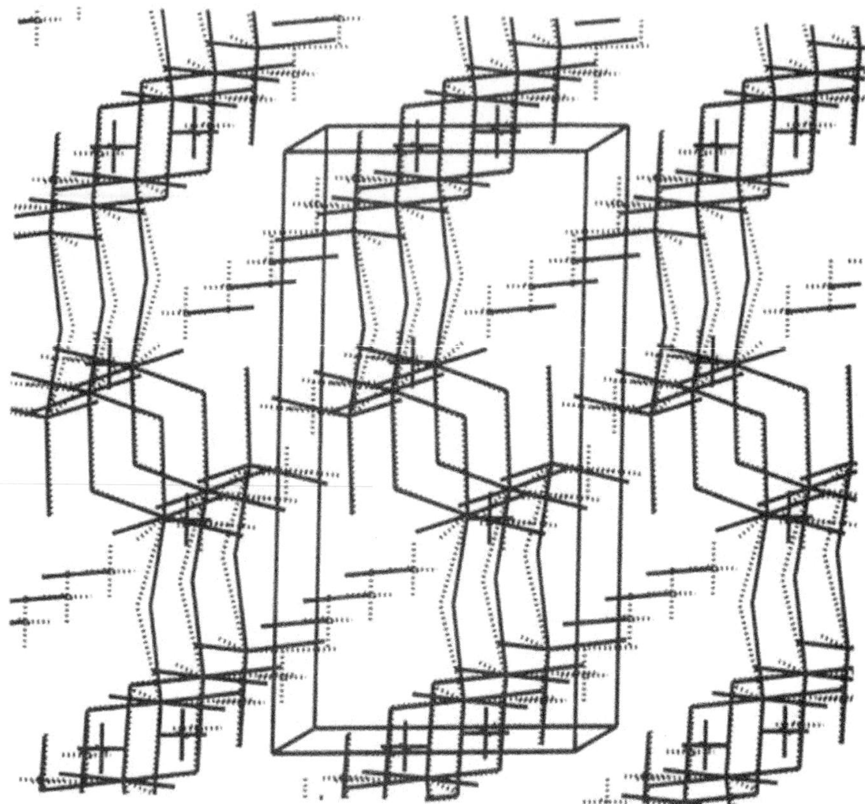

Figure 5.15 Experimental (Hriljac and Toradi, 1993) $BiMo_2O_7OH$ structure (solid line) and structure obtained by simulated annealing driven fitting to the powder diffraction pattern (dashed line).

difficult because of the compounded complexity of the energy hypersurface to be searched. The torsional degrees of freedom of the aspartame molecule (Fig. 5.17) combined with the rotational and translation degrees of freedom of the molecule yield a total of 14 variables which must be optimized, once the unit cell and space group symmetry are determined, to solve the structure. As a test of the simulated annealing methodology for such a system a model calculation, employing synthetic data, for the aspartame molecule is illustrated here. One of the trial structures and its powder diffraction pattern are shown in Fig. 5.18, together with the experimental structure and its diffraction pattern. Although there is superficially good agreement between the patterns at low angle, the calculated structure is clearly very far from the experimental structure. It is only in the higher order reflections that this discrepancy between the patterns becomes obvious. However, the unsatisfactory nature of this particular proposed model is made clear both through visual inspection and energy

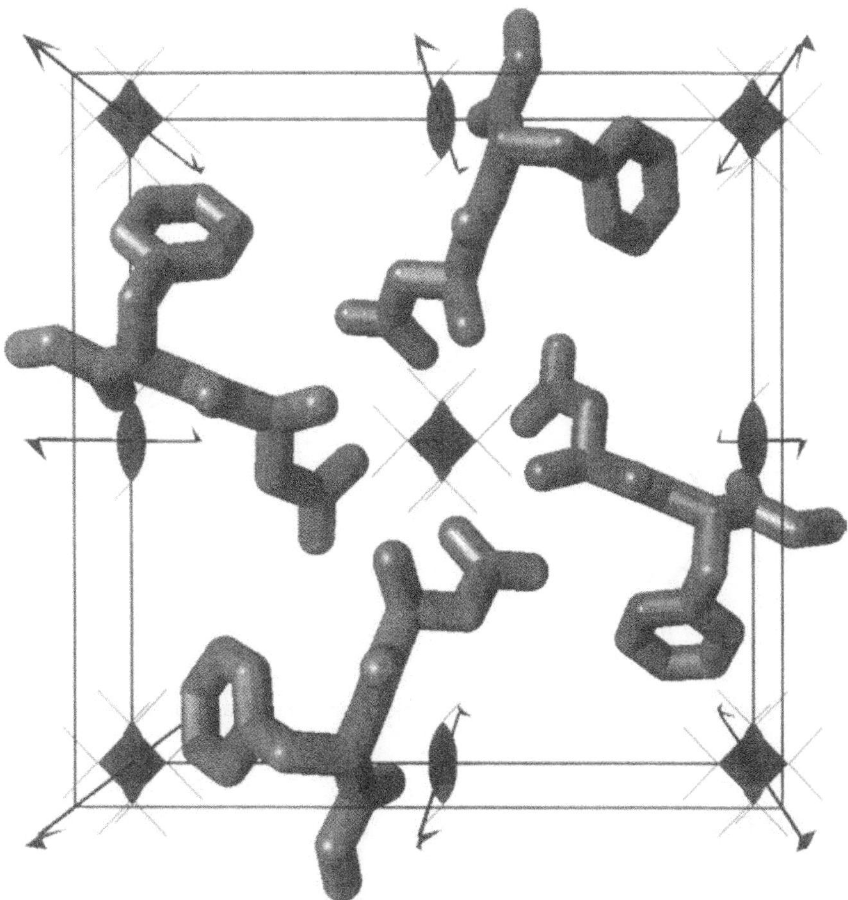

Figure 5.16 Crystal structure of aspartame (Hatada *et al.*, 1985) showing elements of symmetry (four- and two-fold screw axes) within the space group.

calculations. The superficial agreement between the calculated and observed powder patterns illustrated in Fig. 5.18 serves to illustrate the need for the highest possible accuracy in experimental measurement and in the careful scrutiny of proposed models. Subsequent simulated annealing calculations were able to locate structures with lower *R* values in good agreement with the experimental structure. As noted in Section 1 and as illustrated in Fig. 5.1, for complex structures such as aspartame, interactive computer graphics methodologies are a valuable aid in analysing partial structural models (see, e.g., the symmetry element display in Fig. 5.16) and in the completion of the descriptions of the structure through the display of difference Fourier maps, for example, superimposed on a developing model (see Fig. 5.19), allowing the location of missing atoms or groups.

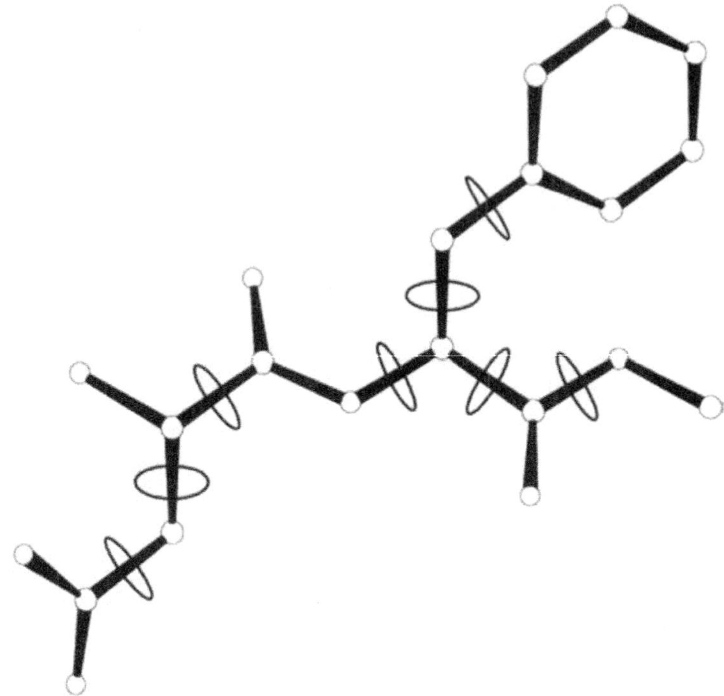

Figure 5.17 Flexible torsional variables within the aspartame molecule.

8 CONCLUSIONS

The growth in computational techniques together with an improved understanding of interatomic interactions ensure that simulated annealing, and other global optimization methods, will continue to have a broad impact on the determination and rationalization of the structural properties of inorganic materials.

It is interesting to note that on occasion the high resolution of experimental data can overwhelm the simulator. Thus, for example, Bartell and coworkers have employed a 'blurring function' in their analysis of powder diffraction data in order to avoid local minima (Bartell *et al.*, 1987). The resulting successful structure solution (Bartell *et al.*, 1987; Cockroft and Fitch, 1988), therefore, is begun at a low resolution, which gradually increases as the model is refined. This procedure in effect mirrors the function distortion illustrated in Fig. 5.5. However, oversimplification can result in solutions that do not capture the true symmetry of a system or obscure important but subtle structural features. We note that smoothing or blurring functions have also been employed to good effect in the comparison of related powder patterns (Karfunkel *et al.*, 1993a,b; Lawton and Bartell, 1994).

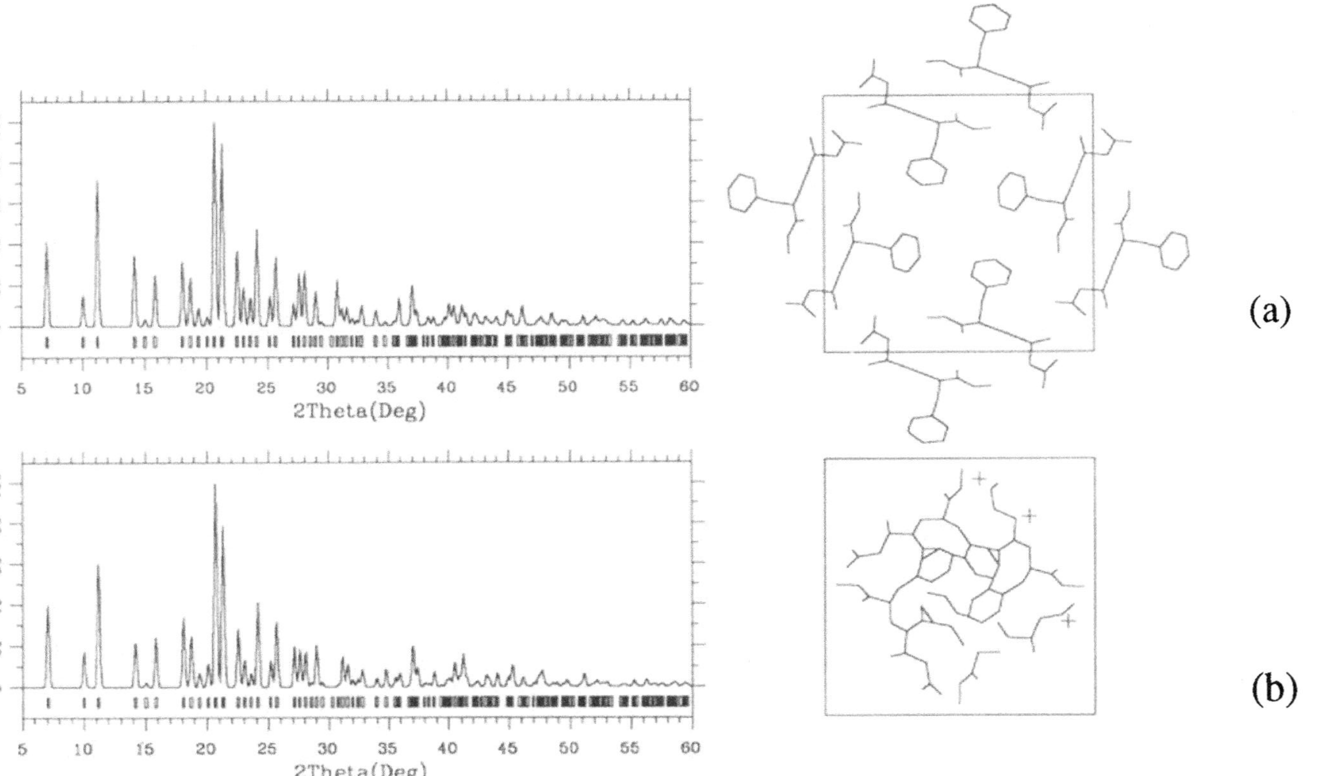

Figure 5.18 (a) Observed (Hatada *et al.*, 1985) aspartame crystal structure and (on the left) its calculated powder X-ray diffraction pattern. This pattern used as the target for a simulated annealing 'solution' of the aspartame crystal structure, generates (among other structures) the configuration shown in (b). The calculated pattern is nominally in good agreement with the pattern of (a). However, the structure is incorrect and sterically infeasible.

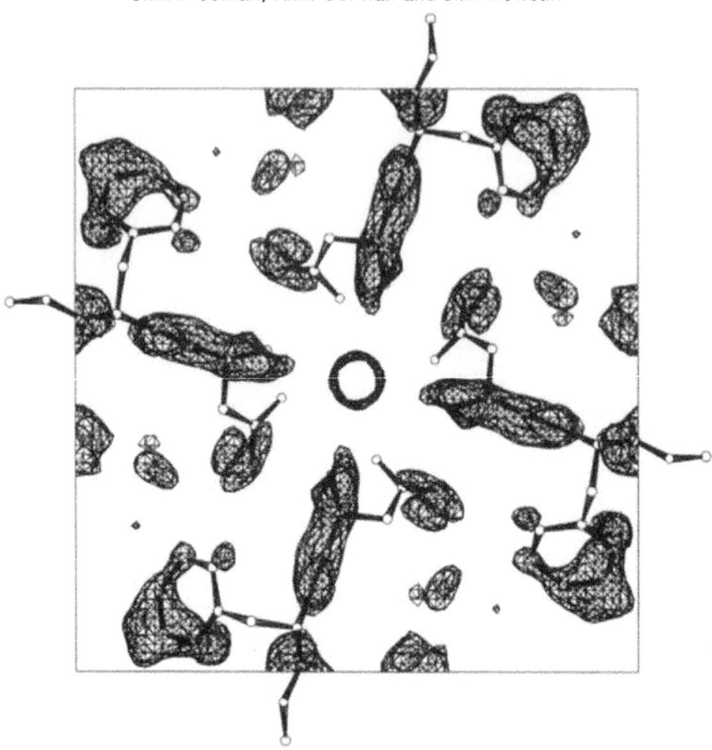

Figure 5.19 Illustration of the use of a difference Fourier map in locating a deficiency in a structural model. In this case the map reveals a region of missing density at the centre of the cell which is occupied in the true structure by a disordered water molecule.

An alternative to experimental resolution reduction is model simplification, e.g. the use of generalized framework atoms to represent tetrahedrally coordinated cations (Deem and Newsam, 1989), or the use of rigid body simulated annealing in the determination of structure from powder diffraction data (Newsam *et al.*, 1992). In practice all tractable molecular simulations use simplification of the details of molecular interactions. Atom-centred atomic charges are, for example, an approximation and, indeed, are impossible to assign unambiguously (Wiberg and Rablen, 1993; Meister and Schwartz, 1994) (although, as discussed in Chapter 1, more sophisticated treatments of electrostatic interactions are possible). In the determination of molecular crystal packing one extension of present methods is the use of generalized, rather than atom based, interaction centres. This approximation significantly reduces the number of points between which the interactions must be computed and also simplifies the energy surface. Two examples of this type of approach are shown in Figs 5.20 and 5.21 where ellipsoidal surfaces encapsulating benzene and crambin (Teeter and Hendrickson, 1979) in the crystalline environments are shown. Ellipsoidal shapes have been employed in the

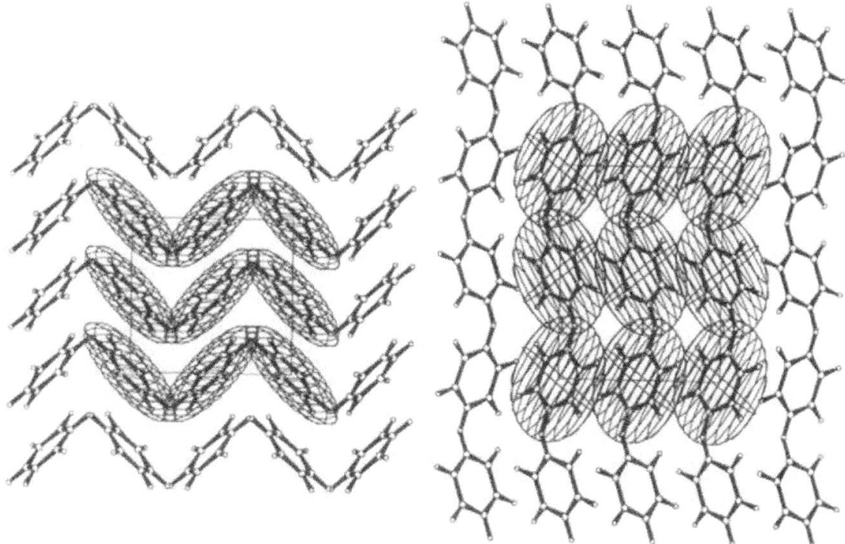

Figure 5.20 Crystal structure of benzene highlighting analysis of its packing with solid ellipsoids.

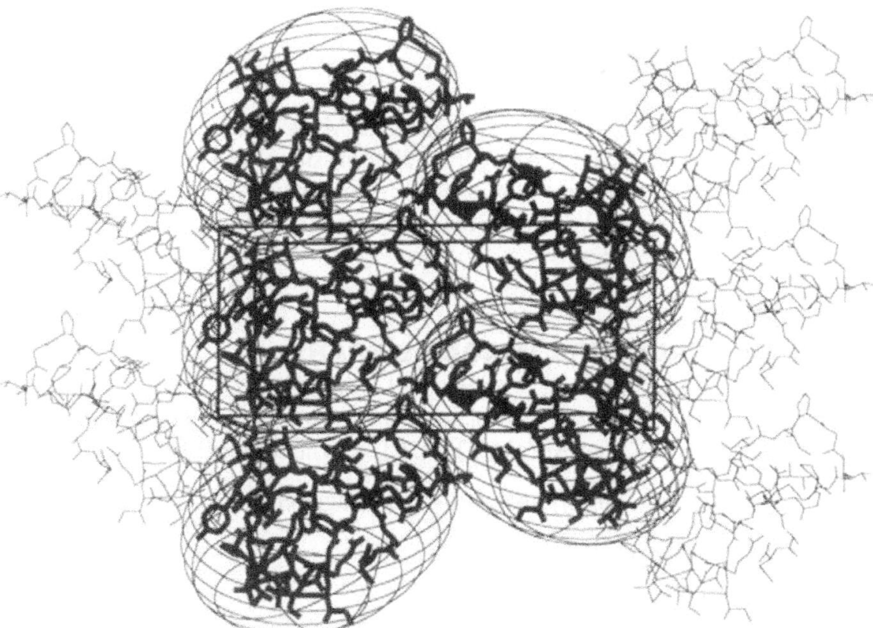

Figure 5.21 Crystal structure of crambin (Teeter and Hendrickson, 1979) highlighting analysis of its packing with solid ellipsoids.

simulation of liquid crystals (Perram and Wertheim, 1985) and the use of Fourier expansions of protein shape has been explored in the analysis of protein form (Taylor *et al.*, 1983; Leicester *et al.*, 1988). The use of such methods in the analysis of organic molecular packing may be of importance in the future.

Improvements in the accuracy of simulations of structure will also undoubtedly stem from the continued evolution of efficient collection, database construction, and analysis of crystal structural information. The Cambridge Crystallographic Database (Allen *et al.*, 1991), for example, presently contains more than 100 000 crystal structures. There is a wealth of information contained within such a compendium and a steady stream of publications appear based on such analyses.

Crystallographic database information can make a significant impact in the derivation and validation of generalized potential models. As noted above, it is the accuracy of the potential model which ultimately governs the reliability of any simulation study. Significant improvements are currently underway in accurate potentials derivation (Hill and Sauer, 1994), multiple least squares fitting procedures (Bush *et al.*, 1994) and the use of *ab initio* quantum mechanically derived energy surfaces (Gale *et al.*, 1992). These improvements, combined with the growing use of direct quantum mechanical methods, will insure that the importance of simulation in structure elucidation will continue to grow.

ACKNOWLEDGEMENTS

We thank Professor C. R. A. Catlow and Drs M. W. Deem, S. M. Levine, B. Vessal, and C. M. Kölmel for many fruitful discussions. The MSI Catalysis and Sorption Project is supported by a consortium of industrial, academic and government institutions. We thank the members for their ongoing support and guidance.

CRYSTAL STRUCTURE IMAGES LIBRARY

A compendium of inorganic crystal structure image files is maintained by the Royal Institution of Great Britain. The URL for the resource is:

http://www.molecular-images.com

REFERENCES

Akporiaye, D.E. and Price, G.D. (1989a) *Zeolites*, **9**, 23.
Akporiaye, D.E. and Price, G.D. (1989b) *Zeolites*, **9**, 321.

Allen, M.P. and Tildesley, D.J. (1987) *Computer Simulation of Liquids*, Oxford University Press, Oxford, UK.

Allen, F.H., Davies, J.E., Galloy, J.J., Johnson, O., Kennard, O., Macrae, C.F., Mitchell, G.F., Smith, J.M. and Watson, D.G. (1991) *J.Chem. Inform. Comput. Sci.* **31**, 187.

Allured, V.S., Kelly, C.M. and Landis, C.R. (1991) *J. Am. Chem. Soc.*, **113**, 1.

Attfield, J.P. (1988) *Acta Crystallogr.*, **B44**, 563.

Baker, J. (1986) *J. Comp. Chem.* **7**, 385.

Barlow, S., Rohl, A.L. and O'Hare, D. (1996a) *J. Chem. Soc. Chem. Commun.*, 257.

Barlow, S., Rohl, A.L., Shi, S., Freeman, C.M. and O'Hare, D. (1996b) *J. Am. Chem. Soc.*, **118**, 7578.

Bartell, L.S., Caillat, J.C. and Powell, B.M. (1987) *Science*, **236**, 1463.

Bell, R.G., Jackson, R.A. and Catlow, C.R.A. (1990) *J. Chem. Soc. Comm.*, 782.

Berg, B.A. (1993) *Nature*, **361**, 708.

Binder, K. (Ed.) (1987) *Applications of the Monte Carlo Method in Statistical Physics*, 2nd edn (Topics in Current Physics Vol. 36), Springer-Verlag, Berlin.

Boisen, M.B., Gibbs, G.V. and Mukowinski, M.S.T. (1994) *Phys. Chem. Mineral.*, **21**, 269.

Brennan, D., Bell, R.G., Catlow, C.R.A. and Jackson, R.A. (1994) *Zeolites*, **14**, 650.

Brodmeier, R. and Pretsch, E. (1994) *J. Comp. Chem.*, **15**, 588.

Brunger, A.T. and Rice, L.M. (1995) *Simulated Annealing Applied in Crystallographic Structure Refinement, in Adaption of Simulated Annealing to Chemical Optimization Problems* (ed. J. H. Kalivas) Elsevier, Amsterdam, p. 259.

Brunger, A.T., Leahy, D.J., Hynes, T.R. and Fox, R.O. (1991) *J. Mol. Biol.*, **221**, 239.

Bull, L.M., Henson, N.J., Cheetham, A.K., Newsam, J.M. and Heyes, S.J. (1993) *J. Phys. Chem.*, **97**, 11776.

Burkett, U. and Allinger, N.L. (1982) *Molecular Mechanics* (ACS monograph No. 177), American Chemical Society, Washington, DC.

Bush, T.S., Gale, J.D., Catlow, C.R.A. and Battle, P.D. (1994) *J. Mater. Chem.*, **4**, 831.

Bush, T.S., Catlow, C.R.A. and Battle P.D. (1995) *J. Mater. Chem.*, **5**, 1269.

Car, R. and Parinello, M. (1985) *Phys. Rev. Lett.*, **55**, 2471.

Castonguay, L.A. and Rappe, A.K. (1992) *J. Am. Chem. Soc.*, **114**, 5832.

Catlow, C.R.A. (1986) *Ann. Rev. Mater. Sci.*, **16**, 517.

Catlow, C.R.A. and Cormack, A.N. (1987) *Int. Rev. Phys. Chem.*, **6**, 227.

Catlow, C.R.A. and Mackrodt, W.C. (Eds.) (1982) *Computer Simulation of Solids*, Lecture Notes in Physics 166, Springer-Verlag, Berlin.

Catlow, C.R.A., James, R., Mackrodt, W.C. and Stewart, R.F. (1982) *Phys. Rev. B*, **25**, 1006.

Catlow, C.R.A., Thomas, J.M., Freeman, C.M., Wright, P.A. and Bell, R.G. (1993) *Proc. R. Soc. Lond. A*, **442**, 85.

Cheetham, A.K. and Taylor, J.C. (1977) *J. Sol. State Chem.*, **21**, 22.

Ciccotti, G., Frenkel, D. and McDonald, I.R. (1987) *Simulation of Liquids and Solids—Molecular Dynamics and Monte Carlo Methods in Statistical Mechanics,* North-Holland Personal Library, Amsterdam.

Cockroft, J.K. and Fitch, A.N. (1988) *Z. Kristallogr.*, **184**, 123.

Crippen, G.M. and Havel, T.F. (1988) *Distance Geometry and Molecular Conformation*, Research Studies Press, New York.

Cvijovic, D. and Klinowski, J. (1995) *Science*, **267**, 664.

David, W.I.F. (1987) *J. Appl. Crystallogr.*, **20**, 316.

Deem, M.W. and Newsam, J.M. (1989) *Nature*, **342**, 260.

Deem, M.W. and Newsam, J.M. (1992) *J. Am. Chem. Soc.*, **114**, 7189.

DeGuzman, R.N., Shen, Y.-F., Neth. E.J., Suib, S.L., O'Young, C.-L., Levine, S.M. and Newsam, J.M. (1994) *J. Mater. Chem.*, **6**, 815.

den Ouden, C.J.J., Jackson, R.A., Catlow, C.R.A. and Post, M.F.M. (1990) *J. Phys. Chem.*, **94**, 5286.

Desiraju, G.R. (1989) *Crystal Engineering, The Design of Organic Solids*, Elsevier, Amsterdam.

Dinur, U. and Hagler, A.T. (1991) In *Reviews in Computational Chemistry*, Vol 2 (eds K.B. Kipkowitz and D.B. Boyd) VCH Publishers, New York, p. 99.

Discover 3.2 (1995) Extensible Systematic Forcefield, ESFF, A Component of Discover 95, developed and distributed by Molecular Simulations Inc., 9685 Scranton Road, San Diego, CA 92121, USA.

Dunitz, J.D. and Bernstein, J. (1995) *Acc. Chem. Res.*, **28**, 193.

Estermann, M. and Gramlich, V. (1991) Abstracts of 1991 Swiss Crystallographic Society, p. 22.

Fletcher, R. and Powell, M.J.D. (1963) *The Computer J.*, **6**, 163.

Freeman, C.M. and Catlow, C.R.A. (1992) *J. Chem. Soc. Chem. Commun.*, 89.

Freeman, C.M., Catlow, C.R.A., Thomas, J.M. and Brode, S. (1991) *Chem. Phys. Lett.*, **186**, 137.

Freeman, C.M., Newsam, J.M., Levine, S.M. and Catlow, C.R.A. (1993) *J. Mater Chem.*, **3**, 531.

Gale, J.D., Catlow, C.R.A. and Mackrodt, W.C. (1992) *Modelling Simul. Mater. Sci. Engng.*, **1**, 73.

Gavezzotti, A. (1991) *J. Am. Chem. Soc.*, **113**, 4622.

Gavezzotti, A. (1994) *Acc. Chem. Res.*, **27**, 309.

Gay, D.H. and Rohl, A.L. (1995) *J. Chem. Soc. Faraday Trans.*, **91**, 925.

Gdanitz, R.J. (1992) *Chem. Phys. Lett.*, **190**, 391.

Gibson, K.D. and Scheraga, H.A. (1995) *J. Phys. Chem.*, **99**, 3752.

Gill, P.E., Murray, W. and Wright, M.H. (1981) *Practical Optimization*, Academic Press, London.

Gilmore, C.J., Henderson, K. and Bricogne, G. (1991) *Acta Crystallogr.*, **A47**, 830.

Harris, K.D.M., Tremayne, M., Lightfoot, P. and Bruce, P.G. (1994) *J. Am. Chem. Soc.*, **116**, 3543.

Hartke, B. (1993) *J. Phys. Chem.*, **97**, 9976.

Hatada, M., Jancarik, J., Graves, B. and Kim, S.H. (1985) *J. Am. Chem. Soc.*, **107**, 4279.

Hill, J.-R. and Sauer, J. (1994) *J. Phys. Chem.*, **98**, 1238.

Holden, J.R., Du, Z. and Ammon, H.L. (1993) *J. Comp. Chem.*, **14**, 422.

Hriljac, J.A. and Torardi, C.C. (1993) *Inorganic Chemistry*, **32**, 6003.

Hwang, M.J., Stockfisch, T.P. and Hagler, A.T. (1994) *J. Am. Chem. Soc.*, **116**, 2515.

Jentys, A., Grimes, R.W., Gale, J.D. and Catlow, C.R.A. (1993) *J. Phys. Chem.*, **97**, 13535.

Kanatzidis, M.G., Baenziger, N.C., Coucouvanis, D., Simopoulos, A. and Kostikas, A. (1984) *J. Am. Chem. Soc.*, **106**, 4500.

Karfunkel, H.R. and Gdanitz, R.J. (1992) *J. Comp. Chem.*, **13**, 1171.

Karfunkel, H.R., Leusen, F.J.J. and Gdanitz, R.J. (1993a) *J. Comput. Aided Mater. Des.*, **1**, 177.

Karfunkel, H.R., Rohde, B., Leusen, F.J.J., Gdanitz, R.J. and Rihs, G. (1993b) *J. Comp. Chem.*, **14**, 1125.

Keen, D.A. and McGeevy, R.L. (1990) *Nature*, **344**, 423.

Kerr, I.S. (1974) *Z. Kristallogr.*, **139**, 186.

Khan, A.A. (1976) *Acta Crystallogr.*, **A32**, 11.

Kirkpatrick, S., Gelatt, C.D. and Vecchi, M.P. (1983) *Science*, **220**, 671.

Kitaigorodskii, A.I. (1973) *Molecular Crystals and Molecules*, Academic Press, New York.
Lacorre, P. and Pannetier, J. (1987) *J. Mag. Mat. Mater.*, **71**, 63.
Laszkur, Z.A., Jones, R.H., Waller, D., Catlow, C.R.A. and Thomas, J.M. (1993) *J. Phys. Chem.*, **97**, 426.
Lawton, S.L. and Bartell, L.S. (1994) *Powder Diffraction*, **9**, 124.
Leicester, S.E., Finney, J.L. and Bywater, R.P.J. (1988) *J. Mol. Graphics*, **5**, 104.
Leusen, F.J.J., Pinches, M.R.S. and Lovell, R. (1995) Personal communication.
Lewis, D.W., Catlow, C.R.A., Sankar, G. and Carr. S.W. (1995) *J. Phys. Chem.*, **99**, 2377.
Maple, J.R., Hwang, M.J., Stockfisch, T.P., Dinur, U., Waldman, M., Ewig, C.S. and Hagler, A.T. (1994) *J. Comp. Chem.*, **15**, 162.
McGreevy, R.L. and Pusztai, L. (1988) *Mol. Simul.*, **1**, 359.
McGreevy, R.L. and Pusztai, L. (1990) *Proc. R. Soc. (Lond.)*, **A430**, 241.
Meier, W.M. and Villiger, H. (1969) *Z. Kristallogr.*, **129**, 411.
Meister, J. and Schwartz, W.H.E. (1994) *J. Phys. Chem.*, **98**, 8245.
Metropolis, N., Rosenbluth, A.W., Rosenbluth, M.N., Teller, A.H. and Teller, G. (1953) *J. Chem. Phys.*, **21**, 1087.
Morris, R.E., Harrison, W.T.A., Nicol, J.M., Wilkinson, A.P. and Cheetham, A.K. (1992) *Nature*, **359**, 519.
Nenoff, T.M., Harrison, W.T.A., Stucky, G.D., Nicol, J.M. and Newsam, J.M. (1993) *Zeolites*, **13**, 506.
Newsam, J.M. (1986) *J. Chem. Soc. Chem. Commun.*, 1295.
Newsam, J.M. (1992) In *Proc. of the Ninth International Zeolite Conference* (eds R.B. von Ballmoos, J.B. Higgins and M.M.J. Treacy), Butterworth-Heinemann, Amsterdam.
Newsam, J.M. and Li, Y.S. (1995) *Catalysis Today*, **23**, 325.
Newsam, J.M., Deem, M.W. and Freeman, C.M. (1992) In *Accuracy in Powder Diffraction II* (NIST Special Publication No. 846) (eds E. Prince and J.K. Stalick) National Institute of Standards and Technology, Bethesda, MD, p. 80.
Newsam, J.M., Freeman, C.M., Gorman, A.M. and Vessal, B. (1996) *J. Chem. Soc. Chem. Comm.*, 1945.
Pannetier, J., Bassas-Alina, J., Rodriguez-Carvajal, J. and Caignaert, V. (1990) *Nature*, **346**, 242.
Perlstein, J. (1994) *J. Am. Chem. Soc.*, **116**, 455.
Perram, J.W. and Wertheim, M.S. (1985) *J. Comp. Phys.*, **58**, 409.
Pertsin, A.J. and Kitaigorodskii, A.I. (1987) *The Atom–Atom Potential Method*, Springer, Berlin.
Piela, L., Kostrowicki, J. and Scheraga, H.A. (1989) *J. Phys. Chem.*, **93**, 3339.
Pluth, J.J. and Smith, J.V. (1980) *J. Am. Chem. Soc.*, **102**, 4704.
Powell, M.J.D. (Ed.) (1982) *Nonlinear Optimization*, Academic Press, New York.
Press, W.H., Flannery, B.P., Teukolsky, S.A. and Vetterling, W.T. (1989) *Numerical Recipes, The Art of Scientific Computing*, Cambridge University Press, Cambridge.
Price, G.D., Wood, I.G. and Akporiaye, D.E. (1992) In *Modelling of Structure and Reactivity in Zeolites* (ed. C.R.A. Catlow) Academic Press, p. 19.
Rabinovick, D. and Schmidt, G.M.J. (1966) *Nature*, **211**, 1392.
Ramdas, S. and Thomas, J.M. (1978) *Chem. Phys. Solids Surf.*, **7**, 31.
Ramprasad, D., Pez, G.P., Toby, B.H. and Markley, T.J. (1995) *J.Am. Chem. Soc.*, **117**, 10694.
Rappe, A.K., Casewit, C.J., Colwell, K.S., Goddard, W.A. and Skiff, W.M. (1992) *J. Am. Chem. Soc.*, **114**, 10024.
Rietveld, H.M. (1969) *J. Appl. Crystallogr.*, **2**, 65.
Santilli, D.C., Harris, T.V. and Zones, S.I. (1993) *Microporous Mater.*, **1**, 329.

Scales, L.E. (1985) *Introduction to Non-linear Optimization*, Macmillan, London.
Sedgewick, R. (1988) *Algorithms* (2nd edn), Addison-Wesley, Reading, Mass.
Shannon, M.D., Casci, J.L., Cox, P.A. and Andrews, S.J. (1991) *Nature*, **353**, 417.
Smith, J.V. (1988) *Chem. Rev.*, **88**, 149.
Soukoulis, C.M. (1984) *J. Phys. Chem.*, **88**, 4898.
Talamo, A., Catlow, C.R.A. and Day, P. (1995) in press.
Tasker, P. W. (1990) *J. Chem. Soc. Faraday Trans.*, **86**, 1311.
Taylor, W.R., Thornton, J.M. and Turnell, W.G. (1983) *J. Mol. Graphics*, **1**, 30.
Teeter, M.M. and Hendrickson, W.A. (1979) *J. Mol. Biol.*, **127**, 219.
Tillmanns, E., Gebert, W. and Baur, W.H. (1973) *J. Solid State Chem.*, **7**, 69.
Vessal, B. (1995) Potentials developed by the Catalysis and Sorption Project supported by a consortium of Industrial, Government and Academic organizations.
Wiberg, K.B. and Rablen, P.R. (1993) *J. Comp. Chem.*, **14**, 1504.
Williams, D.E. (1967) *J. Chem. Phys.*, **47**, 4680.
Xiao, Y.L. and Williams, D.E. (1994) *J. Phys. Chem.*, **98**, 7191.
Young, R.A. (Ed.) (1993) *The Rietveld Method*, IUCr, Oxford University Press, Oxford.

6

Reverse Monte Carlo Methods for Structural Modelling

R.L. McGreevy

1 INTRODUCTION

The technique of Rietveld refinement (Rietveld, 1969; see also Chapter 5) revolutionized crystallography. The essential philosophy of this method is that, rather than attempting to determine individual Bragg peak intensities from a powder diffraction pattern and then attempting to determine the crystal structure from these intensities, a structural model is refined by fitting to the pattern (i.e. the experimental data) as a whole (including background, resolution, etc.). It would be difficult to imagine how much 'routine' crystallography would now be done if this method were not available.

The philosophy of Reverse Monte Carlo (RMC) modelling (McGreevy and Pusztai, 1988) is similar to that of Rietveld refinement, except that the method is applicable to both ordered and disordered structures and is not simply a refinement technique. In the case of Rietveld refinement the 'structural model' is defined in crystallographic terms, i.e. space group, lattice parameters, atom positions in the basis, temperature factors, etc. In RMC the model is simply a collection of atomic coordinates, the number of atoms being large, and so it can in principle be applied to any condensed matter system. From the first study of the structure of liquid argon, using neutron diffraction data (McGreevy and Pusztai, 1988), the range of applications has grown rapidly to include other elemental liquids (Nield *et al.*, 1991; Howe *et al.*, 1993), molecular liquids (Howe, 1990a,b), multicomponent liquids (McGreevy and Pusztai, 1990; Howe and McGreevy, 1991a; Maret *et al.*, 1994), network and metallic glasses (Keen and McGreevy, 1990; Newport *et al.*, 1991; Iparraguire *et al.*, 1993; Pusztai and Svab, 1993; Gereben and Pusztai, 1994), fast ion conducting crystals and glasses (Keen *et al.*, 1990a,b; Börjesson *et al.*, 1992; Nield *et al.*, 1992, 1993, 1994; Karlsson and McGreevy, 1995; Wicks *et al.*, 1995), polymers (Rosi-Schwartz and Mitchell, 1995), disordered magnetic materials (Keen and McGreevy, 1991; Keen *et al.*, 1995), disorder in molecular crystals (McKenzie *et al.*, 1992; Nield *et al.*, 1995a,b) and crystal structure refinement (Montfrooij *et al.*, 1996). The method can be applied to neutron diffraction (McGreevy and Pusztai, 1988),

COMPUTER MODELLING IN INORGANIC CRYSTALLOGRAPHY
ISBN 0-12-164135-X

electron diffraction (McKenzie *et al.*, 1992), X-ray diffraction (Keen and McGreevy, 1990) and Extended X-ray Absorption Fine Structure (EXAFS) data (Gurman and McGreevy, 1990), simultaneously if necessary (Wicks *et al.*, 1995), and constraints based on other information (experimental or otherwise) can be included in the modelling. Over a hundred papers using the RMC method have been published and only a selection of references is given here.

One important point that should be stressed right from the start is that RMC is a method of structural modelling. Despite the close analogy to 'normal' Monte Carlo simulation, RMC has no fundamental theoretical basis, and has much in common with the simulated annealing methods discussed in Chapter 5. There have been many criticisms of the RMC method and these have generally completely misunderstood this point. RMC is extremely effective, as will be seen, particularly in areas for which other methods are unsuitable, but it is essentially a technique for fitting experimental data. The main disadvantage of RMC modelling is that it is computationally expensive. However, the rapid increase in the availability of low cost computing power will certainly make this irrelevant within a few years. This chapter describes the method in some detail, including the use of different types of experimental data and constraints. We then give a few selected examples of applications related to fast ion conducting crystals and glasses, silicate glasses and polymers. For application to the study of liquids, metallic glasses and magnetism the reader is referred elsewhere (McGreevy *et al.*, 1990; McGreevy and Howe, 1992; McGreevy, 1992, 1995).

2 RMC METHOD

2.1 Basic algorithm

RMC is a variation of the standard Metropolis Monte Carlo (MMC) method (Metropolis *et al.*, 1953; see also Chapters 1 and 5). The principle is that we wish to generate an ensemble of atoms, i.e. a structural model, which corresponds to a total structure factor (set of experimental data) within its errors. These are assumed to be purely statistical and to have a normal distribution. Usually the level and distribution of statistical errors in the data is not a problem, but systematic errors can be. We shall initially consider materials that are macroscopically isotropic and that have no long range order, i.e. glasses, liquids and gases. The basic algorithm, as applied to a monatomic system with a single set of experimental data, is as follows:

1. N atoms are placed in a cell with periodic boundary conditions, i.e. the cell is surrounded by images of itself. Normally a cubic cell is used but other geometries may also be chosen. For a cube of side L the number density $\rho = N/L^3$ must equal the required density of the system. The positions of

the N atoms may be chosen randomly, they may have a known crystal structure or they may be a configuration from a different simulation or model.

2. Calculate (C), the radial distribution function for this old (o) configuration

$$g_{\rm o}^{\rm C}(r) = \frac{n_{\rm o}^{\rm C}(r)}{4\pi r^2 {\rm d}r\rho} \tag{1}$$

where $n(r)$ is the number of atoms at a distance between r and $r + {\rm d}r$ from a central atom, averaged over all atoms as centres. The configuration size L should in principle be sufficiently large for there to be no correlation across the cell, i.e. $g(r > L/2) = 1$; $g(r)$ is only calculated for $r < L/2$ and the nearest image convention is used to determine the atomic separations.

3. Transform to the total structure factor

$$A_{\rm o}^{\rm C}(Q) - 1 = \rho \int_0^\infty 4\pi r^2 (g_{\rm o}^{\rm C}(r) - 1) \frac{\sin Qr}{Qr} {\rm d}r \tag{2}$$

where Q is the momentum transfer.

4. Calculate the difference between the measured total structure factor, $A^{\rm E}(Q)$, and that determined from the configuration, $A_{\rm o}^{\rm C}(Q)$,

$$\chi_{\rm o}^2 = \sum_{i=1}^{m} [A_{\rm o}^{\rm C}(Q_i) - A^{\rm E}(Q_i)]^2 / \sigma^2(Q_i) \tag{3}$$

where the sum is over the m experimental points and $\sigma(Q_i)$ is the experimental error. In practice a uniform σ is normally used, since the distribution of systematic errors is normally unknown.

5. Move one atom at random. Calculate the new (n) radial distribution function, $g_{\rm n}^{\rm C}(r)$, and total structure factor, $A_{\rm n}^{\rm C}(Q)$, and

$$\chi_{\rm n}^2 = \sum_{i=1}^{m} [A_{\rm n}^{\rm C}(Q_i) - A^{\rm E}(Q_i)]^2 / \sigma^2(Q_i) \tag{4}$$

6. If $\chi_{\rm n}^2 < \chi_{\rm o}^2$ then the move is accepted and the 'new' configuration becomes the 'old' configuration. If $\chi_{\rm n}^2 > \chi_{\rm o}^2$ the move is accepted with probability $\exp[-(\chi_{\rm n}^2 - \chi_{\rm o}^2)/2]$. Otherwise it is rejected.
7. Repeat from step 5.

As this procedure is iterated χ^2 will initially decrease until it reaches an equilibrium value about which it will fluctuate. The resulting configuration should be a three-dimensional structure that is consistent with the experimental total structure factor within experimental error. Statistically independent configurations may then be collected. In MMC, configurations are normally assumed to be independent if separated by N accepted moves, but in practice we normally use at least $5N$ moves.

The difference between RMC and MMC is simply that in RMC the difference between calculated and measured total structure factors (χ^2) is sampled (minimized), while in MMC the potential energy (U/kT) is sampled. Otherwise the two algorithms are identical. It is particularly important that RMC uses a proper Markov chain, so that the final structure should be independent of the initial configuration. This makes the method an *ab initio* structural determination akin to those discussed in Chapter 5 rather than a refinement. However, in some circumstances the method is deliberately used as a refinement. This involves only accepting moves that decrease χ^2 in step (6) and corresponds to setting $T = 0$ in MMC. The use of the RMC algorithm in practice has been discussed in detail elsewhere (McGreevy *et al.*, 1990, 1994).

In Fig. 6.1 we show a simple two-dimensional example of the course of an RMC run. In this case we start with a configuration that is a square lattice, though any other could have been chosen. The 'data' comes from an MMC simulation using a Lennard–Jones potential. As the RMC modelling progresses the $g(r)$ of the configuration gradually gets closer to that of the data until, in equilibrium, a good fit is obtained. The deviations are purely statistical errors due to the small simulation size in this example. A 'real' RMC model contains typically 4000 atoms and configurations as large as 30 000 atoms have been used.

Two of the most common systematic errors that occur in experimental total structure factors are small normalization errors in the form of additive and multiplicative constants, particularly for X-ray data because of the Q dependence of the form factor. Such errors can be accounted for within the RMC algorithm, since the RMC structure factors are correctly normalized (given the correct density, atomic composition, etc.). The required multiplicative factor which minimizes χ^2 is given by

$$\alpha = \frac{\sum_{i=1}^{m} A^{\mathrm{E}}(Q_i)A^{\mathrm{C}}(Q_i)}{\sum_{i=1}^{m}(A^{\mathrm{E}}(Q_i))^2} \tag{5}$$

and the additive factor by

$$\beta = \frac{1}{m}\sum_{i=1}^{m}(A^{\mathrm{E}}(Q_i) - A^{\mathrm{C}}(Q_i)) \tag{6}$$

When both factors are used the expressions are more complex (Howe, 1989). If the factors are large then clearly the original data or the model parameters need to be checked.

Figure 6.1 Example of RMC modelling of a simple test system. On the right is the configuration (two dimensional) and on the left $g(r)$ calculated from the configuration (solid curve) compared to the 'data' (dotted curve). The starting configuration (square lattice) is at the top and an equilibrium configuration at the bottom, with intermediate configurations in between.

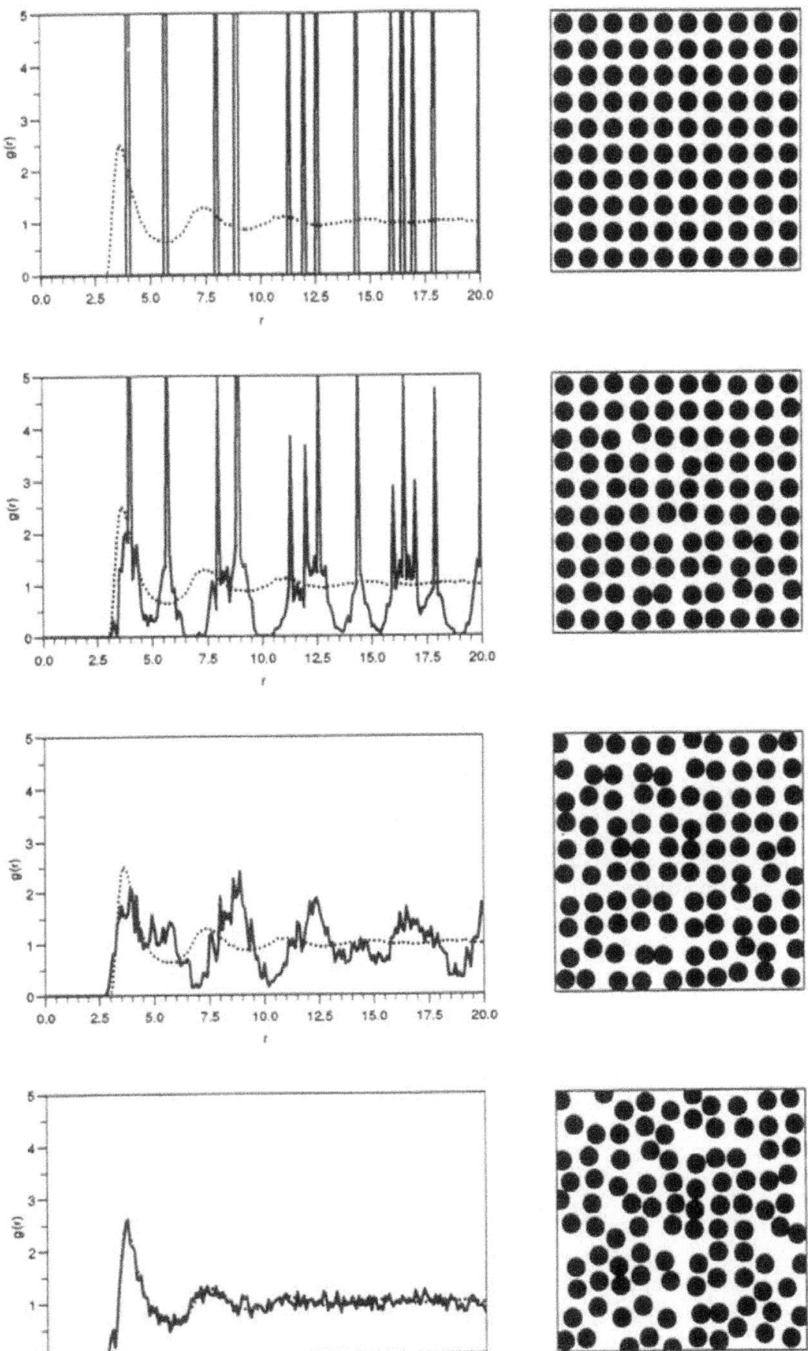
g(r)
r
0.0
2.5
5.0
7.5
10.0
12.5
15.0
17.5
20.0

2.2 Multiple data sets

The algorithm described above is specifically for modelling a single set of diffraction data which could be obtained using either X-rays, neutrons or electrons. The fit may be either to the structure factor or to the radial distribution function, though the former is recommended because the distribution of errors in the latter may be highly non-uniform. In practice a fit is normally made first to the radial distribution function, then to a subset of the total structure factor points, and finally to all the structure factor points. This considerably reduces the time required.

The RMC method is more general than this simple algorithm in that *any* set(s) of data that can be directly related to the structure can be modelled. It can be applied to isotopic substitution in neutron diffraction, or equivalently to anomalous scattering in X-ray diffraction, to EXAFS (multiple edges) and possibly to NMR data. All data sets can be modelled simultaneously simply by adding the respective χ^2 values.

For a multicomponent system studied by neutron (ne) diffraction

$$\chi^2_{\mathrm{ne}} = \sum_k \sum_{i=1}^{m} [F^{\mathrm{C}}_{\mathrm{ne}_k}(Q_i) - F^{\mathrm{E}}_{\mathrm{ne}_k}(Q_i)]^2/\sigma^2_{\mathrm{ne}_k}(Q_i) \tag{7}$$

where

$$F_{\mathrm{ne}_k}(Q_i) = \sum_{\alpha,\beta} c_\alpha c_\beta b_{\alpha_k} b_{\beta_k}[A_{\alpha\beta}(Q_i) - 1] \tag{8}$$

c_α is the concentration and $b_{\alpha k}$ the coherent neutron scattering length for species α in (isotopically substituted) sample k. $A_{\alpha\beta}(Q)$ are the partial structure factors. For X-ray (xr) diffraction

$$\chi^2_{\mathrm{xr}} = \sum_k \sum_{i=1}^{m} [F^{\mathrm{C}}_{\mathrm{xr}_k}(Q_i) - F^{\mathrm{E}}_{\mathrm{xr}_k}(Q_i)]^2/\sigma^2_{\mathrm{xr}_k}(Q_i) \tag{9}$$

where

$$F_{\mathrm{xr}_k}(Q_i) = \sum_{\alpha,\beta} c_\alpha c_\beta f^*_{\alpha_k}(Q_i) f^*_{\beta_k}(Q_i)(A_{\alpha\beta}(Q_i) - 1) \tag{10}$$

and

$$f^*_{\alpha_k}(Q_i) = F_{\alpha_k}(Q_i)/\sum_\alpha c_\alpha f^2_{\alpha_k}(Q_i) \tag{11}$$

$f_{\alpha k}(Q_i)$ is the Q-dependent form factor for X-ray wavelength λ_k. The normalized value $f^*_{\alpha_k}(Q_i)$ is used in the definition of $F_{\mathrm{xr}k}$ for the usual case where the scattered intensity is measured with constant statistical error. Note that clearly there is no requirement for the Q_i points in the neutron data (equation (5)) or the X-ray data (equation (8)) to be the same, or even for the Q ranges to be the same. For EXAFS (ex) data the normal definition of a fit index has been used, i.e.

$$\chi^2_{\rm ex} = \sum_\alpha \left(\sum_{i=1}^{m} [Q_i^p (F^{\rm C}_{{\rm ex}_\alpha}(Q_i) - F^{\rm E}_{{\rm ex}_\alpha}(Q_i))]^2 \Big/ \sum_{i=1}^{m} [Q_i^p \sigma_{{\rm ex}_\alpha}(Q_i) F^{\rm E}_{{\rm ex}_\alpha}(Q_i)]^2 \right) \tag{12}$$

The Q_i^p weighting is normally used with $p = 3$ to give all points in the spectrum $F_{{\rm ex}_\alpha}(Q_i)$ approximately equal amplitude. The subscript α here refers to the spectrum measured at the absorption edge of species α. This spectrum is related to the partial radial distribution functions, $g_{\alpha\beta}(r)$ by

$$F_{{\rm ex}_\alpha}(Q_i) = \sum_\beta 4\pi\rho \int_0^\infty r^2 [g_{\alpha\beta}(r) - 1] f_\beta(Q_i, r)\,{\rm d}r \tag{13}$$

where $f_\beta(Q_i,r)$ is the contribution to the EXAFS of a single atom of type β at distance r. This is calculated by one of the standard EXAFS data analysis packages such as the SERC Daresbury package EXCURV.

For simultaneous fitting of the data sets obtained by different experimental techniques the quantity

$$\chi^2 = \chi^2_{\rm ne} + \chi^2_{\rm xr} + \chi^2_{\rm ex} \tag{14}$$

is then minimized in step 4 of the RMC algorithm. The relative weighting of the data sets is taken care of by the choice of the various σ values. Clearly the required computer time increases significantly if multiple data sets are fitted.

2.3 Constraints

Other information that cannot be used directly can be made use of in the form of constraints; this may include NMR, Electron Paramagnetic Resonance (EPR), Raman scattering and 'chemical' knowledge. The most commonly used constraint is on the closest approach distance of two atoms. Because of systematic errors in the experimental data, and often because of the limited data range, the data would not 'forbid' some atoms from coming very close together. However, we know that this is physically unrealistic so an excluded volume is defined. Often realistic values for the closest approach distances can be determined from direct Fourier transformation of the measured total structure factors. If an unsuitable choice is made then this is usually obvious since spurious sharp 'spikes' occur in $g(r)$ at low r. While the closest approach constraint is very simple, it is very powerful when used in conjunction with a fixed density. For many materials the dominant effect determining the structure is packing, and hence to include information on atomic sizes (these are minimum sizes rather than, e.g., ionic radii) implicitly in the model severely limits the number of structures that are consistent with the data.

The second most commonly used constraint is on the coordination of atoms. A coordination number $n_{\alpha\beta}$ is defined as the number of atoms of type β within two fixed distances of one of type α. Normally the lower fixed distance is the closest approach of atoms α and β. If we define the proportion of atoms type α in the configuration with the 'correct' coordination as f_{RMC}, and the required proportion with such a coordination as f_{req}, then a term

$$\chi^2_{\mathrm{co}} = (f_{\mathrm{req}} - f_{\mathrm{RMC}})^2/\sigma^2_{\mathrm{co}} \tag{15}$$

is added to the overall χ^2. Multiple coordination constraints can be applied to a model by adding the appropriate χ^2. The 'error' σ_{co} in this case simply acts as a weighting factor for the constraint relative to the data. If $\sigma_{\mathrm{co}} \approx 0$ then it is effectively impossible for atoms with the 'correct' coordination to change it; this can be used to 'mimic' the effect of covalent bonding. In many cases hard sphere Monte Carlo simulation with such coordination constraints, i.e. RMC with no data, can be used to produce structures with suitable topology prior to fitting to the data.

Other constraints that can be used include the average coordination number

$$\langle n_{\alpha\beta}\rangle = \int_{r_1}^{r_2} 4\pi r^2 \rho\,[g_{\alpha\beta}(r) - 1]\,\mathrm{d}r \tag{16}$$

which can be obtained from EXAFS data if it is not to be used directly. Three-body information such as bond angle constraints can also be included if the relative angles of chemical bonds are known from, e.g., crystallographic studies. This has been done in a study of amorphous silicon (Kugler *et al.*, 1993). Finally semi-rigid molecules can be included explicitly in the model if appropriate. In this case the molecules are defined by the atomic centre, the coordinates of the atoms relative to the centre, two Euler angles and a Debye–Waller factor (to account for molecular vibrations). In step 5 of the algorithm the molecules are then moved and rotated randomly (Howe, 1990a).

2.4 Crystalline materials

Crystalline materials may be either macroscopically isotropic (powders) or anisotropic (single crystals). One may be interested either in the equilibrium (average) crystal structure, normally referred to as 'the' crystal structure, or in the deviations from that structure, i.e. structural disorder. The RMC method may be applied in different ways to these different cases, but before describing the method it is instructive to consider precisely what is measured in a diffraction experiment.

The total structure factors that have been referred to above are obtained by measuring the total scattered intensity at a particular angle, i.e. including both elastic and inelastic scattering. They are related to the van Hove (1954) correlation function $G(\mathbf{r},t)$ by

$$A(\mathbf{Q}) = \int A(\mathbf{Q}, \omega)\,\mathrm{d}\omega = \int G(\mathbf{r}, t = 0)\exp(i\mathbf{Q}\cdot\mathbf{r})\,\mathrm{d}\mathbf{r} \tag{17}$$

where $g(\mathbf{r}) = G(\mathbf{r},t = 0)$. The total structure factor is therefore related to the *instantaneous* structure and this is what is being modelled by the RMC method that has been described.

In 'normal' diffraction studies of crystalline materials the structure factor that is measured is only elastic (Bragg) scattering, though often total scattering is actually measured and it is simply assumed that all sharp features are elastic scattering and all broad features are not; broad features are then simply subtracted from the measured scattering pattern. Here the elastic structure factor will be referred to as $S(\mathbf{Q})$ to distinguish it from the total structure factor $A(\mathbf{Q})$.

$$S(\mathbf{Q}) = A(\mathbf{Q}, \omega = 0) = \int\int G(\mathbf{r}, t)\exp(i\mathbf{Q}\cdot\mathbf{r})\,\mathrm{d}\mathbf{r}\mathrm{d}t \tag{18}$$

and so it is related to the *time average* structure. This cannot strictly be calculated from a structural model, in the sense that a static model of atomic coordinates represents an instantaneous structure, unless it is assumed that a space average, e.g. over an appropriately defined unit cell, is equivalent to a time average. Even if one has a molecular dynamics model, i.e. $G(\mathbf{r},t)$, to strictly calculate the elastic scattering requires integration to $t = \infty$ which is impossible.

The elastic scattering, being related to a time average, does not directly contain information on local disorder. The distinction between imposing a model of the local disorder and showing consistency with the (elastic scattering) data, and deriving the model from the data, is often forgotten in normal crystallographic analysis. Information on local structural correlations is provided by the diffuse scattering, which may be either due to static disorder (elastic diffuse scattering) or dynamic disorder (thermal diffuse scattering). As the level of disorder increases, so the intensity of the Bragg peaks decreases and the intensity of the diffuse scattering increases. In such cases the local structural correlations are best studied by RMC modelling of the total structure factor, i.e. *both* Bragg and diffuse scattering.

For all RMC methods of modelling crystalline materials the configuration must consist of an integral number of unit cells along each axis, i.e. it must be a supercell. The unit cell must therefore be predetermined by conventional crystallographic techniques.

The RMC methods that are described in detail below have so far been successfully used for crystal structure refinement. There have been some attempts at *ab initio* structure determination, i.e. starting the model from random atomic positions within the unit cells that make up the configuration. Best results are found (for simple structures) if $g(r)$ is modelled first. However, it is clear that the minimization of χ^2 is much more difficult than for disordered structures; χ^2 is a multidimensional function of the $3N$ atomic coordinates and for disordered materials this function is relatively smooth, so while there may be problems involving local minima these can usually be overcome in some empirical manner, e.g. by using σ as a variable. This is equivalent to raising the temperature in an MMC simulation. However, for crystalline materials there are numerous deep local minima, so if RMC is to be used as an *ab initio* method it must be combined with some form of simulated annealing approach (see, e.g. Chapter 5 and Pannetier *et al.*, 1990). This will be computationally expensive.

2.5 Powder diffraction

Because of the existence of long range order there will still be significant oscillations in $g^C(r)$ for $r > L/2$, leading to truncation errors in the transform to $A^C(Q)$ (equation (2)). Since the experimental data $A^E(Q)$ are measured in Q space they do not contain such errors and cannot be compared directly to $A^C(Q)$. There are a number of ways round this problem.

2.5.1 Transform of $A^E(Q)$

If the system is sufficiently disordered (i.e. a high Debye–Waller factor), the diffractometer resolution is sufficiently low and the Q range is sufficiently large, then the experimental structure factor can be transformed directly to $g^E(r)$ and the comparison made with $g^C(r)$ (Keen *et al.*, 1990b). However, this procedure is not applicable in many cases, i.e. there are still oscillations in $A^E(Q)$ at the maximum Q measured. In addition the low Q resolution leads to a loss of real space resolution in the RMC model, i.e. a broader distribution of atoms around their crystal sites.

2.5.2 Convolution of $A^E(Q)$

The radial distribution function $g_L^C(r)$ calculated from a (finite size) RMC model is a section of the 'complete' radial distribution function $g^C(r)$, multiplied by a step function

$$\begin{aligned} m(r) &= 1 \quad r < L/2 \\ m(r) &= 0 \quad r > L/2 \end{aligned} \tag{19}$$

$A_L^C(Q)$ is therefore the convolution of the 'true' structure factor $A^C(Q)$ and the transform of $m(r)$

$$m(Q) = \frac{\sin(QL/2)}{Q} \tag{20}$$

i.e.

$$A_L^C(Q_j) = \frac{1}{\pi}\int A^C(Q_i)\left(\frac{\sin(|Q_i - Q_j|L/2)}{|Q_i - Q_j|} - \frac{\sin(|Q_i + Q_j|L/2)}{|Q_i + Q_j|}\right)\mathrm{d}Q_i \tag{21}$$

If the experimental structure factor $A^E(Q)$ is similarly convoluted with $m(Q)$,

$$A_L^E(Q_j) = \frac{1}{\pi}\int A^E(Q_i)\left(\frac{\sin(|Q_i - Q_j|L/2)}{|Q_i - Q_j|} - \frac{\sin(|Q_i + Q_j|L/2)}{|Q_i + Q_j|}\right)\mathrm{d}Q_i \tag{22}$$

then it can be compared directly to $A_L^C(Q)$. While the *shape* of the structure factor has been modified by the convolution, all the *intensity* information is maintained so a quantitative comparison of experiment and simulation can be made (Nield *et al.*, 1992). If $g^E(r)$ is already flat at $r = L/2$ then the convolution will not alter $A^E(Q)$. The problem with this method is that the convolution redistributes errors as well as 'real' data, with an unknown effect.

2.5.3 Inverse method

The $g^E(r)$ may be calculated from $A^E(Q)$ using an inverse method, MCGR (Monte Carlo g of r) (Soper, 1992; McGreevy, 1994), rather than a direct transform. MCGR is similar to a 'one-dimensional' version of RMC. A numerical $g^E(r)$ is modified by a Monte Carlo method until its transform is consistent with $A^E(Q)$. Because $g^E(r)$ can be generated out to any desired r value, the effect of truncation can be avoided. In practice this means that the maximum r value is chosen to match the experimental Q resolution. The resolution could in principle be included in the calculation. However, the required r range increases as the Q resolution improves so this produces a requirement for an enormous array of $Q \times r$ points and computer memory becomes a limiting factor (recent developments overcome this problem). This method, in contrast to those described above, can be applied to systems with any degree of structural disorder. Figure 6.2 shows results for $YBa_2Cu_3O_{6.95}$ (Karlsson and McGreevy, 1995).

2.5.4 Direct calculation

The structure factor can be calculated directly from the atomic positions. For neutron diffraction

$$A(\mathbf{Q}) = \frac{1}{2\pi N'}\left\langle \sum_{i=1}^{N'}\sum_{j=1}^{N'} b_i b_j \exp\left(i\mathbf{Q}\cdot(\mathbf{r}_i - \mathbf{r}_j)\right)\right\rangle \tag{23}$$

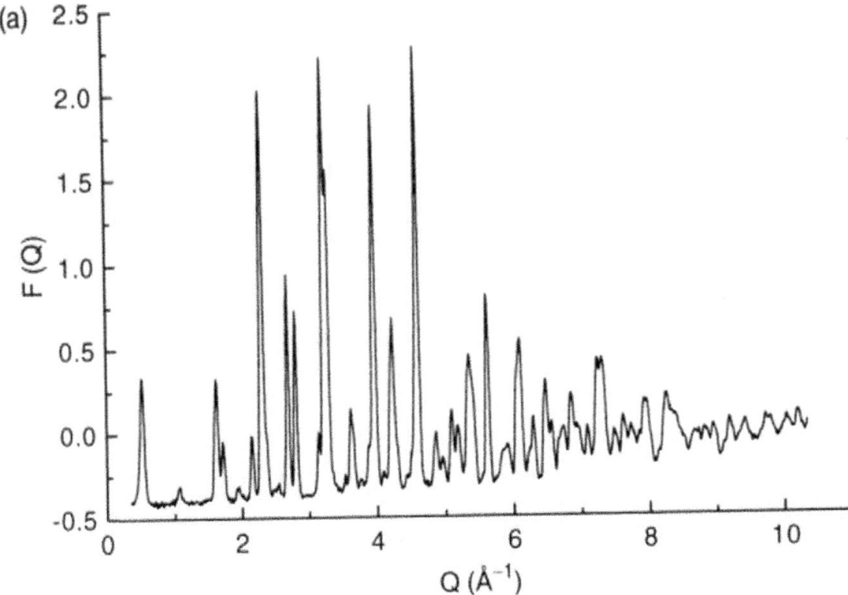

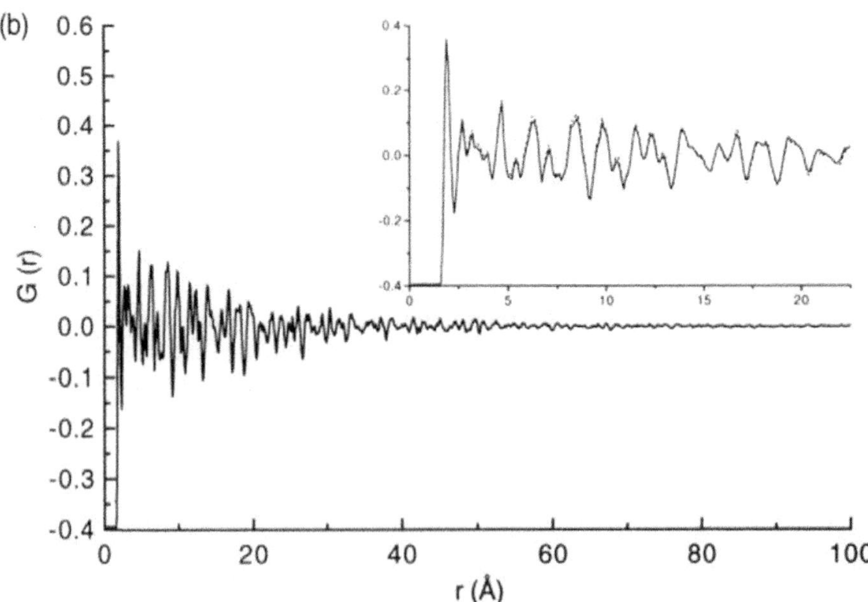

Figure 6.2 (a) Experimental total structure factor for $YBa_2Cu_3O_{6.95}$, measured by neutron diffraction (solid curve) and MCGR fit (dashed curve) the fit is so good that this curve can hardly be distinguished. (b) Corresponding total radial distribution function over the whole *r* range for which it has been generated. The inset shows the low *r* section subsequently used for RMC modelling (solid curve) and the RMC fit (broken curve).

where the summation runs over the N' atoms at positions $\mathbf{r}_j$ in the simulation cell and all the (infinite number of) images. Since the simulation cell is a supercell of the unit cell this expression reduces to

$$A(\mathbf{Q}) = \frac{1}{2\pi N}\left|\sum_{j=1}^{N} b_j \exp(i\mathbf{Q}\cdot\mathbf{r}_j)\right|^2 \tag{24}$$

for $\mathbf{Q}$ values corresponding to standing waves in the simulation cell, i.e. $\mathbf{Q} = 2\pi(h,k,l)/L$ for integers h, k, l (the Bragg peaks form a subset of these points). The summation is now only over the atoms in the simulation cell and so can be performed. $A(\mathbf{Q})$ must then be convoluted with the experimental resolution function before comparison with the data. This method is best in principle, but the computational cost is greater than for methods based on a $g(r) \leftrightarrow A(\mathbf{Q})$ transform. In cases where the diffuse scattering is relatively featureless it is possible to use only a few points in addition to the Bragg peaks and calculate the diffuse background by interpolation. Figure 6.3 shows results for $YBa_2Cu_3O_{6.95}$ (Montfrooij *et al.*, 1996). The agreement with the data is as good as that achieved with conventional Rietveld refinement, but in this case the diffuse background has been calculated from the RMC model rather than fitting to a polynomial. This means that the diffuse scattering and thermal factors from the

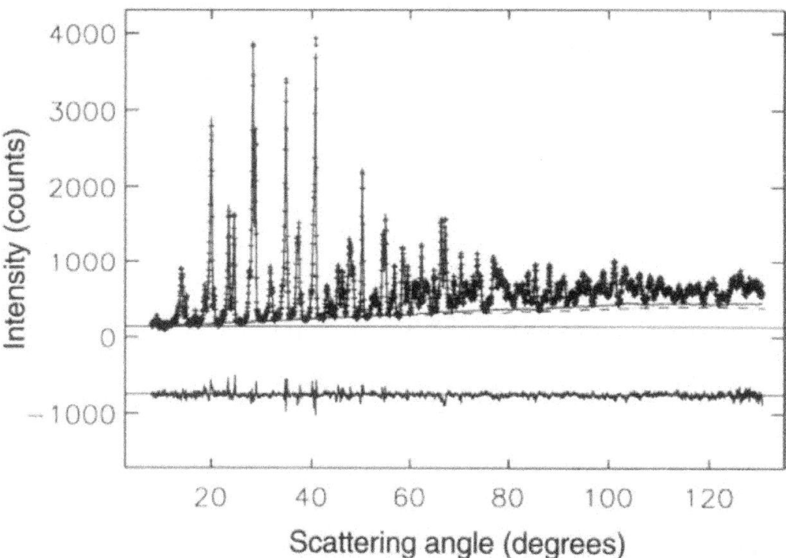

Figure 6.3 Experimental total structure factor for $YBa_2Cu_3O_{6.95}$ (+), direct RMC fit (solid curve) and the difference (lowest solid curve). The solid horizontal line shows the incoherent scattering contribution. The sloping solid line is the diffuse scattering calculated from the RMC model and the sloping dashed line is the background fitted by Rietveld refinement.

RMC model must be consistent, whereas there is no such requirement in Rietveld refinement. The result is that the RMC refinement produced more physically sensible values for some anisotropic temperature factors, in good agreement with single crystal results. However, it took much, much longer!

2.5.5 Calculation of Bragg scattering only

It might be thought that this would be the easiest calculation, but in fact it is not. While the total structure factor calculated directly, as in the previous section, is normally dominated by elastic Bragg scattering at Bragg peak positions, it nevertheless does contain diffuse scattering which may be significant for very disordered systems. Strictly, elastic Bragg scattering can only be calculated directly from a dynamic model, e.g. molecular dynamics, where the time-averaged positions of the atoms can be determined. To calculate them from a static model requires an assumption that space and time averages are equivalent, which is in fact the normal assumption made when powder diffraction data are analysed.

We denote the position of the jth atom in the ith unit cell by $\mathbf{r}_{i_j}$ where $\mathbf{r}_{i_j} = \mathbf{r}'_{i_j} + \mathbf{T}_i$ and $\mathbf{T}_i$ is the origin of the ith unit cell. The average position of each atom in the unit cell $\langle \mathbf{r}'_{i_j} \rangle$ must then be calculated. These will determine the equilibrium structure factor

$$S_e(\mathbf{Q}) = \frac{1}{2\pi N} \left| \sum_{i_j} b_{i_j} \exp(i\mathbf{Q} \cdot \mathbf{r}_{i_j}) \right|^2 \tag{25}$$

The effect of disorder must now be considered. For a harmonic cubic crystal with isotropic displacements we can simply calculate the mean square displacement $\langle \mathbf{u}_{i_j}^2 \rangle$, where

$$\mathbf{u}_{i_j} = \mathbf{r}'_{i_j} - \langle \mathbf{r}'_{i_j} \rangle \tag{26}$$

The structure factor is then

$$S(\mathbf{Q}) = S_e(\mathbf{Q}) \exp[-2W(\mathbf{Q})] \tag{27}$$

where the Debye–Waller factor is

$$W(\mathbf{Q}) = \tfrac{1}{2} \langle (\mathbf{Q} \cdot \mathbf{u}_{i_j})^2 \rangle = \tfrac{1}{6} Q^2 \langle \mathbf{u}_{i_j}^2 \rangle \tag{28}$$

For anharmonic crystals this can be extended to

$$W(\mathbf{Q}) = \tfrac{1}{2} \langle (\mathbf{Q} \cdot \mathbf{u}_{i_j})^2 \rangle - \tfrac{1}{6} \langle (\mathbf{Q} \cdot \mathbf{u}_{i_j})^3 \rangle - \tfrac{1}{24} \langle (\mathbf{Q} \cdot \mathbf{u}_{i_j})^4 \rangle - 3 \langle (\mathbf{Q} \cdot \mathbf{u}_{i_j})^2 \rangle^2 + \ldots \tag{29}$$

For anisotropic displacements the expressions are yet more complex. However, it can be seen that since the relevant information to calculate the expansions for $W(\mathbf{Q})$ to any order is available from the RMC model (in practice there may be statistical limitations), the elastic structure factor can be calculated.

This algorithm is currently being tested. For structural refinement it has the advantage over the direct method, outlined in Section 2.5.4, that it can be applied to data that have not had the diffuse background properly measured (which is normally the case for crystalline materials). There is a price paid for this in the assumptions that must be made and it must be judged in individual cases whether these are important.

2.6 Single crystal diffraction

It might initially be thought that diffraction measurements from single crystals would necessarily be preferable to those from powders, since the latter only provides an angular average of the information available from the former. However, the situation is in practice not quite so simple. Because of time and other practical restrictions, the number and range of $\mathbf{Q}$ points measured in a single crystal study of Bragg and diffuse scattering is often considerably smaller than in a powder study and the statistical accuracy (except at Bragg peaks) is often worse. It is then not clear that the information content of the single crystal diffuse scattering is greater than that of the powder pattern. What is clearly required is a combination of the two techniques; this is possible with RMC modelling and such an approach is currently being developed.

Because of the small size of the configuration (even if it contains many thousands of atoms) the statistical accuracy of the three-dimensional distribution $g(\mathbf{r})$ is too poor to enable direct transformation to the single crystal structure factor $A(\mathbf{Q})$. In addition there are truncation problems due to long range order, as mentioned earlier for powders. We therefore use essentially the same method as in Section 2.5.4, e.g. equation (24). It is, if possible, sensible to choose the $\mathbf{Q}$ points at which the measurement is to be made to be appropriate to the model, otherwise it is necessary to interpolate the data onto these points. For example, if the model is cubic and contains $10 \times 10 \times 10$ unit cells then $\mathbf{Q} = 0.1(h\mathbf{A}, k\mathbf{B}, l\mathbf{C})$ where $\mathbf{A}$, $\mathbf{B}$ and $\mathbf{C}$ are the reciprocal lattice vectors and h, k, l are integers.

For modelling of single crystal data there is a problem with fitting Bragg and diffuse scattering simultaneously due to the different effects of the resolution function when cutting through the Ewald sphere. So far only diffuse scattering has been successfully fitted (Nield *et al.*, 1995b). For the Bragg scattering it is necessary to calculate the peak intensities by normal integration, with the diffuse background subtracted, and then fit them as a separate data set simultaneously with the diffuse scattering data. This method is currently under development. An RMC fit to a single plane of diffuse scattering data for ice Ih is shown in Fig. 6.4 (Nield *et al.*, 1995a). The algorithm can cope with multiple planes and is now being extended to off-symmetry directions.

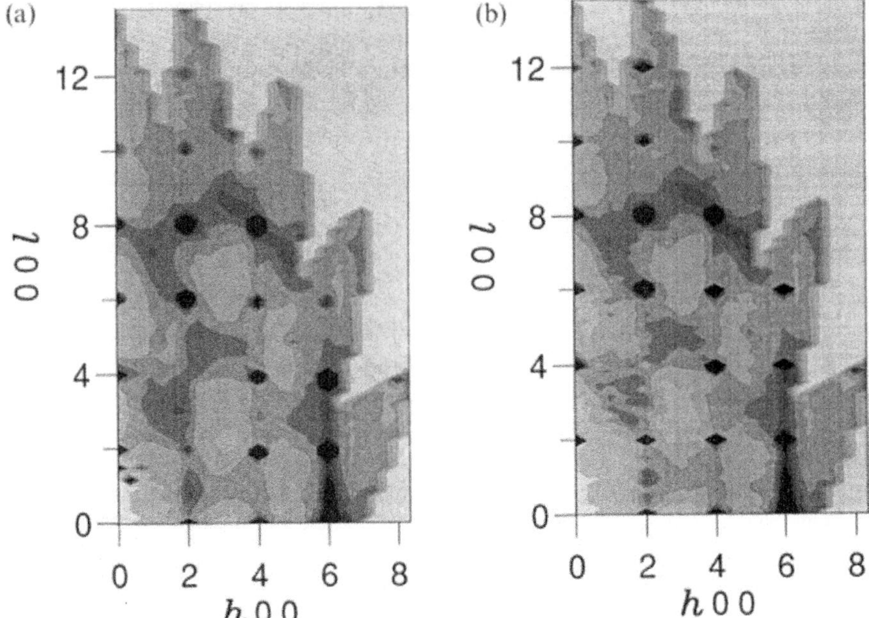

Figure 6.4 (a) Single crystal diffuse neutron scattering data measured in the (orthorhombic) $h0l$ plane of ice Ih and (b) RMC fit.

2.7 Uniqueness

One question that is continually raised with regard to the RMC method concerns the uniqueness of the structures generated, though the word 'unique' is often used mistakenly for the word 'correct'. The RMC model *may be* correct, in the sense that it is truly representative of the real structure, even though it is *never* unique in the sense that it is the only model that is consistent with the available information. In fact there has never been any claim that RMC models are in practice either unique or correct, but then this is true of *all* models, however they have been obtained. The non-uniqueness of RMC is not a disadvantage; it is an advantage. Constraints can be used in an empirical fashion to 'map out' the range of possible structures that are consistent with the data, and these can then be considered in the light of other available information or some possibilities can be discarded if new information becomes available. This point has been discussed in more detail elsewhere (McGreevy, 1993).

It has been shown (Evans, 1990) that in the particular case of a system described by pairwise additive potentials the one-dimensional structure factor of a macroscopically isotropic system does in fact contain all the information necessary to define the three-dimensional structure uniquely. Tests have shown that for such systems RMC (without constraints) reproduces the three body correlations correctly, i.e. the algorithm 'works' (Howe and McGreevy, 1991b). In systems where the potentials are definitely not pairwise additive it may nevertheless be possible to use constraints to obtain the 'correct' answers. However this should not be taken to mean that RMC can always produce the correct answer; often we may not know suitable constraints or they may not even be definable. It should also be noted that if inappropriate constraints are applied in the case of a system with pairwise additive potentials then it may be possible to fit the data, and yet the 'wrong' structure will be obtained.

2.8 Advantages of RMC

RMC modelling has many advantages as a general method of structural modelling.

1. Full use is made of all available data, not just particular features, in a quantitative rather than a qualitative manner. Many other models are based on particular structural features, such as peak positions and average coordination numbers from radial distribution functions, which can be misleading (McGreevy, 1991).
2. No interatomic potential is required.
3. The model is self-consistent and corresponds to a possible physical structure. This is not true of all models, particularly those which only concern short range structural correlations.
4. The macroscopic density is correct. There are some modelling techniques which can produce satisfactory agreement with experimental data, but only if a significantly wrong density is used.
5. Many different types of data can be combined simultaneously.
6. RMC is general and easily adapted to different physical problems.

2.9 Comparison with simulation methods using potentials

It has been stated above that the fact that no interatomic potential is required is one advantage of RMC. The determination of suitable potentials that will *accurately* reproduce the experimental diffraction data on even comparatively simple systems may be a complex task, as discussed elsewhere in this book. However, if it has been achieved then structural models produced this way are clearly equally as possible as those produced by RMC. The availability of potentials is then advantageous because (a) different thermodynamic states of the same system can be investigated (though potentials may not be transferable across, e.g., a metal–insulator transition), whereas RMC would require separate data sets and (b) some parts of potentials are transferable between systems.

Molecular dynamics also enables the calculation of dynamical correlations, whereas Monte Carlo procedures only provide structural information (however, some dynamical information can be inferred, as will be illustrated later).

We should refer also to the methods of *ab initio* molecular dynamics simulation (e.g. Car and Parrinello, 1985). We recall from Chapter 1 that by a combination of conventional molecular dynamics and density functional theory a full quantum mechanical treatment can be given to the electrons, and hence to the interatomic forces. Despite the importance of the method, it is extremely computationally expensive and has so far largely been applied to light atoms and small models (hundreds of atoms at most). Even then the quantum mechanical treatment is only applied to the bonding electrons, and core electrons are treated with conventional pseudopotentials. The results are dependent on the choice of pseudopotential and in some cases this choice is empirical. There is, therefore, still much scope for alternative techniques. In fact it is possible to compare the results of *ab initio* simulations and RMC modelling in the case of liquid metal alloys such as molten KPb. It was proposed that 'Zintl ions', e.g. tetrahedral Pb_4^{4-} ions, were a significant feature of the structure of such liquids. RMC modelling (Howe and McGreevy, 1991a) suggested that, rather than well-defined complex ions, a more network-like structure within which there were many three-coordinated Pb with tetrahedral bond angles was a better representation. This caused some controversy at the time, but later *ab initio* simulations of similar materials such as KSi (Galli and Parrinello, 1991), NaSn (Seifert *et al.*, 1992) and CsPb (de Wijs, 1995) came to exactly the same conclusion.

It is worth noting at this point that the results of simulations are usually judged, as is only natural, by comparison with experimental data. However, the comparison is often judged to be good when it is in fact only qualitative, i.e. well outside the experimental errors. Since the structures of many materials, particulary disordered ones, are determined to a large extent by packing considerations, even simple hard sphere models should agree qualitatively with the data. The success of a structural model should therefore be judged by how much *better* it does than a simple, e.g. hard sphere, model. Detailed analysis of the model can only be justified when comparison with the data is good in *this* context. However, it is also fair to say that it is often difficult for simulators to judge the true accuracy of experimental data.

3 APPLICATIONS

3.1 Structure and Fast Ion Conduction in crystalline materials

In this section we will illustrate the application of the RMC method to the study of disorder in crystals, with special emphasis on solids with high ionic conductivity, as discussed in Chapter 1 and 4.

3.1.1 Silver bromide

Silver bromide has the fcc rock salt structure. It is not normally classified as a fast ion conductor, since it undergoes no obvious transition, but at temperatures close to the melting point the ionic conductivity increases anomalously and is only a factor of three smaller than that of AgI (the 'archetypal' fast ion conductor) at the melting point. In a series of experiments (Keen *et al.*, 1990a,b; Nield *et al.*, 1992) the total structure factor of AgBr has been measured at temperatures to within 1 K of the melting point (701 K). At elevated temperatures the thermal factors are high and few Bragg peaks are visible in the diffraction pattern, making a conventional crystallographic study difficult. Typical data and an RMC fit to the convoluted structure factor (i.e. using the method outlined in Section 2.5.2) are shown in Fig. 6.5. The agreement is excellent.

The RMC model consisted of 8 × 8 × 8 unit cells, i.e. 4096 ions, with the lattice parameter (density) being determined directly from the data. The model has been extended to 16 × 16 × 16 unit cells, 32 768 ions, but no significant differences were found in the results (Keen *et al.*, 1990b). The starting

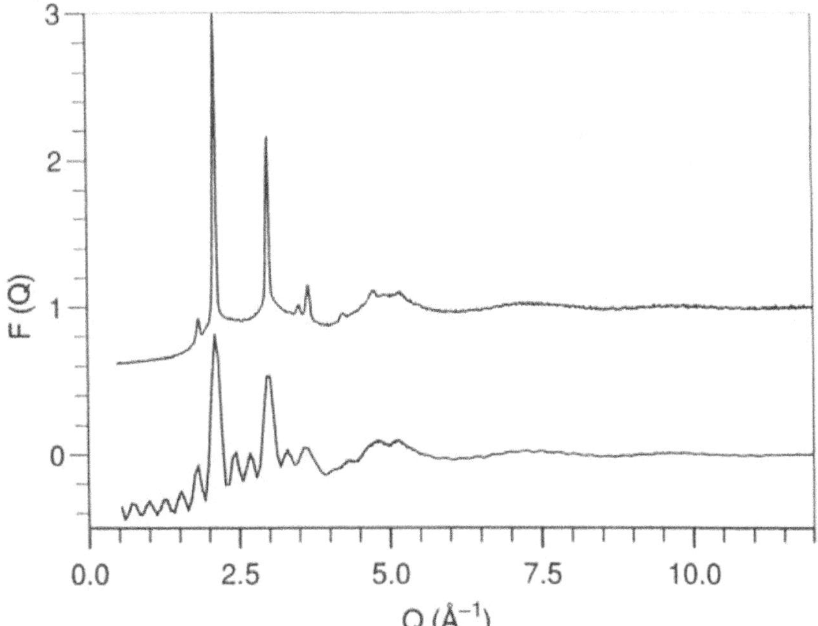

Figure 6.5 Total structure factor, $F^{E}(Q)$, for AgBr at 699 K (upper solid curve, displaced vertically) and comparison between $F_{L}^{E}(Q)$ (see Section 2.4, lower solid curve) and RMC fit, $F_{L}^{C}(Q)$, (dashed curve). The fit is so good that the dashed curve is almost invisible.

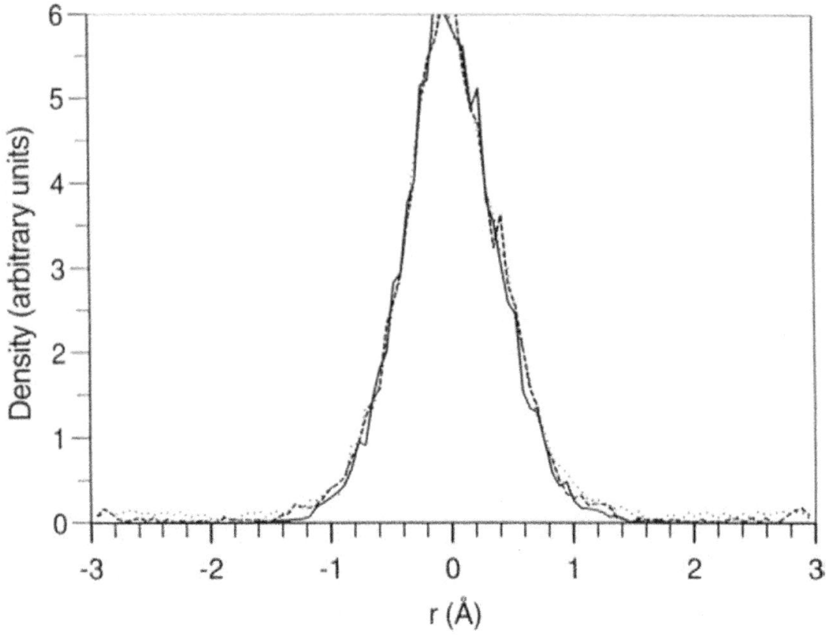

Figure 6.6 Density profile across Ag^+ lattice sites in ⟨100⟩ (solid curve), ⟨110⟩ (dashed curve) and ⟨111⟩ (dotted curve) directions in AgBr at $T = 669$ K.

configuration for modelling was the perfect crystal structure. The final models can be used to determine the average distribution of ions within the unit cell, usually averaged over about 10 independent configurations to improve statistics. By examining this distribution in detail, or by examining the model(s) directly, one can obtain considerable information on the local structural correlations. Figure 6.6 shows the density distribution of Ag^+ ions along the principal symmetry directions at 669 K. The distribution is largely harmonic (Gaussian), but there are small 'tails' in the ⟨110⟩ and ⟨111⟩ directions indicating some anharmonic motion. At higher temperatures the ⟨111⟩ 'tail' becomes a small peak (Fig. 6.7) which indicates partial occupation of the $(\frac{1}{4}, \frac{1}{4}, \frac{1}{4})$ interstitial sites. By making some empirical judgement of a dividing line between the lattice and interstitial sites the area of the peak, and hence the number of interstitials, can be determined. This increases rapidly close to the melting point (701 K) with $\sim$4% of the Ag^+ ions being in interstitial positions at 699 K.

Conventionally it is considered that it is not possible to derive dynamical information, and hence determine the mechanism of ionic conduction, from Monte Carlo simulations. However, by examining the density distribution along different possible pathways for conduction, one can determine whether such processes are 'active' and make some estimate of their relative probabilities (Fig. 6.8). Clearly in this case the dominant pathway will be via the interstitial

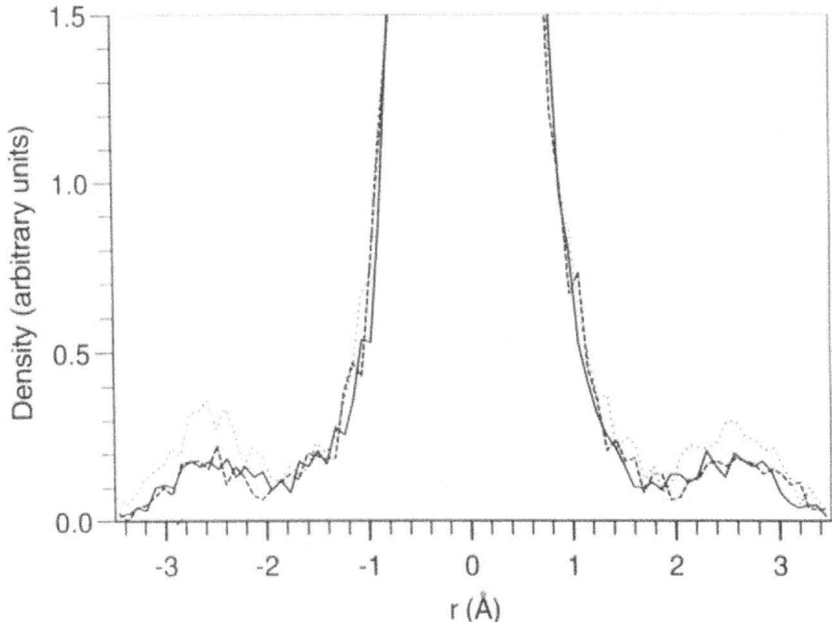

Figure 6.7 Density profile across Ag^+ lattice sites in ⟨111⟩ directions in AgBr at T = 669 K (solid curve), 689 K (dashed curve) and 699 K (dotted curve).

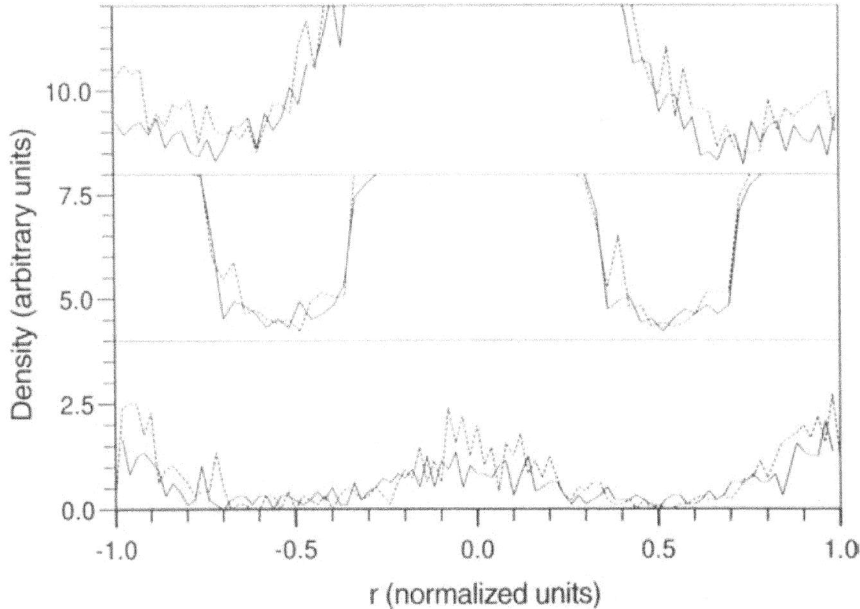

Figure 6.8 Density profiles along possible pathways for ionic conduction in AgBr at 669 K (solid curve) and 699 K (dashed curve). Top: from $(-\frac{1}{4}, -\frac{1}{4}, -\frac{1}{4})$ to $(\frac{1}{4}, \frac{1}{4}, \frac{1}{4})$ through (0, 0, 0) (interstitialcy jump). Centre: from $(-\frac{1}{2}, -\frac{1}{2}, 0)$ to $(\frac{1}{2}, \frac{1}{2}, 0)$ through (0, 0, 0) (direct vacancy jump). Bottom: from $(-\frac{1}{4}, \frac{1}{4}, \frac{1}{4})$ to $(\frac{3}{4}, \frac{1}{4}, \frac{1}{4})$ through $(\frac{1}{4}, \frac{1}{4}, \frac{1}{4})$ (direct interstitial jump).

site, but even then there are a number of different possible mechanisms including colinear and non-colinear interstitialcy jumps. Five possible mechanisms were examined and it was concluded that the interstitialcy jump dominates but that the other mechanisms are possible close to melting (Keen *et al.*, 1990b). The colinear jump was found to be most important by examining the distribution of interstitial–lattice site–vacancy angles in the RMC model.

By making accurate measurements of the Bragg scattering within 1 K of the melting point it was confirmed that the melting transition in AgBr is first order, despite the rapid increase in disorder (Nield *et al.*, 1992). Examination of the Ag–Br–Ag and Br–Br–Br bond angle distributions in the RMC model (Fig. 6.9) then led to the conclusion that AgBr is undergoing a second order transition to a structure in which the Br^- ions maintain their fcc positions, but the Ag^+ ions are distributed between the octahedrally coordinated (previously lattice) and tetrahedrally coordinated (previously interstitial) sites. This would be a 'true' fast ion conducting phase with rapid motion of Ag^+ between the alternative sites, similar to a proposed high pressure phase of AgI (Tallon 1986). However, melting occurs before this second order transition is complete, as had previously been suggested on the basis of theoretical arguments (Andreoni and Tosi, 1983).

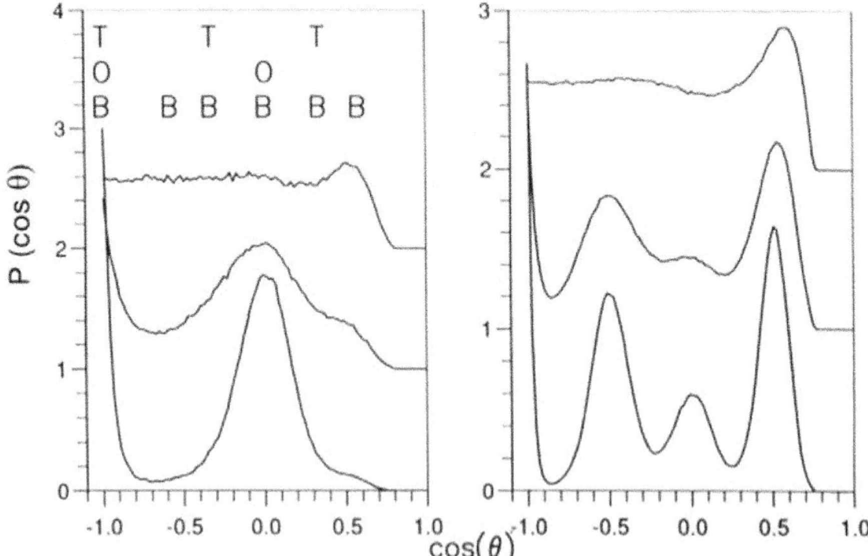

Figure 6.9 Ag–Br–Ag (left) and Br–Br–Br (right) bond angle distributions at 490, 699 and 703 K in AgBr (T_m = 701 K). Successive temperatures are displaced vertically by 1.0. The cosines which would be obtained for various fcc crystals are given, where T indicates four Ag^+ randomly occupying the eight tetrahedral sites, O indicates occupation of the four octahedral sites and B random occupation of the T and O sites; in all cases the Br^- occupy a perfect fcc sublattice.

3.1.2 Silver iodide

Silver iodide undergoes a first order structural phase transition at 420 K from the β-phase (hexagonal Wurtzite structure), which is metastable with respect to the γ-phase (cubic sphalerite structure), to the α-phase where the I^- ions occupy a bcc lattice within which the Ag^+ ions jump rapidly between a number of possible sites. The ionic conductivity is very high; upon melting it actually decreases. AgI is probably the most widely studied fast ion conductor, with much of the work concentrating on determination of the exact distribution of Ag^+ sites and conduction pathways.

Good fits to the total structure factor were obtained with six different RMC models using a variety of starting configurations and convergence criteria (Nield *et al.*, 1993). Four of these produced essentially the same result, with Ag^+ ions occupying tetrahedrally coordinated sites and conduction occurring via trigonal sites, and to a lesser extent via octahedral sites (Fig. 6.10). By fitting the density

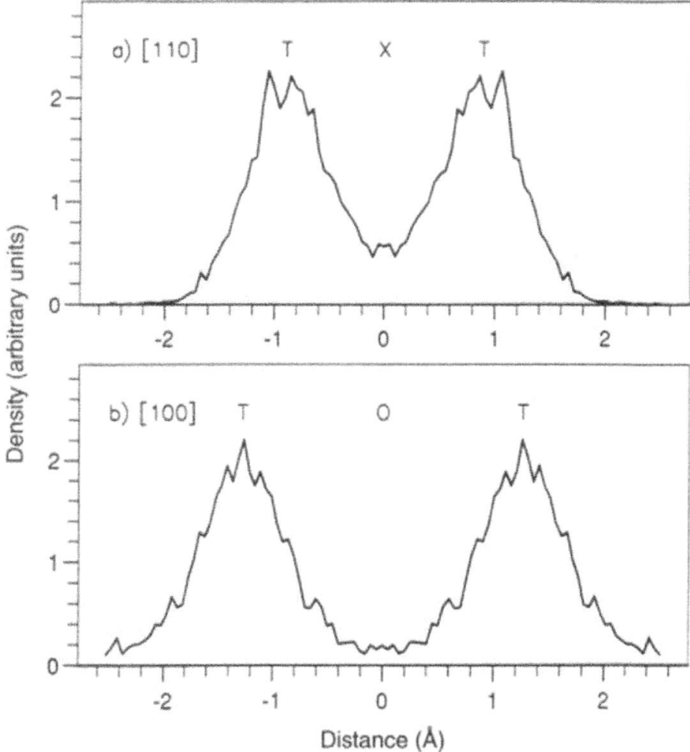

Figure 6.10 The Ag^+ density distribution in the average unit cell of α-AgI at 420 K. Top: ⟨110⟩ direction connecting two tetrahedral (T) sites and centred on a trigonal (X) position. Bottom: ⟨100⟩ direction connecting tetrahedral sites and centred on an octahedral (O) position.

distribution $\sim$75% of Ag^+ ions could be assigned to the tetrahedral sites, i.e. be considered to be vibrating about these sites, with $\sim$21% diffusing via trigonal sites and $\sim$4% via octahedral sites.

In the other two models the structures were deliberately forced into local minima with maximum occupation of tetrahedral and octahedral sites, respectively. In the latter model the density distribution was found to peak between, rather than at, the symmetry sites and so was considered to be unlikely. The models could also be compared by calculation of the effective single site potential, $V(\mathbf{r})$. If one makes the assumption that the Ag^+ density distribution, relative to the equilibrium site, is given by

$$\rho(\mathbf{r}) = A \exp[-V(\mathbf{r})/kT] \tag{30}$$

where A is a constant and k is Boltzmann's constant, then the potential is given by

$$\frac{V(\mathbf{r})}{k} = -T \ln[\rho(\mathbf{r})] + T \ln(A) \tag{31}$$

$V(\mathbf{r})$ was found to be temperature independent, in agreement with equation (30); it is shown in Fig. 6.11 for two directions. The energy barrier for conduction via the trigonal site is $\sim$600 K for the 'normal' RMC model and $\sim$1500 K for maximum tetrahedral site occupancy. The barriers via the octahedral site are 1300 K and infinity, respectively. For the model with maximum octahedral occupancy there is no barrier for conduction via trigonal sites, which confirms that this model is unrealistic. The energy barrier determined from the temperature dependence of the ionic conductivity is $\sim$1200 K. The model with maximum tetrahedral occupancy is therefore too ordered, while the 'normal' model is possibly too disordered. However, as there is no qualitative difference between these models, the conclusions that the equilibrium sites are tetrahedral and that conduction occurs dominantly via trigonal sites are valid.

3.1.3 CuBr

A similar study has been made of CuBr (Nield *et al.*, 1994) which has transitions from γ- to β- to α-phases. Interstitial Cu^+ ions are found in the empty tetrahedral sites in the β-phase by a similar method to that described above for AgBr. The results for the α-phase are almost identical to those for α-AgI in terms of the relative distribution of the Cu^+ and Ag^+ ions, despite the fact that the experimental data are different because of the different combinations of scattering lengths and ionic sizes.

3.1.4 Li_2SO_4 and $LiNaSO_4$

While many of the results of the RMC modelling studies described above are not new, in that they have been suggested on the basis of other work, it should

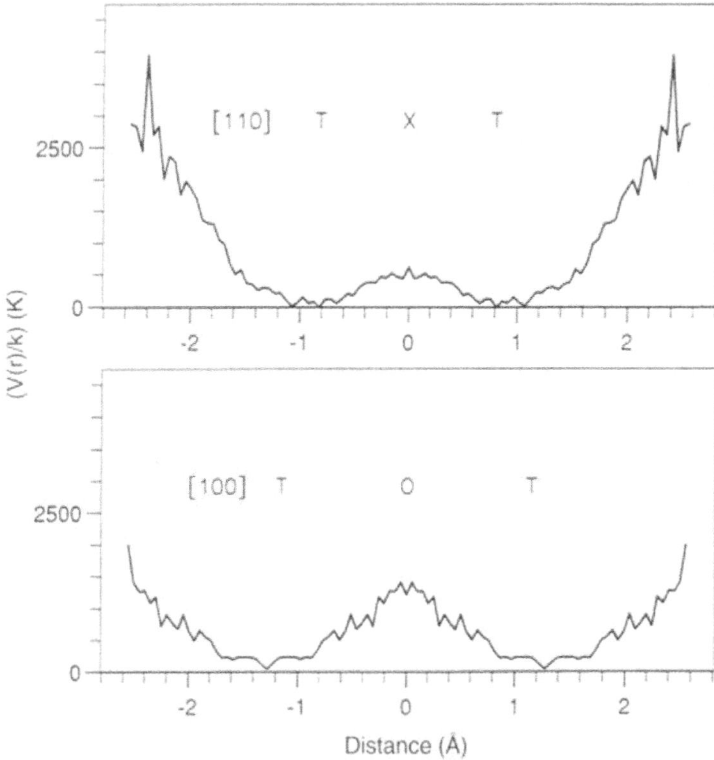

Figure 6.11 Effective single site potential, $V(\mathbf{r})/k$, in α-AgI at 420 K along ⟨110⟩ (top) and ⟨100⟩ (bottom) directions. The vertical offset is arbitrary.

be noted that they are all obtained self-consistently from the results of a single, relatively simple, experiment. The models are microscopic, and hence detailed, rather than being purely descriptive. RMC modelling can, as illustrated, be used to examine different possibilities rather than just to determine that a single model fits the data; if it fails to distinguish these possibilities directly, then they can be judged on the basis of other information. Another important point to note, with regard to modelling of fast ion conductors, is that the RMC model intrinsically includes all disorder, both thermal and static. What is immediately clear is that it is, in fact, impossible to draw an unambiguous dividing line between the two. The correlation between ionic vibrations and diffusion is important to the conduction mechanism yet many other models, particularly those based on elastic scattering, implicitly separate the two.

The above point is illustrated well by work on Li_2SO_4 and $LiNaSO_4$. Two models existed to describe the ionic conduction in such materials and no consensus had been reached despite a vehement debate over almost 20 years. The 'paddle wheel' model suggested that diffusive motion of Li^+ (or Na^+) must be highly correlated to rotations of the sulphate ions, whereas the 'percolation' model proposed that the motions were essentially uncorrelated, simply occurring because of the local free volume made available by large anion vibrations and/or rotations. RMC modelling (Karlsson and McGreevy, 1995) showed that both types of process could be identified, and indeed all sorts of processes in between. Rotational motions of the SO_4^{2-} groups were clearly important in enhancing the ionic conductivity, but should be considered in the context of large thermal vibrations, not as if in a rigid lattice.

3.2 Fast ion conduction in glasses

If 'traditional' glass formers such as $AgPO_3$ or $(Ag_2O)_x(B_2O_3)_{1-x}$ are doped with Ag or Li halides it is possible to produce glasses with high ionic conductivity (10^{-2} $(\Omega\ \mathrm{cm})^{-1}$) at ambient temperature. These are of technological interest as battery or sensor materials. Despite a considerable amount of work there is as yet no consensus on the structure, and hence the conduction mechanisms, of such glasses. One set of models proposes that AgI tends to form interconnected microclusters with a local structure similar to that of the superionic conductor α-AgI. Diffusion of Ag^+ ions then takes place along the connections between the microclusters. Neutron diffraction experiments show a considerable change in intermediate range order upon introduction of AgI into the base glass, characterized by a sharp diffraction peak at anomalously low Q values (0.7–0.8 $Å^{-1}$) for the doped glass which is not observed in the undoped glass (Börjesson *et al.*, 1989a,b) (see Fig. 6.12). The peak height increases approximately proportionately to the level of doping and to the ionic conductivity (Börjesson and Howells, 1991).

It might be thought that RMC modelling of a single structure factor for a multicomponent system would provide little reliable information. However, this is a good example of how constraints can be used in a very powerful way. From 'chemical knowledge' and crystallography it is believed that the structure of $AgPO_3$ glass consists of tetrahedral PO_4 units which form a random 'polymeric' network by sharing two corners (bridging oxygens). Two of the corners are not shared (non-bridging oxygens) and cross-linking of the phosphate chains occurs by ionic bonding of Ag^+ and one single bonded non-bridging oxygen. It is unlikely that the covalently bonded phosphate chain is disrupted by doping with salt, so the following method has been used to model the structures of $(AgI)_x(AgPO_3)_{1-x}$ glasses (Wicks *et al.*, 1995).

1. A Hard Sphere Monte Carlo (HSMC) simulation is run for P atoms at densities appropriate to the doped and undoped glasses, with the constraint that each P atom is two-fold coordinated. This creates a –P–P–P–P– chain structure that is ~99% perfect; to produce a perfect

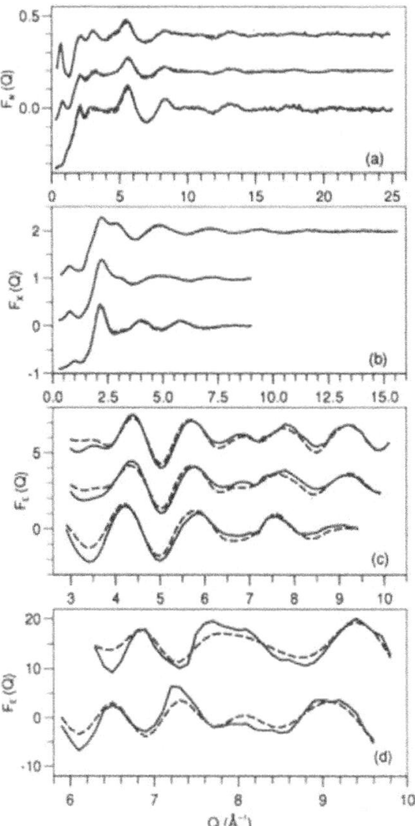

Figure 6.12 Experimental data (solid curves) and RMC fits (dashed curves) for $(AgI)_x(AgPO_3)_{1-x}$ glasses with $x = 0$ (bottom), 0.3 (centre) and 0.5 (top). (a) Neutron diffraction, (b) X-ray diffraction, (c) Ag K edge EXAFS and (d) I L_{III} edge EXAFS (no data for $x = 0$).

structure would require an inordinate amount of computer time and in many cases would probably not correspond to reality.

2. O atoms are added, initially at the centre of each P–P bond, and the HSMC is run with the constraint of two-fold P–O coordination. This creates a non-linear –O–P–O–P–O–P–O– chain structure.
3. Two more O atoms are added to each P atom, though not now at bond centres. HSMC is again run with a constraint of four-fold P–O coordination. This creates a network of PO_4 units that share two O atoms. By suitable choice of P–O and O–O separations these can be made approximately tetrahedral (though this information is in fact contained in the structure factor).
4. The appropriate numbers of Ag and I (depending on x) are added at random. RMC is run, fitting to neutron diffraction data, with the

constraint that the existing four-fold P–O coordination is maintained, i.e. the network can change its conformation but not its connectivity.

The origin of the low Q peak was found quite easily. Even before fitting to the data at step 4 the PO_4 network created by HSMC showed a distinct peak at $Q \approx 0.8$ Å^{-1} in the partial structure factor $A_{PP}(Q)$ at the lower PO_3 density (0.0309 Å^{-3}) corresponding to $x = 0.5$, but not at the higher density (0.059 Å^{-3}) corresponding to $x = 0$. For the network to maintain its connectivity, i.e. covalent bonding, with a significantly lower density the phosphate chains must 'stretch', leaving large gaps in the structure which are filled by the dopant salt AgI. It is the density fluctuations between the 'normal' glass regions and the gaps that produce the low Q peak. The low Q peak does not occur in Ag and I partial structure factors.

This prediction was contrary to current thinking, which presumed that the peak must be due to AgI scattering because its intensity was proportional to the dopant level. The RMC model prediction was easily tested by X-ray diffraction. If the peak was due to AgI scattering then it would be very large in the X-ray structure factor, which is dominated by Ag and I scattering, but if it was due to the phosphate network it would be very small. The experimental X-ray structure factors confirmed the RMC model. However, the model did not predict the X-ray structure factor correctly at all Q, because of the differing information content of the neutron structure factor, so the models were then fitted to both neutron and X-ray data. Finally the models were fitted to neutron diffraction, X-ray diffraction, Ag K edge EXAFS and I L_{III} edge EXAFS data as shown in Fig. 6.12 (Wicks *et al.*, 1995).

The series of RMC models could then be investigated in detail, in order to determine the relationship between structure and ionic conductivity. A method was used in which a chosen Ag^+ ion was allowed to 'explore' its local accessible free volume. If this was sufficiently large for a diffusive jump the ion was moved (by a random amount within the free volume) and then one of its (previous) neighbours was chosen. If an ion could not be moved sufficiently then another neighbour was chosen, and so on. In this way 'trees' of conducting ions could be built up. It was found that for $x < 0.3$ there were a number of isolated trees, but for $x > 0.3$ a percolation transition occurred and then the majority of Ag^+ were in a single tree. By using the tree connectivity as an estimate of ionic conductivity good qualitative agreement with experimental results could be obtained. Another method used was to run the RMC program and use the Monte Carlo move as an effective 'time step', calculating the mean square displacement and hence the effective diffusion constant. Again a transition was found between $x = 0.2$ and 0.3. Results are shown in Fig. 6.13.

The models also explained the range of compositions over which glasses could be formed. Ag^+ that 'bridged' between phosphate chains by bonding to O^- could be identified. At low AgI dopings there were many such bridges but the number decreased rapidly with increasing x; it could be predicted that for

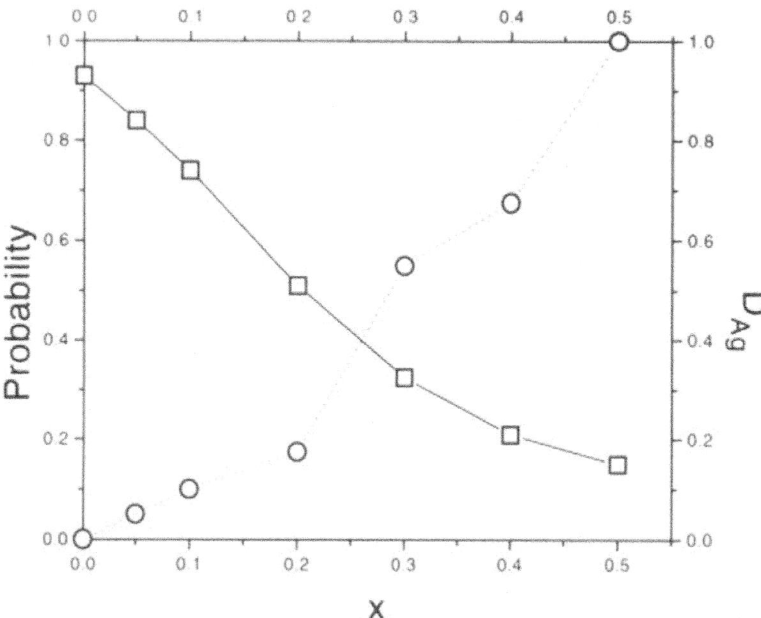

Figure 6.13 Squares and solid curve: probability of Ag^+ cross-linking phosphate chains in $(AgI)_x(AgPO_3)_{1-x}$ glasses as a function of x. Circles and dotted curve: effective Ag^+ diffusion constant, calculated as described in the text. There is a transition between local and macroscopic diffusion between $x = 0.2$ and 0.3.

$x > 0.6$ the number would be insufficient to maintain connectivity between the phosphate chains and a glass would no longer be formed (see Fig. 6.13). Experimentally, the limit is $x \approx 0.55$. Further work on borate glasses (Swenson *et al.*, 1995), where the network is (topologically) two-dimensional, and more recently on tungstate and molybdate glasses which contain molecular ions and not a 'conventional' network (Wicks, 1993; Swenson *et al.*, 1996), has confirmed that free volume and cross-linking are the dominant factors determining the ionic conductivity and glass forming range, respectively.

3.3 Silicate glasses

Complex silicate glasses, discussed also in Chapter 12, are of commercial, geological and academic interest. They can be considered as more of a challenge for RMC than the glasses described above since the network is three-dimensional, thus putting considerably more constraints on the models. The structure of pure vitreous SiO_2 is generally agreed to be a continuous random network of corner sharing SiO_4 tetrahedra (see also Chapters 1 and 9). Each oxygen bridges two tetrahedra. When alkali oxide (known as a modifier) is added, the extra oxygen introduced causes some breaking of the network. Each Si remains four-fold coordinated to O but some of these O are not shared (non-bridging). Tetrahedra with n bridging O are known as Q_n species. The proportions of these can be determined by Magic Angle Spinning ^{29}Si Nuclear Magnetic Resonance (MASNMR) (Murdoch *et al.*, 1985). In the first work of its type, MASNMR data has been used quantitatively in constructing RMC models of $(K_2O)_x(SiO_2)_{1-x}$ glasses (Wicks, 1993). The initial configurations have been constructed in the following manner.

1. Si atoms at the appropriate density are introduced at random and then HSMC is run with constraints on n-fold Si–Si coordination equivalent to the proportions of Q_n species determined from NMR. For instance, in the case of $(K_2O)_{0.2}(SiO_2)_{0.8}$ the ratios of Q_4:Q_3:Q_2 are 0.57:0.35:0.08. The coordination that is achieved is 98–99% perfect.
2. O atoms are added at the mid-point of each Si–Si bond. Single O atoms are then added close to those Si that now have only two or three neighbours (two-fold or three-fold Si–Si coordination). HSMC is run with a constraint on four-fold Si–O coordination. In the 'ideal' modified random network model (Greaves *et al.*, 1981) the ratio of non-bridging to bridging oxygens is given by $2y/(2+y)$ where $y = x/(1-x)$; it can be checked that the resulting model is close to this prediction.
3. Alkali atoms are added at random and RMC is run, keeping the Si–O coordination constraint.

In the same way as for the $(AgI)_x(AgPO_3)_{1-x}$ glasses, the model is now free to change its conformation, but not its topology. However the topological constraint is more severe in this case since it is 'three-dimensional', rather than 'one-dimensional' for the phosphate chain structure. Nevertheless good fits to the data can still be obtained, even in the low Q region of the structure factor which becomes quite complex for intermediate x ($x \approx 0.3$). This result suggests that the structures of such covalent glasses are largely determined by packing considerations (i.e. macroscopic density) and simple topological bonding constraints. Details such as the preferred bond angles then determine the details of the structure.

Further discussion of the application of RMC to amorphous structures is given in Chapter 12.

3.4 Polymers

Polymers are materials of immense technological importance. Much past work has concentrated on macroscopic structures; however, there is now increasing interest in the role of the short and intermediate range structure in determining macroscopic structure and hence physical properties. An important new class of materials are ion conducting polymers (polymer electrolytes) which are at the forefront of thin film battery design. Clearly the application of RMC as described above for studying crystalline and glassy fast ion conductors can be extended to this field.

The first such RMC work was performed on the simplest polymer, molten polyethylene (Rosi-Schwartz and Mitchell, 1994). The initial polymer chain was created with commercial molecular modelling software, though an MC method similar to those described above can also be used. The model was then either relaxed and a short MD simulation run using the best available force fields, or fitted to neutron diffraction data using RMC. While the initial model showed significant differences from the experimental data both the final MD and RMC models were in equally good agreement with it. They also agreed well in all details of the chain correlations both locally and over intermediate range.

This result might seem to suggest that the RMC method has no advantage. However for more complex polymers such as polytetrafluoroethylene (PTFE) (Rosi-Schwartz and Mitchell, 1995) and polymethylmethacrylate (PMMA) (Ward and Mitchell, 1995) it is much more difficult to develop suitable potentials. Polymers with charged groups present an even more severe problem. For RMC the modelling is in principle no more difficult, though construction of the initial chains takes longer and it may take a considerable time to fit the data because motions are hindered by the bulky side groups. It is probably necessary to develop new methods to speed this up. Figure 6.14 shows an RMC fit to neutron diffraction data for molten PTFE (Rosi-Schwartz and Mitchell, 1995) and Plate VI shows the resulting structural model.

4 FUTURE DEVELOPMENTS

As has been illustrated, RMC modelling has already been applied to many different problems. The ability to combine many types of experimental data to produce a single structural model, and to use constraints, makes it a very powerful technique. There are some obvious areas of further application in chemical crystallography. One example would be zeolite or other 'cage' type materials, where the study of the behaviour of guest molecules such as hydrocarbons in the host material is important for the understanding of many chemical processes as discussed in Chapters 1 and 5. Such applications would be directly comparable to the work already discussed on fast ion conductors, if slightly more complex. Another area is the structural study of surfaces, e.g. reflectivity and surface diffraction (which could be combined). Recent developments in the quantitative calculation of electron microscopy images may also make this amenable to RMC modelling within a few years. The first work on magnetic structures of crystalline materials is now being done. In such areas, where the RMC method is already well developed or the development is obvious, it is now in many cases the accuracy or range of experimental data

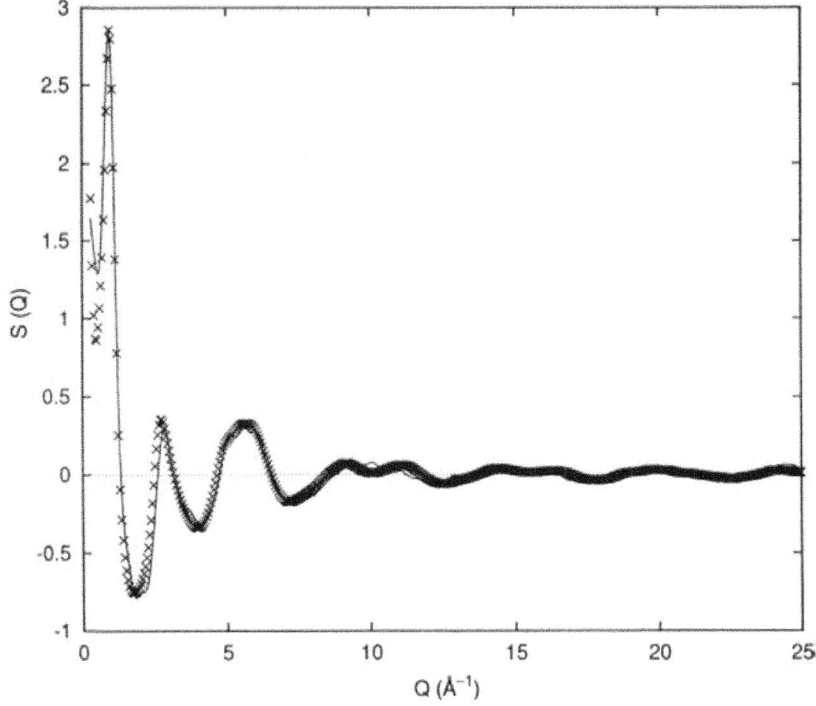

Figure 6.14 Experimental neutron diffraction data (×) and RMC fit (solid curve) for molten PTFE.

that needs improving, or even that suitable data (e.g. with backgrounds properly subtracted and correctly normalized) have not yet been measured.

One of the main areas of development will be in the use of models, since of course the construction of models is not of itself a useful exercise. The major problem here is that it is not always obvious, particularly in the case of disordered materials, how one should 'isolate' the particular correlations of interest from the overwhelming 'background' when one does not usually know which correlations are of interest! It is already clear that methods based on sophisticated graphics (not just 'ball-and-stick' representations) will play a major role.

In a general sense the RMC method can be applied to any physical problem where the relationship between the data measured and the model required is indirect. The most important feature of the method in this context is that it uses a proper Markov chain, i.e. it is not simply a refinement technique. The price paid for this is in computational time, which puts a limit on the model size; however, with the rapidly decreasing price/performance ratio of computers this problem is disappearing. One example of such an extension would be to the structure of macroscopic materials. For example a major goal of work on polymers is to relate the macroscopic and microscopic structures and properties. This is probably some years away, but it is becoming feasible to consider problems such as modelling nanocrystals with 100 000 atoms.

REFERENCES

Andreoni, W. and Tosi, M.P. (1983) *Solid State Ionics*, **11**, 49.

Börjesson, L. and Howells, W.S. (1991) *Solid State Ionics*, **40/41**, 702.

Börjesson, L., Torell, L.M., Dahlborg, U. and Howells, W.S. (1989a) *Phys. Rev. B*, **39**, 3404.

Börjesson, L., Torell, L.M., Dahlborg, U. and Howells, W.S. (1989b) *Phil. Mag. B*, **59**, 105.

Börjesson, L, McGreevy R.L. and Howells, W.S. (1992) *Phil. Mag. B*, **65**, 261.

Car, R. and Parrinello, M. (1985) *Phys. Rev. Lett.*, **55**, 2471.

de Wijs, G.A. (1995) PhD thesis, University of Groningen.

Evans, R.A. (1990) *Mol. Simul.*, **4**, 409.

Galli, G. and Parrinello, M. (1991) *J. Chem. Phys.*, **95**, 7504.

Gereben, O. and Pusztai, L. (1994) *Phys. Rev. B*, **50**, 14136.

Greaves, G.N., Fontaine, A., Lagarde, P., Raoux, D. and Gurman, S.J. (1981) *Nature*, **293**, 611.

Gurman, S.J. and McGreevy, R.L. (1990) *J. Phys.: Condens. Matter*, **2**, 9463.

Howe, M.A. (1989) *Physica B*, **160**, 170.

Howe, M.A. (1990a) *Mol. Phys.*, **69**, 161.

Howe, M.A. (1990b) *J. Phys.: Condens. Matter*, **2**, 741.

Howe, M.A. and McGreevy, R.L. (1991a) *J. Phys.: Condens. Matter*, **3**, 577.

Howe, M.A. and McGreevy, R.L. (1991b) *Phys. Chem. Liq.*, **24**, 1.

Howe, M.A., McGreevy, R.L., Pusztai, L. and Borzsak, I. (1993) *Phys. Chem. Liq.*, **25**, 204.

Iparraguire, E.W., Sietsma, J. and Thijsse, J. (1993) *J. Non-Cryst. Solids*, **156–158**, 969.

Karlsson, L. and McGreevy, R.L. (1995) *Solid State Ionics*, **76**, 301.

Keen, D.A. and McGreevy, R.L. (1990) *Nature*, **344**, 423.

Keen, D.A. and McGreevy, R.L. (1991) *J. Phys.: Condens. Matter*, **3**, 7383.

Keen, D.A., Hayes, W., McGreevy, R.L. and Clausen, K.N. (1990a) *Phil. Mag. Lett.*, **61**, 349.

Keen, D.A., McGreevy, R.L. and Hayes, W. (1990b) *J. Phys.: Condens. Matter*, **2**, 2773.

Keen, D.A., McGreevy, R.L., Bewley, R.I. and Cywinski, R. (1995) *Nucl. Inst. Meth. Phys. Res. A*, **354**, 48.

Kugler, S., Pusztai, L., Rosta, L, Bellisent, R., Menelle, A. and Chieux, P. (1993) *Phys. Rev. B*, **48**, 7685.

Maret, M., Lancon, F. and Billard, L. (1994) *J. Phys.: Condens. Mätter*, **6**, 5791.

McGreevy, R.L. (1991) *J. Phys.: Condens. Matter*, **3**, F9.

McGreevy, R.L. (1992) *Int. J. Mod. Phys. B*, **7**, 2965.

McGreevy, R.L. (1993) *J. Non.-Cryst. Solids*, **156–158**, 949.

McGreevy, R.L. (1994) *Proc. Third European Powder Diffraction Conference* (eds R. Delhez and E.J. Mittemeijer), Trans. Tech. Publications Ltd, Aedermannsdorf, Switzerland. Materials Science Forum, **166–169**, 45.

McGreevy, R.L. (1995) *Nucl. Inst. Meth. Phys. Res. A*, **354**, 1.

McGreevy, R.L. and Howe, M.A. (1992) *Ann Rev. Mater. Sci.*, **22**, 217.

McGreevy, R.L. and Pusztai, L. (1988) *Mol. Simul.*, **1**, 359.

McGreevy, R.L. and Pusztai, L. (1990) *Proc. R. Soc. A*, **430**, 241.

McGreevy, R.L., Howe, M.A., Keen, D.A. and Clausen, K.N. (1990) *IOP Conference Series*, Vol. 107, (ed. M.W. Johnson), p. 165.
McGreevy, R.L., Howe, M.A. and Wicks, J.D. (1994) *RMCA Version 3 Manual.* The manual and the RMCA program are available on the world wide web (http:\\www.studsvik.uu.sc).
McKenzie, D.R., Davis, C.A., Cockayne, D.J.H., Muller, D.A. and Vassallo, A.M. (1992) *Nature*, **355**, 622.
Metropolis, N., Rosenbluth, A.W., Rosenbluth, M.N., Teller, A.H. and Teller, E. (1953) *J. Phys. Chem.*, **21**, 1087.
Montfrooij, W., McGreevy, R.L., Hadfield, R. and Andersen, N.-H. (1996) *J. Appl. Cryst.*, **29**, 285.
Murdoch, J.B., Stebbins, F. and Carmichael, I.S.E. (1985) *Am. Mineral.*, **70**, 332.
Newport, R.J., Honeybone, P.J.R., Cottrell, S.P., Franks, J., Revell, P., Cernik, R.J. and Howells, W.S. (1991) *Surface Coatings and Technol.*, **47**, 608.
Nield, V.M., Howe, M.A. and McGreevy, R.L. (1991) *J. Phys.: Condens. Matter*, **3**, 7519.
Nield, V.M., Keen, D.A., Hayes, W. and McGreevy, R.L. (1992) *J. Phys.: Condens. Matter*, **4**, 6703.
Nield, V.M., Keen, D.A., Hayes, W. and McGreevy, R.L. (1993) *Solid State Ionics*, **66**, 247.
Nield, V.M., Keen, D.A., Hayes, W. and McGreevy, R.L. (1994) *Physica B*, **202**, 159.
Nield, V.M., Li, J.-C., Ross, D.K. and Whitworth, R.W. (1995a) *Physica Scripta* **T57**, 179.
Nield, V.M., Keen, D.A. and McGreevy, R.L. (1995b) *Acta Crystallogr.*, in press.
Pannetier, J., Bassas-Alsina, J., Rodriguez-Carvajal, J. and Caignaert, V. (1990) *Nature*, **346**, 343.
Pusztai, L. and Svab, E. (1993) *J. Non-Cryst. Solids*, **156–158**, 973.
Rietveld, H.M. (1969) *J. Appl. Crystallogr.*, **2**, 65.
Rosi-Schwartz, B. and Mitchell, G.R. (1994) *Polymer*, **35**, 3139.
Rosi-Schwartz, B. and Mitchell, G.R. (1995) *Nucl. Instr. Meth. in Phys. Res. A*, **354**, 17.
Seifert, G., Pastore, G. and Car, R. (1992) *J. Phys.: Condens. Matter*, **4**, L179.
Soper, A.K. (1992) *IOP Conference Series*, Vol. 107 (ed. M. W. Johnson). IOP publishing, Bristol, UK. p. 192.
Swenson, J., Börjesson, L. and Howells, W.S. (1995) *Physica Scripta*, **T57**, 117.
Swenson, J., McGreevy, R.L., Börjesson, L., Howells, W. and Wicks, J.D. (1996) *J. Phys. Condens. Matter*, **8**, 3545.
Tallon, J. (1986) *Phys. Rev. Lett.*, **57**, 2427.
van Hove, L. (1954) *Phys. Rev.*, **95**, 249.
Ward, D.J. and Mitchell, G.R. (1995) *Physica Scripta*, **T57**, 153.
Wicks, J.D. (1993) D. Phil. thesis, University of Oxford.
Wicks, J.D., Börjesson, L., Bushnell-Wye, G., Howells, W.S. and McGreevy, R.L. (1995a) *Phys. Rev. Lett.*, **74**, 726.

7

Defects, Surfaces and Interfaces

J.H. Harding

1 INTRODUCTION

Defects exert a crucial influence on the properties of materials and simulations have made a major contribution to our understanding of defect structures and properties. The state of the art up to 1990 has been summarized by Harding (1990). A more recent review of oxide surfaces has been given by Colbourn (1992). The older review of Duffy (1986) is still the only one devoted to grain boundary simulations.

The purpose of this chapter is twofold. Firstly, we shall discuss the methods of simulating defects and interfaces using classical models. We further restrict attention to static and quasi-harmonic calculations. A discussion of molecular dynamics is given in Chapter 4. Although some details of the methods will be given, the main objective is to indicate what kinds of calculation are possible within this framework and, perhaps equally important, where the problems begin. Second, we shall choose a number of examples to illustrate the kind of work that is being done at present. In the light of this we shall point to areas where further progress is needed.

We now turn to the models underlying the simulations. These are reviewed in greater detail in Chapter 3; this paragraph considers only the basic points necessary for understanding what follows. In the classical model of ionic crystals we assume that the crystal can be described as ions with the full charge corresponding to the oxidation state with short-range interionic potentials acting between the ions. That is, following the approach outlined in Chapter 1, we write the full crystal potential as

$$V(r_1 \cdots r_n) = V^{(0)} + \sum_{i,j=1}^{N}{}' V^{(2)}(|r_i - r_j)|) + \sum_{i,j,k=1}^{N}{}' V^{(3)}(r_i, r_j, r_k) + \sum_{i,j,k,l=1}^{N}{}' V^{(4)}(r_i, r_j, r_k, r_l) + \cdots \quad (1)$$

COMPUTER MODELLING IN INORGANIC CRYSTALLOGRAPHY
ISBN 0-12-164135-X

The first term is the energy required to create the ion in the electronic state that it has in the crystal. This is usually assumed to be constant. Recent work (Harding and Pyper, 1995) has shown that this assumption is untrue for oxides. However, it is possible to absorb most of the effect into the short-range part of the pair potential (the second term). The three-body and higher terms can often be neglected. We define the short-range part of the pair potential as

$$\phi_{ij}(r) = V_{ij}^{(2)}(r) - \frac{q_i q_j}{r_{ij}} \tag{2}$$

This includes the Pauli repulsion and (attractive) dispersion terms. The polarizability of the ions is included using the shell model (Dick and Overhauser, 1964) which, as discussed in Chapter 3, models the polarizability using a massive core linked to a mass-less shell by a spring. The theoretical basis of this model is uncertain, but its practical success has been attested over 20 years. Probably the best way to consider it is as a sensible model for linking the electronic polarizability of the ions to the forces exerted by the surrounding lattice. It is therefore a many-body term, a fact that should be remembered if one wishes to consider three-body potentials in the description of the crystal. A recent development in the field has been the use of quantum calculations. These are discussed in detail elsewhere (Chapter 8) but some results will be compared with the classical simulations in this chapter.

2 BULK DEFECTS

2.1 Preliminary remarks

The idea of a point defect was introduced into solid state physics to explain why crystals can transport mass as rapidly as they do. At a finite temperature elementary thermodynamic considerations show that point defects are inevitable; however high the energy needed to generate a defect, the entropy gained because of the large number of possible configurations of the defective crystal ensures that the concentration is non-zero. There are two basic mechanisms of defect production: *Frenkel defect formation* where an ion moves off its lattice into an interstitial position leaving a vacancy and *Schottky defects* where ions are removed from the lattice to the surface. This *intrinsic* disorder will be present in (hypothetical) perfectly pure crystals. *Extrinsic* disorder, is, however, also universal, but for different reasons, namely that it is impossible to produce absolutely pure crystals; the impurities are often present as point defects where the alien atoms substitute for the host atoms in the lattice; moreover, if the valence of the impurity differs from the host atom it will induce a charge compensatory defect population. Many crystals (particularly transition metal oxides) are non-stoichiometric. The crystal has an excess or deficiency of oxide ions and electroneutrality is maintained by altering the charge state of the cations or anions in the crystal. When the concentration of defects is high they can cluster or order to produce shear planes or, in extreme cases, new three-dimensional structures.

The calculations we shall discuss give the energy of taking an ion or ions out of the lattice and removing them to infinity (or bringing ions into the lattice from infinity). This enables us to calculate the energy of a defect process by a suitable combination of these calculations. Even the Schottky process is not an exception to this. Provided that the crystal surface contains a large number of available sites, the configuration of the surface is unaffected but the crystal volume is increased by a number of lattice sites. Thus the process of bringing back ions from infinity to the surface adds a cohesive energy term to the energy of formation of the defect.

We need to relate these defect energies to defect processes and thus to experimental observables. In the dilute limit, the free energy of formation of a defect is related to its concentration by the law of mass action, as shown by Howard and Lidiard (1964). From here we get the standard parameters h_P and s_P, the constant pressure enthalpy and entropy for the formation of the defect. Diffusion or conductivity experiments yield the defect migration parameters. These can be calculated as the difference between the saddle-point and ground state configurations. Diffusion can also be tackled by the complementary method of molecular dynamics (see Chapter 4). Molecular dynamics is most useful when the energy barriers are small and the concentration of defects is high. If the energies involved are large it may be impractical to use molecular dynamics because of the length of time the simulation needs to be run to get adequate statistics. Then the static methods come into their own.

However, most calculations are not constant pressure, but constant volume calculations (strictly speaking constant lattice parameter but the distinction is not important here) performed on a static lattice. The relationship between constant pressure and constant volume defect parameters has been discussed extensively in the literature. The results are summarized by Catlow *et al.* (1981). The three most important are:

$$g_P = f_V \tag{3}$$

$$h_P = u_V - T\left(\frac{\partial V}{\partial T}\right)_P \left(\frac{\partial f_V}{\partial V}\right)_T \tag{4}$$

$$s_P = s_V - \left(\frac{\partial V}{\partial T}\right)_P \left(\frac{\partial f_V}{\partial V}\right)_T \tag{5}$$

where g_P is the defect Gibbs energy at constant pressure and f_V the defect Helmholtz energy at constant volume. These results are true for low defect concentrations (to first order in the defect volume). Gillan (1981) and Harding (1985) have pointed out that the identification of the enthalpy of a defect at finite temperature with the internal energy at zero temperature relies on the cancellation of two terms. Thus $h_P(T) \approx u_V(0)$. There is no similar cancellation for the entropy. Calculations of the defect entropy at constant volume can give qualitatively different answers to the experimental values at constant pressure.

2.2 Methods of calculating defect parameters

Most calculations of the internal energy of a defect process (hereafter referred to simply as the energy) use classical potential models. Two kinds of approach have been used to calculate defect processes: first, that based on the Mott–Littleton approximation and second, the supercell method (closely related to the lattice minimization methods discussed in Chapter 3).

In his original, simple calculations Jost (1933) showed that it was essential to calculate the polarization response of the lattice to a charged defect if the energy was to be even qualitatively correct. This insight was developed by Mott and Littleton (1938) who suggested that the lattice should be divided into two regions: an inner region where the explicit relaxation of the ions had to be calculated and an outer region where a continuum approximation would be adequate. For the outer region, they showed that (for a cubic crystal) the polarization energy for a defect of charge Q with respect to the lattice is

$$E_{\mathrm{ML}} = -\frac{Q^2 V_{\mathrm{m}}}{8\pi\varepsilon_0}\sum_i \frac{M_i}{|r_i|^4} \tag{6}$$

where V_{m} is the molar volume, r_i is the distance of the ith ion from the defect and M_i is the Mott–Littleton parameter for that sublattice, given by

$$M_i = \left(\frac{\alpha_i}{\sum_j \alpha_j}\right)\left(1 - \frac{1}{\varepsilon}\right) \tag{7}$$

Here α_i is the polarizability of the ith sublattice and ε is the static dielectric constant.

Formally, we can write methods of calculating defect energies based on this approach by writing the energy of the defective crystal as

$$E = E_1(\mathbf{r}) + E_2(\mathbf{r},\xi) + E_3(\xi) \tag{8}$$

where $E_1(\mathbf{r})$ is the energy of the inner region, $E_3(\xi)$ the energy of the outer region and $E_2(\mathbf{r},\xi)$ the interaction energy between the two regions. The displacements of ions in the inner region (calculated explicitly) are denoted by $\mathbf{r}$; the displacements of ions in the outer region are denoted by ξ. If the inner region is large enough, we can assume that the outer region consists of perfect crystal with harmonic displacements. The formal solution of equation (8) under these conditions is given in a number of places (e.g. Catlow and Mackrodt (1982) which also gives a full discussion of the practical implementation of this method). The minimization of the ion positions in the inner region is performed by one of the standard techniques discussed in Chapter 3. Three general-purpose computer codes (HADES (Norgett, 1974), CASCADE (Leslie, 1982) and GULP (Gale, 1993)) have been written to implement this approach.

The supercell method is discussed in some detail in Chapter 3. When applied to defect calculations, the defect is introduced into the supercell to which periodic boundary conditions are applied. The main conceptual difficulty occurs when charged defects are present since, if the cell is charged, the Coulomb summation diverges. The problem can be overcome by introducing a uniform background of charge density to compensate for the net charge of the cell. If we wish to compare the results with calculations on isolated defects we must also subtract the interaction between the defects. These questions are discussed by Leslie and Gillan (1985) who show that the corrections are

$$\text{background correction} = Q^2/8V_m\varepsilon_0\eta^2 \tag{9}$$

$$\text{interaction correction} = \alpha_M Q^2/8\pi\varepsilon_0\varepsilon L \tag{10}$$

where Q is the net charge of the supercell, V_m the volume of the primitive unit cell of the crystal, η the parameter defining the partition between the real and reciprocal space parts of the Ewald summation, α_M the Madelung constant and L the spacing between the defects. In using this method for defects, one must remember that an artificial ordering is imposed on the defect population by the periodic boundary conditions. The interaction correction removes most of the effect of this, but for small cells not all of it. In some cases, the effect may be desirable and the correction is not wanted. In such a case, one is really modelling a special type of structure as in Chapter 3. And indeed structures based on ordered arrangements of defects are well known in non-stoichiometric compounds.

Calculations of the entropy of formation of defects have also been performed. The calculation of defect entropies at constant volume is essentially a calculation of the effect of the defect on the lattice phonon spectrum. Again, this problem may be approached either by a supercell method or by a two-region approach. (A third method involving Green functions is too clumsy to act as the basis of a general computer code.) All these methods assume that a static lattice calculation has already been performed and a set of ion positions in the relaxed configuration is available. The methods then assume that the defective lattice can be treated within the harmonic approximation in exactly the same way as the perfect lattice. The supercell calculation is performed as a standard calculation in lattice dynamics. The two-region approach allows the inner region to vibrate, but holds the outer region static. The two methods converge to similar values in the limit of large numbers of ions. Details are given in Gillan and Jacobs (1983), Harding (1985) and Allan *et al.* (1987).

An example of the kinds of result obtainable can be seen in the work of Harding (1985) on calcium fluoride. The calculated defect parameters for the formation enthalpy and entropy of the anion Frenkel defect are 2.81 eV and 5.4k, respectively. This compares with the experimental values of Jacobs and

Ong (1976) of 2.71 eV and 5.5k where k is Boltzmann's constant. (The constant volume numbers are 2.71 eV for the defect energy and $-1.65k$ for the defect entropy, underlining the points made above about constant pressure and constant volume calculations). The agreement here is exceptionally close; the interionic potentials for calcium fluoride are well established. However, there is no difficulty in principle in carrying out the calculations for more complex systems. The problem is often the lack of reliable experimental data for comparison.

2.3 Recent studies of point defects

It should be pointed out that these methods have now been used far beyond the simple ionic systems where they began. For a discussion of some of these applications the reader should consult the chapters on superconducting ceramics (Chapter 10) and zeolites (Chapter 9).

Two examples illustrate current practice. The study of Ilett and Islam (1993) on the defect chemistry of the catalyst La_2O_3 calculates the diffusion processes, substitution reactions and redox reactions for this material. The calculations show that the defect population is dominated by anion Frenkel defects but the concentrations are low. A detailed search for possible migration pathways suggests that the oxygen diffusion is dominated by vacancy migration within the layers. The calculations predict a migration energy of 0.6 eV in agreement with the experimental range of 0.6–0.7 eV. Calculations of the energy required to dope the oxide with various cations show the sensitivity to ion size first observed by Butler *et al.* (1983) in doped CeO_2. The closer the size of the dopant ion to that of the host cation, the lower the interaction energy. The calculations of the redox reactions require estimates of the various electronic terms (ionization energies, electron affinities). The results are in qualitative agreement with experiment. However, the difficulties inherent in simply combining electronic terms with classical results have led workers to investigate the possibility of writing hybrid quantum/classical codes such as ICECAP (see e.g. Harding *et al.*, 1985).

An illustration of the use of both classical and quantum calculations is the study of Stefanovich *et al.* (1994) on the stabilization of the cubic phase of ZrO_2. Hartree–Fock calculations are used for the individual phases, defect energies and heats of mixing are calculated classically. The results show that the stabilization is a complex balance of effects: the cubic phase is favoured by a combination of the oxide vacancies imposing a negative pressure, the lowering of the dielectric constant in the cubic phase (which increases the (attractive) defect–defect interaction) and changes in electronic structure when highly ionic cations like Ca^{2+} and Mg^{2+} are present. Even this is not all: a study of the entropy terms is still needed and segregation effects to grain boundaries should probably be considered. Nevertheless, this kind of study shows how computer simulation can be used to tackle intrinsically complex problems in materials chemistry.

A number of recent calculations have compared the classical result with quantum mechanical calculations. In many cases, the results from the latter techniques confirm those from classical calculations with a gratifying accuracy. However, one topic on which there is continuing controversy is the nature of the polarons in transition metal oxides. Since the classical method subsumes all the quantum mechanics of the problem into the potential function, it can only tackle problems of electronic structure in a few specific cases, the most common example of which is in non-stoichiometric oxides. Here the question is the location of the electronic hole when the system is metal deficient. The only way such a problem can be tackled by classical methods is to use the small polaron approximation and assume that the hole resides on an ion to produce a new (in effect substitutional) ion with an extra positive charge. This can be successful; and the use of the small polaron approximation in crystals is discussed in detail by Shluger and Stoneham (1993). However, all calculations on the first-row transition metal oxides have assumed that the extra charge resides on the metal ion. Recent quantum calculations (Towler *et al.*, 1994) have thrown doubt on this assumption, suggesting that the hole is on the oxide ion. Moreover, the question of whether the hole is a small polaron for all these oxides is, at present, quite uncertain. Further discussion is given in Chapter 8.

3 SURFACES AND OTHER INTERFACES

To the surrounding ions, an interface is a giant defect. They will relax towards or away from it. Such relaxations may be so large that the resulting structure is quite different from a simple termination of the bulk. It is then usual to refer to reconstructions. When considering interfaces, it is useful to consider the crystal as a stack of planes. The creation of a surface then consists of breaking the stack in half; the creation of grain boundaries is twisting or tilting the top of the stack relative to the bottom and the creation of hetero-interfaces is the bringing of two stacks of different materials together. From this picture, it is clear that the three types of calculation have much in common and can be performed with the same general computer program.

Even without performing calculations, some qualitative insights can be obtained. The most important of these concerns polar directions. These are normals to stacks of planes where the planes are all charged and the charges alternate in sign. The simplest example is the [111] direction in the rock-salt structure. It is impossible to create a simple surface using such a stack. This is because the stack has a dipole moment normal to the surface that is proportional to the number of pairs of planes in the stack. Such a surface, with a divergent moment, cannot exist unless it is stabilized by defects. Kummer and Yao (1964) pointed out that a dipole layer with charge density $\sigma/2$ and spacing

na will cancel the crystal potential produced by a stack of n planes of alternating charge density $\pm\sigma$ and interplanar spacing a. More complex cases can easily be treated (Sayle *et al.*, 1996). These dipole layers should be considered as representations of a far more complicated surface rearrangement. They may indicate the presence of large concentrations of charged species or alternatively facetting of the surface.

The calculation of interfaces uses the two-region strategy discussed before. Planes close to the interface are explicitly relaxed. Planes further away are held rigid, but the stack is allowed to move as a block; this permits surface dilatation. The two-dimensional Coulomb sums are performed using the method first suggested by Parry (1975, 1976); an alternative derivation is given in Heyes *et al.* (1977). Two computer programs have been written to calculate the properties of interfaces: MIDAS (Tasker, 1979; Harding, 1988) and MARVIN (Gay and Rohl, 1995).

3.1 Surfaces

The surface energy, Υ, is defined as

$$\Upsilon = \frac{\text{(energy of crystal with surface)} - \text{(energy of same number of bulk ions)}}{\text{(surface area)}} \tag{11}$$

In liquids, this is equivalent to the surface tension, since a liquid surface cannot support a stress. When it is stretched, ions from the bulk move to the surface to maintain the same structure. However, solid surfaces do support stresses and so if a surface is stretched reversibly, we may define (Shuttleworth, 1950)

$$\gamma = \frac{\mathrm{d}}{\mathrm{d}A}(\Upsilon A) = \Upsilon + A\frac{\mathrm{d}\Upsilon}{\mathrm{d}A} \tag{12}$$

where γ is the surface stress. Calculations using the techniques described above are static calculations and the results are internal energies at zero temperature. However, the constraint of constant lattice parameter now corresponds to constant surface area rather than to constant volume.

At first, calculations were performed on simple surfaces such as the (100) surface of NaCl or MgO. The MgO (100) surface has now been calculated by *ab initio* quantum methods (both LDA and Hartree–Fock) and remarkable agreement with the classical methods found. These calculations are discussed in Chapter 8 in this book. The level of agreement can be seen in Table 7.1.

A more remarkable case of agreement between the classical simulations is provided by the basal plane of alumina, where recent LDA calculations support the prediction of large displacements of the surface cations made by the classical simulation (Manassidis and Gillan, 1994).

Table 7.1 Surface energies and surface structure for MgO (Pugh and Gillan, 1994).

Parameter	Experiment	Shell model	Hartree–Fock	LDA
Surface energy (J/m^2)	1.04 ± 1.2	1.07	1.16	1.03
Rumple (%)	2.0 ± 2.0	3.0	2.5	1.7
Relaxation (%)	0.0 ± 0.75	1.0	−0.7	0.7

Table 7.2 Madelung potentials (reduced units) at surface sites in MgO (after Colbourn, 1992).

Site	Unrelaxed	Relaxed anion	Relaxed cation
Bulk	0.852	0.852	0.852
(100) Surface	0.820	0.820	0.826
Kink	0.926	0.791	0.791
Corner	0.666	0.818	0.828
Ledge	0.846	0.875	0.869

If one can calculate the energies of surfaces, it is possible to calculate the equilibrium crystal morphology. This has been done for a number of cases (e.g. Mackrodt *et al.*, 1987; Parker *et al.*, 1992). The morphology may, of course, depend on kinetic rather than thermodynamic factors. A number of attempts have been made to overcome this problem, often based on the ideas of Hartman (1980) and Hartman and Bennema (1980) but none can yet be said to be satisfactory.

Most surfaces contain steps, kinks and corners, which are important since they are likely to be the high energy sites where gas/solid reactions occur. A number of simulations have been performed for the (100) surface of MgO. These show large relaxations from the simple geometrical constructions, demonstrating that the latter should be used with caution. Also, the spread in Madelung potentials for the various sites is greatly reduced, as shown in Table 7.2.

These results suggest that the electronic structure of the ions may not be altered by the presence of the surface, a point supported by the recent calculations of Abarenkov and Frenkel (1991).

We now turn to recent work of Shluger *et al.* (1994) in modelling the interactions of an atomic force microscope with an ionic surface. Here by calculating the distortion of the surface caused by the tip, the simulations can show what the best distance is for atomic scale resolution (3–5 Å). The calculations also suggest that it should be possible to make images of charged impurities. More work along these lines is in progress; the results, particularly when combined with suitable experiments, will be of considerable interest.

3.2 Grain boundaries

A grain boundary is, formally at least, two surfaces back-to-back. The two crystallites joined by the boundary are rotated with respect to each other. If the axis of rotation is parallel to the interface, the boundary is called a *tilt*; if perpendicular it is called at *twist*. Most boundaries are combinations of these extremes. All simulations of grain boundaries have been of highly symmetric structures because of computational limitations.

Calculations of grain boundaries can first attempt to calculate the structure and compare the results with microscopy. A recent example of this kind of study is that of Lee *et al.* (1993). Twin boundaries were obtained by chemical vapour deposition of TiO_2 onto a sapphire substrate. HREM (High Resolution Electron Microscopy). Images of the (101) and (301) twin boundaries were compared with an atomistic simulation by computing an image from the calculated structure of the boundaries. The calculated images are consistent with experiment. Slight distortions of the compared structure with respect to the HREM image can be ascribed to the inadequacies of the TiO_2 potential, which gives a c/a ratio for the perfect structure that is too large.

3.3 Hetero-interfaces

The formation of a hetero-epitaxial interface involves matching two incommensurate lattices. The inevitable misfit can be accommodated by compression or expansion of either or both of the component materials. The consequent straining of the materials destabilizes the interface. Sometimes the interface is stabilized by incorporating defects into the structure. A good example of what simulation can achieve is provided by the work of Sayle and co-workers (Sayle *et al.*, 1993, 1994) on the BaO/MgO interface. Using a Near-Coincidence Site lattice theory, they first generated interfaces of low misfit (less than 3%) and demonstrated that these are more stable than interfaces of high misfit, as one would expect. Calculations on the growth of BaO on the (100) surface of MgO predict the growth of three-dimensional islands rather than monolayers, which is consistent with the experiments of Cotter *et al.* (1988).

3.4 Metal–ceramic interfaces

An important class of hetero-interface is that between metals and ceramics. This occurs in such areas as metal oxidation and catalyst support. It is also important in understanding such phenomena as wetting. Stoneham and Tasker (1985) argued for the importance of image terms in the adhesion of these interfaces. Simple image theory predicts that the adhesion energy of a

Table 7.3 Comparison of image calculations with those undertaken with LDA techniques.

	Binding energy (J/m^2)		Dilatation (Å)	
Silver position	LDA	Image	LDA	Image
Over O^{2-}	0.54, 1.54	0.58	3.22, –	2.52
Over Mg^{2+}	–, 0.81	0.46	–, 2.51	2.52
Interstitial	–, 1.05	0.87	–, –	2.74

In the LDA column the first calculations are those of Li *et al.* (1993) and the second those of Schönberger *et al.* (1992). The image calculations are from Duffy *et al.* (1994).

metal–ceramic interface must be of the order of the cohesive energy of the ceramic. However, this cannot be a complete account since this theory breaks down at distances of the order of the atom–ion separation. Two models have been developed that correct this difficulty by removing the divergences which otherwise occur at the image plane. Finnis (1991, 1992) modelled the metal as a discrete lattice of polarizable spheres; the basic idea of image theory is retained by constraining the spheres to be equipotential. An alternative approach is that of Duffy *et al.* (1993) who removed the divergence by cutting off the Fourier expansion of the induced charge distribution at the Fermi wave vector.

Calculations have been performed on the Ag/MgO(100) interface with quantum (Local Density Approximation (LDA)) calculations, as shown in Table 7.3. As can be seen, the only significant difference between the LDA calculations and the image theory concerns whether the silver atoms sit over the oxygen ions (LDA) or the magnesium (image theory). A high resolution transmission electron microscopy study has been performed by Trampert *et al.* (1992). Both configurations can be seen in the experimental micrographs. The experimental binding energy is 0.45 ± 0.1 J/m^2.

More recently, calculations have been performed on the Nb/Al_2O_3 interface by Kruse *et al.* (1993) (also discussed in Finnis and Kruse (1994)) and Duffy *et al.* (1996). The first set comprise LDA calculations on a monolayer of Nb atoms replacing the top aluminium layer in the (0001) surface of Al_2O_3. The results indicate that the bonding is largely ionic. This is clearly not an adequate representation of the bulk of niobium on alumina, and further calculations are in progress at the time of writing. The second set is based on the models discussed above with the Nb(111)/Al_2O_3 interface being chosen. The cases of both aluminium and oxygen termination were considered. Only the second case was bound, with the niobium atoms sitting in the vacant aluminium sites. The structure is in good agreement with HREM images of Nb films grown on the (0001) face of Al_2O_3 by molecular beam epitaxy (Mayer *et al.*, 1992).

4 DEFECTS AND INTERFACES

The presence of an interface can have profound effects on the defect population of a crystal. Such effects will not necessarily be short ranged. *Local* charge neutrality need not be conserved at interfaces. Charged interfaces and space charge layers are common. Many interfaces also have a discontinuity in the dielectric constant, which will occur for surfaces and hetero-interfaces (in particular the metal–ceramic interfaces). There is therefore an interaction between such an interface and a charged defect. The interaction energy can be calculated using the method of images:

$$E = \frac{1}{4\pi\varepsilon_0} \frac{Q^2}{4R\varepsilon_A} \frac{\varepsilon_A - \varepsilon_B}{\varepsilon_A + \varepsilon_B} \tag{13}$$

where positive R is into the medium with the larger dielectric constant and A, B label the media on either side of the interface. Equation (13) shows that the defect is attracted into the material of higher dielectric constant. However, the detailed atomic structure modifies, and can even reverse, this result.

A computer code, CHAOS (Duffy and Tasker, 1983), has been written to calculate the energy of defects at interfaces. It uses the structure of the relaxed interface as a starting point and then applies the Mott–Littleton method discussed above. The presence of the interface complicates the calculation. Details are given in the reference above; the essential difference is that part of the polarization calculation must be done as a sum over planes.

The most important long-range effect of the atomic structure on charged defects is that due to rumpling. Cations and anions relax differently at a surface. Seen from a long distance (at the atomic level), the effect of rumpling is to replace one neutral plane by two closely spaced planes of opposite charge, which gives rise to a potential V calculated as:

$$V = (\sigma_c u_c + \sigma_a u_a)/2\varepsilon_0 \tag{14}$$

where σ_c, σ_a are the charge densities on the cation and anion subplanes and u_c, u_a the displacements of the charged planes with respect to the mean dilatation. This potential is of infinite range and causes a shift of the defect energy with respect to the calculation performed in an infinite crystal. At long distances this correction is given by QV where Q is the charge of the defect with respect to the lattice. Clearly, the term cancels out when neutral combinations of defects are considered.

Most calculations of defects at interfaces have been carried out for surfaces, where one of the fundamental issues is that of segregation. Here Mackrodt and Tasker (1989) have shown that it is important not to confuse the experimental heat of segregation (the Arrhenius slope of the coverage plot) with the calculated energy of segregation of an ion to the surface. This identification is true in the case of very low coverage when the Langmuir isotherm applies.

However, it is possible to obtain Langmuir-like plots even at high coverages. In this case, the heat of segregation is given by

$$H = \Delta h + x_s(x_s + 1)\frac{d\Delta h}{dx_s} \tag{15}$$

where H is the heat of segregation, Δh is the energy of segregation of an ion to the surface and x_s is the site fraction of impurity ions at the surface.

Beyond this, one can attempt to calculate the defects at boundaries and thus consider grain boundary processes such as diffusion, where much less has been done. Duffy and Tasker (1984) have discussed segregation and space charge effects. Their calculations suggest that any defect will segregate to grain boundaries since, in so complex a structure, there is always a site more favourable than the bulk. Attempts have been made to study grain boundary diffusion (Duffy and Tasker, 1986). The results correctly show that diffusion down the boundary is easier than in the bulk. However, such calculations are greatly complicated by the problems of finding the saddle-point of the diffusion pathway and technical problems of calculating the true minimized structure.

5 CONCLUSIONS

The derivation of a suitable model which is fundamental to the field of classical simulations becomes ever more important as simulations are used to predict the behaviour of complex systems where the properties of the perfect lattice are not known. For simple systems, the use of quantum mechanical calculations has often validated the classical model; and it is clear that in some cases it will replace it. Calculations on surfaces are inherently more complex. Here one hope is that it will be possible to apply such calculations to understanding phenomena such as catalysis, where the classical simulations are often useful in revealing details of the surface structure; however, any detailed study of the reaction inevitably requires a quantum mechanical calculation. The field of grain boundaries and hetero-interfaces, after a long period of neglect, has begun to see renewed activity. Calculations have been of special interfaces in special orientations, a constraint which was imposed by computation limitations. With the ever-increasing computer power available (including the ability to display the resulting structures), rapid progress should now be possible in such calculations.

The simulations of the type we have discussed are limited insofar as they are *static* simulations and one must choose the starting configuration. Since they are *classical* simulations, they can only consider electronic structure through the potential model. However, as argued in Chapter 1 and elsewhere in this book, provided these limitations are understood and respected, they are frequently the method of choice. They are often useful in simple systems where a large number of cases must be considered. Above all they can be used to study large, complex systems that can be investigated in no other way.

REFERENCES

Abarenkov, I.V. and Frenkel, T. Yu (1991) *J. Phys. Condens. Mater.*, **3**, 3471.
Allan, N.L., Mackrodt, W.C. and Leslie, M. (1987) *Advances in Ceramics,* Vol. 23 (eds C.R.A. Catlow and W.C. Mackrodt), American Ceramic Society Inc., Ohio, p. 257.
Butler, V., Catlow, C.R.A., Fender, B.E.F. and Harding, J.H. (1983) *Solid State Ionics*, **8**, 109.
Catlow, C.R.A. and Mackrodt, W.C. (1982) *Computer Simulation of Solids* (eds C.R.A. Catlow and W.C. Mackrodt), Springer, Berlin, p. 1.
Catlow, C.R.A., Corish, J., Jacobs, P.W.M. and Lidiard, A.B. (1981) *J. Phys. C*, **14**, L121.
Colbourn, E.A. (1992) *Surf. Sci. Rep.*, **15**, 281.
Cotter, M., Campbell, S., Egdell, R.G. and Mackrodt, W.C. (1988) *Surf. Sci.*, **197**, 208.
Dick, B.G. and Overhauser, A.W. (1964) *Phys. Rev.*, **164**, 90.
Duffy, D.M. (1986) *J. Phys. C*, **19**, 4393.
Duffy, D.M. and Tasker, P.W. (1984) *J. Appl. Phys.*, **56**, 971.
Duffy, D.M. and Tasker, P.W. (1986a) Harwell Report AERE-R 11059. (A revised manual is in preparation.)
Duffy, D.M. and Tasker, P.W. (1986b) *Phil. Mag. A*, **54**, 759.
Duffy, D.M., Harding, J.H. and Stoneham, A.M. (1993) *Phil. Mag. A*, **67**, 865.
Duffy, D.M., Harding, J.H. and Stoneham, A.M. (1994) *J. Appl. Phys.*, **76**, 2791.
Duffy, D.M., Harding, J.H. and Stoneham, A.M. (1996) **44**, 3293.
Finnis, M.W. (1991) *Surf. Sci.*, **241**, 61.
Finnis, M.W. (1992) *Acta Metall.*, **40**, S25.
Finnis, M.W. and Kruse, C. (1994) *Mater. High Temp.*, **12**, 189.
Gale, J.D. (1993) General Utility Lattice Program, Imperial College, London.
Gay, D.H. and Rohl, A.L. (1995) *Chem. Soc. (Faraday II)*, **91**, 935.
Gillan, M.J. (1981) *Phil. Mag. A*, **43**, 301.
Gillan, M.J. and Jacobs, P.W.M. (1983) *Phys. Rev. B*, **28**, 759.
Harding, J.H. (1985) *Phys. Rev. B*, **32**, 6861.
Harding, J.H. (1988) Harwell Report AERE-R 13127 (A revised manual is in preparation: Duffy, D.M. and Harding, J.H. (1996).)
Harding, J.H. (1990) *Rep. Prog. Phys.*, **53**, 1403.
Harding, J.H. and Pyper, N.C. (1995) *Phil. Mag. Lett.*, **71**, 113.
Harding, J.H., Harker, A.H., Keegstra, P.B., Pandey, R., Vail, J.M. and Woodward, C. (1985) *Physica*, **131B**, 151.
Hartman, P. (1980) *J. Cryst. Growth*, **49**, 157, 166.
Hartman, P. and Bennema, P. (1980) *J. Cryst. Growth*, **49**, 145.
Heyes, D.M., Barber, M. and Clarke, J.H.R. (1977) *J. Chem. Soc. (Faraday II)*, **10**, 1485.
Howard, R.E. and Lidiard, A.B. (1964) *Rep. Prog. Phys.*, **27**, 197.
Ilett, D.J. and Islam, M.S. (1993) *J. Chem. Soc. (Faraday)*, **89**, 3833.
Jacobs, P.W.M. and Ong, S.H. (1976) *J. Phys. (Paris) Colloq.*, **37**, C7, 331.
Jost, W. (1933) *J. Chem. Phys.*, **1**, 466.
Kruse, C., Finnis, M.W., Milman, V., Payne, M.C., Vita, A.D. and Gillan, M.J. (1993) *J. Amer. Ceram. Soc.*, **77**, 431.
Kummer, T. and Yao, Y. Y. (1964) *Canad. J. Chem.*, **45**, 421.
Lee, W.-Y., Bristowe, P.D., Gao, Y. and Merkle, K.L. (1993) *Phil. Mag. Lett.*, **68**, 309.
Leslie, M. (1982) SERC Daresbury Laboratory Report DL-SCI-TM31T, CCL Daresbury Laboratory, Warrington, WA4 4AD.
Leslie, M. and Gillan, M.J. (1985) *J. Phys. C*, **18**, 973.
Li, C., Wu, R., Freeman, A.J. and Fu, C.L. (1993) *Phys. Rev. B*, **48**, 8317.

Mackrodt, W.C. and Tasker, P.W. (1989) *J. Amer. Ceram. Soc.*, **72**, 1576.
Mackrodt, W.C., Davey, R.J., Black, S.N. and Docherty, R. (1987) *J. Cryst. Growth*, **80**, 441.
Manassidis, I. and Gillan, M. J. (1994) *J. Amer. Ceram. Soc.*, **77**, 335.
Mayer, J., Gutrkunst, G., Möbus, G., Dura, J., Flynn, C.P. and Rühle, M. (1992) *Acta Metall. Mater.*, **40**, S217.
Mott, N.F. and Littleton, M.J. (1938) *Trans. Faraday Soc.*, **34**, 485.
Norgett, M.J. (1974) Hartwell Report AERE-R 7650, AEA Technology, Harwell Laboratory, Didcot, Oxon OX11 0RA.
Parker, S.C., Lawrence, P.J., Freeman, C.M., Levine, S.M. and Newsam, J.M. (1992) *Catalysis Lett.*, **15**, 123.
Parry, D.E. (1975 *Surf. Sci.*, **49**, 433; note also the correction, Parry, D.E. (1975) *Surf. Sci.*, **54**, 195.
Pugh, S. and Gillan, M.J. (1994) *Surf. Sci.*, **320**, 331.
Sayle, T.X.T., Catlow, C.R.A., Sayle, D.C., Parker, S.C. and Harding, J.H. (1993) *Phil. Mag. A*, **68**, 565.
Sayle, D.C., Parker, S.C. and Harding, J.H. (1994) *J. Mater. Chem.*, **4**, 1883.
Sayle, D.C., Gay, D.H, Rohl, A.L., Catlow, C.R.A., Harding, J.H., Perrin, M.A. and Nortier, P. (1996) *J. Mater. Chem.*, **6**, 633.
Schönberger, U., Andersen, O.K. and Methfessel, M. (1992) *Acta Crystallogr.*, **40**, S1.
Shluger, A.L. and Stoneham, A.M. (1993) *J. Phys. Condens. Mater.*, **5**, 3049.
Shluger, A.L., Rohl, A.L., Gay, D.H. and Williams, R.T. (1994) *J. Phys. Condens. Mater.*, **6**, 1825.
Shuttleworth, R. (1950) *Proc. Phys. Soc. A*, **63**, 444.
Stefanovich, E.V., Shluger, A.L. and Catlow, C.R.A. (1994) *Phys. Rev. B*, **49**, 11560.
Stoneham, A.M. and Tasker, P.W. (1985) *J. Phys. C*, **18**, L543.
Tasker, P.W. (1979) *Phil. Mag. A*, **39**, 119.
Towler, M.D., Allan, N.L., Harrison, N.M., Saunders, V.R., Mackrodt, W.C. and Aprà, E. (1994) *Phys. Rev. B*, **50**, 5041.
Trampert, A., Ernst, F., Flynn, C.P., Fischmeister, H.F. and Rühle, M. (1992) *Acta Metall. Mater.*, **40**, S227.

8

Electronic Structure

N.M. Harrison

1 INTRODUCTION

The rapid growth of *ab initio* quantum mechanical (QM) simulations of condensed matter means that a comprehensive review of theoretical approaches and applications would in itself occupy a book. In this chapter the emphasis will be placed on QM studies of silicate and oxide systems. The key technologies will be identified and a critique of the possibilities and inadequacies of current theory presented. Although we discuss the technical details of implementation of QM methods, and some fundamental issues with regard to the description of electron interactions, the intention is to provide a general reference for non-experts in this field. Recent work based on semi-empirical approaches to QM simulations will not be reviewed (e.g. LaFemina, 1992; Goniakowski *et al.*, 1993).

The first section of the chapter deals with various theoretical models. The computational realization of these models will also be discussed in the context of modern computer technology. Applications are discussed in terms of complex physical structures (silicate crystals, defects, surfaces) where the problems are largely technical, and complex electronic structure (Mott–Hubbard insulators) where fundamental questions about electron interactions are relevant.

2 PERIODIC QUANTUM THEORY

On a modern computer the single particle Schrödinger equation

$$H\psi_i = (-\nabla^2 + V_{en} + V_{eff})\psi_i = \varepsilon_i\psi_i$$

where V_{en} is the electron–nuclear interaction and ε_i are the single-particle eigenvalues, can be solved quickly and reliably for periodic materials with several hundred atoms per unit cell. There are a bewildering array of techniques used to achieve this. With a few notable exceptions (Green's function multiple

COMPUTER MODELLING IN INORGANIC CRYSTALLOGRAPHY
ISBN 0-12-164135-X

scattering theory, for example (Wang *et al.*, 1995)), the orbitals, ψ_i, are expanded in a basis set yielding a problem in linear algebra which is most often solved using diagonalization or direct minimization. The various techniques may therefore be distinguished by the basis set employed. Approaches also differ in the choice of the mean field potential V_{eff}. Usually this involves one of the local (LDA) or gradient corrected (GGA) approximations to Density Functional Theory (DFT) which are based on a detailed knowledge of the homogeneous electron gas (Kohn and Sham, 1965; Colle and Salvetti, 1975; Ceperley and Alder, 1980; Perdew and Zunger, 1981; Perdew, 1991) or direct application of Hartree–Fock (HF) theory where V_{eff} is replaced by the exact exchange interaction. Recently attempts to combine DFT and HF approaches have met with some success (Zupan and Causà, 1994a,b, 1995). In many calculations a pseudopotential is introduced as an approximation to the interaction between the core and valence electrons. This approximation means that the basis set is required to describe pseudovalence orbitals only and, if the pseudopotential is chosen suitably, these will be much smoother than the real valence functions. The most important consequence of this approximation is that one can use a very simple basis set consisting of plane waves (PW) commensurate with the underlying lattice (referred to in what follows as the PW-LDA method). The consequences of the use of a pseudopotential depend strongly on the system studied. The potential itself is not unique and depends on a number of parameters (e.g. the core radius r_c) which must be chosen carefully to minimize errors in the energy expression. In 'weak scattering' systems (alkali metals, silicon, germanium, etc.) the approximation is very well justified and consistent results are obtained. When strong scattering is present (oxides, transition metal ions, etc.) more care must be taken in defining the potential and often it is the main source of error.

One can identify several distinct trends in recent work. The first is based on the dramatic breakthrough of Car and Parrinello (1985). For calculations based on a plane wave basis set, and thus 'soft' pseudopotentials, they demonstrated that the diagonalization bottleneck could be avoided by use of direct minimization (originally phrased in terms of fictitious electron dynamics). This has made possible PW-LDA calculations using several million basis functions and thus up to several hundred atoms per cell. The calculation still scales as N_{atoms}^3 but is far quicker than diagonalization when using such large basis sets. The method requires atoms to be treated accurately with 'soft' pseudopotentials and thus much recent effort has been dedicated to developing such potentials for heavier atoms (Vanderbilt, 1990; Troullier and Martins, 1991). Several excellent reviews of this methodology have been published (Gillan, 1991; Payne *et al.*, 1992).

The second trend has been in attempts to recognize and exploit the local nature of chemical interactions, particularly in semi-conducting and insulating materials. One of the goals of this research is to avoid the N_{atoms}^3 scaling of the computational complexity of current algorithms (Wang *et al.*, 1995) which has obvious advantages for studies of large systems. For reasons that will become apparent below, the combination of local basis set and HF approximation have been used extensively in studies of silicate and oxide materials (Dovesi *et al.*, 1992 and references therein). Powerful screening algorithms have been developed which exploit the locality of the orbitals to allow matrix elements of H to be computed efficiently. The compact basis set (10–20 functions per atom) allows standard diagonalization techniques to be applied in large systems. For HF theory the natural choice for the local basis set is atom centred Gaussian functions multiplied by real spherical harmonics (linear combination of atomic orbitals, LCAO). The use of such orbitals has the advantage that matrix elements are analytically determined which removes the need to perform numerical integrals and thus the precision of such calculations is very high allowing the total energy and small energy differences to be computed with confidence. The main approximation in this approach is the choice of functions to include in the basis set. The total energy is variational with respect to the flexibility of the basis set and care must be taken to ensure convergence. The local basis function approach is also a convenient framework for the implementation of embedding techniques in which the calculation concentrates on the electronic and structural relaxation of a small region of the material 'embedded' in a static description of the rest of the crystal. This technology has recently been implemented and used to study bulk and surface defects in insulating materials (Pisani, 1993; Pisani *et al.*, 1994).

A wide range of methods for solving the Schrödinger equation has been developed over the years, differing in the basis set used, the computational implementation and the underlying one electron potential. Worthy of special note are: LAPW (Linear Augmented Plane Waves (Liu *et al.*, 1994a)), PAW (Projected Augmented Wave (Blöchl, 1993)) and local Slater orbitals (Philipsen *et al.*, 1994). The PW-LDA and LCAO-HF have been extensively used on silicate and oxide materials and suffice to exemplify the effects of a varying basis set and one electron potential in these systems. The remainder of this chapter will focus on applications of these techniques. First, materials with complex geometrical structures will be considered (silicates and simple oxides) where the problems associated with the application of the theory are confined to the scale of the system. This discussion will serve to illustrate the current 'state of the art' and to document the accuracy to be expected of these techniques. Second, recent attempts to compute the ground state properties of materials with complex electronic structure, mostly transition metal oxides, will be reviewed. In these systems the detailed nature of the QM description is still unclear.

3 COMPLEX PHYSICAL STRUCTURES

3.1 Silicates

As is clear from other chapters in this book, the importance of materials based on the SiO_2 unit can hardly be overemphasized. Silica is a major component of the Earth's crust, and is widely used in glass, ceramics, optical fibres and microelectronics industries, stimulating an extensive range of studies of the material. Of the 40+ polymorphs, most are based on corner sharing tetrahedral SiO_4 units, the structurally simplest exception being the mineral stishovite in which silicon is sixfold coordinated with oxygens arranged in a distorted octahedron.

The relatively few QM studies of these materials reflect both the structural complexity (including low symmetry) and also the strong scattering associated with the oxygen potential, which makes the generation of reliable and efficient pseudopotentials for PW-LDA calculations difficult. There has been a concerted effort to solve this problem over the last few years which has yielded a new family of ultra-soft pseudopotentials with which oxygen has been described with plane wave cut-offs as low as 500 eV (Vanderbilt, 1990). In the all electron LCAO-HF method the major approximation is associated with the choice of basis set. This has been studied in some detail and reliable basis sets developed (Nada *et al.*, 1990). In the absence of analytic forces, the use of this method to optimize fully the geometry of such complex structures is rare.

In order to document the reliability of these calculations we consider recently published data for the α-quartz and stishovite phases (Tables 8.1 and 8.2, respectively). In each case there are data from a number of PW-LDA calculations and from the LCAO-HF technique. The literature references in the

Table 8.1 Various theoretical calculations of V_0 (Å^3), c/a ratio and B (GPa) for α-quartz. The percentage error is shown in parenthesis.

	V_0	c/a	B
Expt	113.1	1.10	34–37
PW-LDA (a)	113.1 (0.0)	1.10 (0.0)	37
PW-LDA (b)	113.0 (−0.08)	1.10 (0.0)	–
PW-LDA (c)	107.4 (−5.04)	1.13 (3.0)	38.1
LCAO-HF (d)	114.2 (1.00)	1.09 (−0.9)	–

See text for references (a)–(d).

Table 8.2 Various theoretical calculations of V_0 (Å^3), c/a ratio, B (GPa) and stability with respect to α-quartz, dE(eV/SiO_2) for stishovite. The percentage error is shown in parenthesis.

	V_0	c/a	B	dE
Expt	46.51	0.638	31	0.52
PW-LDA (a)	46.18 (−0.7)	0.641 (0.47)	28.2 (−9)	0.07 (−87)
PW-LDA (b)	47.15 (1.4)	0.612 (−4.1)	– –	–
PW-LDA (c)	45.64 (−1.8)	0.645 (1.1)	29.2 (−5.8)	0.1 (−81)
LCAO-HF (d)	46.69 (0.39)	0.643 (0.78)	– –	0.33 (−37)

tables are: (a) Liu *et al.* (1994b); (b) Allan and Teter (1987); (c) Chelikowsky *et al.* (1990, 1991) and Keskar *et al.* (1991); and (d) Nada *et al.* (1990).

If we consider the structural data in the light of the fact that the input to these calculations is simply the atom types and lattice structure, this data establishes the QM approach as a predictive tool for studying the structure of silicates. On the whole, observed bond lengths are reproduced to better than 1% and elastic properties within 10%. The variation in results published by different authors is rather large and, for the PW-LDA calculations, is apparently due mainly to the different choices of pseudopotential. A meaningful comparison of different approximations to DFT or of those with the HF approximation requires errors in the energy expression to be confined to a few meV per formula unit.

The errors in the HF approach may be understood in quite general terms. The correlation energy could, in principle, be calculated using configuration interaction (CI) theory. This mixes the various excited states of the system into the single determinant description of the ground state. The contribution decreases as one considers states of higher energy. In wide band gap insulators one therefore expects correlation effects to be small and dominated by the self-correlation of an ion. There will be a consequent tendency to overestimate the size of ions which results in bond lengths being 1–2% too long and volume dependent elastic constants being 5–10% too high (Dovesi *et al.*, 1992). The data here are consistent with these expectations.

The energy difference between the two structures is rather poorly described (Table 8.2). It is usual for binding energies within the HF approximation to be underestimated by 20–30% as correlation effects contribute significantly to the stability of a material (Dovesi *et al.*, 1992); correction terms based on density functional theory improve these results to within 5% (Causa *et al.*, 1991). Such calculations have not yet been performed for these materials. In contrast, LDA estimates of binding energies are usually somewhat overestimated (15–20%) with errors in relative stabilities being small ($\pm 5\%$). Indeed the relative stability of the α- and β-phases of quartz and cristabolite are well described (Liu *et al.*, 1994b). It is difficult to attribute the origins of the remaining errors in the relative stability seen in this case. They may be due to the fact that we are comparing materials with significantly different structural topologies.

The PW-LDA approach has been used to calculate structural properties and their pressure dependence for many of the smaller silica structures (Liu *et al.*, 1994b). At a pressure of ~ 30 GPa α-quartz undergoes a gradual transition to an amorphous phase. Near this transition there are subtle changes in bond lengths, Si–O–Si bond angles and the packing of the SiO_4 tetrahedra. Studies based on empirical force fields become increasingly unreliable as the transition is approached. PW-LDA studies of this high pressure phase have been performed (Chelikowsky *et al.*, 1990, 1991; Keskar *et al.*, 1991). At pressures of up to 10 GPa the structural changes are in excellent agreement with recent experiments – the theoretical calculations were performed up to 20 GPa.

Amorphous silica has also been studied recently using first principle molecular dynamics (Sarnthein *et al.*, 1995). Such simulations are currently restricted to rather small unit cells (24 SiO_2 units) and short time spans (~ 10 ps), unlike the simulations using interatomic potentials described in Chapter 9. Nevertheless, the direct simulation of the liquid phase followed by a rapid quench produces an amorphous state with structural and electronic properties in reasonable agreement with neutron diffraction and spectroscopic studies. QM simulations are able to treat a wide range of structural environments on an equal footing while providing information on both structural and electronic properties.

Periodic LCAO-HF calculations have been used to examine the electronic structure and related properties of a number of silaceous structures such as zeolite cages (White and Hess, 1993; Anchell *et al.*, 1994; Nicholas and Hess, 1994) and clay-like minerals (Hess and Saunders, 1992). The computational cost of such studies restricts structural optimization to calculations using minimal basis sets. Nevertheless, important properties of these structures such as the diffusion of cations through the zeolite framework (Anchell *et al.*, 1994) and the position of the hydrogen atoms in the mineral kaolinite (Hess and Saunders, 1992) have been studied.

To a large extent interatomic potential based modelling techniques may be used to describe the structural energetics of these frameworks. While it is clear that QM theory has an important role to play in understanding structural energetics and determining model parameters, it is in situations where the local electronic structure is strongly perturbed that parameterization of models becomes difficult and such techniques are essential. Extending the periodic techniques to studies of realistic defects and chemical reactions within silicate frameworks is of primary importance. This area has been studied extensively using QM calculations on small clusters – often embedded in an electrostatic or mechanical environment to mimic the response of the infinite crystal (e.g. Greatbanks *et al.*, 1995; Maseras and Morokuma, 1995). Such methods must be implemented with great care in order to quantify (and minimize) the effects of finite cluster size and the boundary conditions on calculated properties. In this regard, self-consistent embedding schemes are of particular interest. These schemes aim for a consistent description of the 'chemically interesting' region with a cluster-QM approach and the surrounding lattice within a periodic-QM theory. This approach allows one to bring the techniques of cluster-QM methods to bear in a reliable and efficient manner on an essentially solid state problem. An LCAO-HF embedding method (the perturbed cluster approach) has been implemented (Pisani, 1993; Pisani *et al.*, 1994). Early results for silicates have been published. The energetics of both charged and neutral defects in α-quartz (Corà and Pisani, 1994) have been examined with small basis sets, demonstrating that the local structure and charge density may be computed reliably. The properties of isolated and embedded cluster calculations in studies of the oxygen vacancy in α-quartz have been examined in a recent publication (Sulimov *et al.*, 1994).

3.2 Simple oxides

Non-transition metal (TM) bearing oxides, or those in which the *d*-shell is essentially empty, may be classified as 'simple' oxides. The ground state electronic structure of these systems is well known and the application of standard QM techniques leads to reliable and well-understood trends with respect to observation. The system often used to exemplify this class of materials is MgO. It has been by far the most popular system for both theory and experiment and thus the one for which the most detailed information is available. The simple rock-salt structure allows detailed calculations of the lattice structure, elastic constants and phonon vibrations to be performed. A selection of results from different methods are collected in Table 8.3 and compared to experimental data extrapolated to the athermal limit (McCarthy and Harrison, 1994). As for the silicate materials described above, the lattice properties are reproduced well. The PW-LDA results for the lattice constant (De Vita *et al.*, 1992; Langel and Parrinello, 1994; Refson *et al.*, 1995) range from slightly too short to ~2% too long. For the bulk modulus there is a tendency to be rather soft (−8% to −3%). Again the variation in published results is probably attributable to the various pseudopotentials used. The HF data (McCarthy and Harrison, 1994 and references therein) follows the trends described above for ionic systems, the lattice constant being very slightly too long and the bulk modulus too stiff (+10%). Density functional estimates of the correlation correction to the Hartree–Fock energy surfaces (HF + DFT in Table 8.3) provide an extra binding term which appears to behave essentially as an additive two-body potential reducing the lattice constant and increasing the bulk modulus (McCarthy and Harrison, 1994). Calculations can therefore be extended to structurally complex systems with some confidence. It is the defect properties and surface chemistry of the oxide materials which underpin their importance in such a wide range of applications and it is on these topics that recent calculations have focused.

Table 8.3 The lattice constant (a), bulk modulus (B) and frequency of the transverse optic phonon at the zone centre (TO-Γ) for MgO from PW-LDA and LCAO-HF calculations compared to observations extrapolated to the athermal limit.

Method	*a* (Å)	*B* (GPa)	TO-Γ (THz)
PW-LDA	4.17–4.27	154–162	12.4
LCAO-HF	4.195–4.20	181–186	11.92
HF + DFT	4.086	205	11.46
Experiment	4.19	167	11.8–12.23

3.3 Bulk defects

There has been an increasing focus on QM studies of defects in alkaline-earth oxides in recent years. An early review of this area was given by Henderson (1980). Interest has been stimulated by the growing power of QM techniques, the availability of well-characterized samples (highly crystalline and with low impurity concentrations) and the important technological applications of these materials which are derived from the high melting point and refractive properties. The study of defects within a QM formalism raises many problems, some of which are purely technical while others involve genuine conceptual issues. First, in order to calculate accurate defect energies in ionic crystals one must pay close attention to the convergence of the total energy with respect to system size and computational parameters. This is true whether one is using QM theory or model potentials (see also Chapter 7) but, due to the extra computational cost, it is often a more serious problem in the former. It is also true for supercell, embedding and cluster approaches. Second, the treatment of charged defects has been the subject of much debate. In periodic supercell studies a mechanism must be found to remove the excess charge in each unit cell. Almost invariably this is achieved by adding a homogeneous charged background which is, conveniently, an automatic consequence of treating the Coulomb interactions within the Ewald approximation. The best mechanism for removing the residual interaction between the charge defects and the subsequent convergence of defect formation energies with system size is the subject of current research (see below). We will consider some recent calculations of the properties of neutral defects before examining this issue.

3.3.1 Neutral defects

The substitution of divalent cations for Mg^{2+} ions in MgO has been extensively studied within the ionic model (Mackrodt and Stewart, 1979). Recently these systems have been the subject of a series of systematic studies within the LCAO-HF formalism in which the highly symmetric nature of the system is exploited to lower computational costs and to simplify the geometrical relaxation (Freyria-Fava *et al.*, 1993; Dovesi and Orlando, 1994; Orlando *et al.*, 1994a). These studies contain the only published attempt to demonstrate the convergence of the supercell method for the treatment of defects within a QM formalism in ionic systems. The defect formation energy (E_D) of, e.g., substitution of Ca for Mg ions in MgO may be referenced to the energies of the isolated ions ($E_{Mg^{2+}}$ and $E_{Ca^{2+}}$), i.e.

$$E_D = E^{S_n}_{Ca^{2+}} - \frac{n}{2} E^{bulk}_{MgO} + E_{Mg^{2+}} - E_{Ca^{2+}}$$

where $E^{S_n}_{Ca^{2+}}$ is the energy of the Ca substituted n-ion supercell. This definition allows for the maximum cancellation of errors in the description of the ions due

Table 8.4 Defect formation energy when Ca is substituted for Mg in MgO as a function of supercell size. The number of shells of atoms neighbouring the defect is shown in the second column.

Supercell size	Shells of neighbours relaxed	E_D (eV)
8	0	7.20
16	1	6.56
32	2	6.25
64	3	6.30
128	4	6.28

to the use of a finite basis set. The variation of E_D with supercell size is given in Table 8.4 (Dovesi and Orlando, 1994).

The defect formation energy has converged to within 0.1 eV in the 32-ion supercell. If geometrical relaxations are neglected the convergence is even more rapid – a 16-ion cell being sufficient. The structural relaxation lowers E_D by some 0.8 eV. A rapid convergence of the charge distribution and electronic structure is also observed. Within an ionic model based on empirical two-body potentials which also describes dipolar distortions of the ions, E_D is computed as 6.25 eV (Freyria-Fava *et al.*, 1993). The energetics and detailed geometrical relaxations within this model agree remarkably well with the *ab initio* calculations.

These techniques have been extended to studies of defects in less ionic systems. In a recent study of carbon substitution in silicon (Orlando *et al.*, 1994b) it was observed that the strong local perturbation of the electronic structure was screened rapidly (first nearest neighbours) by electronic polarization. Upon optimization of the lattice geometry important relaxations were observed out to the fifth neighbours and thus a converged defect energy again required a 32-atom unit cell. In this study the correction to the defect formation energy from correlation effects was estimated using density functional theory and found to be small (0.24 eV).

3.3.2 Charge defects

As stated above the treatment of charged defects in periodic supercell calculations requires the use of a technique to restore the charge neutrality of the system. The dominant feature in the convergence of these calculations with supercell size is then the interaction of the charged defects in the presence of the neutralizing background. The cation and anion vacancies in MgO have been the subject of much recent study. The first periodic QM treatment of this problem was performed with the PW-LDA technique (De Vita *et al.*, 1992). The formation energy for the Schottky pair (E_s, the energy to extract Mg^{2+} and O^{2-} ions forming isolated charged vacancies) was reported. The residual interaction of the defects was estimated as

$$\Delta E = \frac{-\frac{1}{2}\alpha q^2}{\varepsilon L} \tag{1}$$

where q is the net defect charge, ε was taken to be the experimental static dielectric constant (9.86) and α and L are the Madelung constant and lattice parameter of the supercell lattice, respectively. This sort of correction, based on the simple physical argument that the charged defects have Coulombic interactions across a medium characterized by ε, has been widely used to accelerate the convergence of defect energies in studies of alkali-halide vacancies with semi-classical shell model potentials. De Vita *et al.* demonstrated that, using an empirical model, supercell calculations converged to the isolated Schottky defect formation energy for a cell containing 128 ions. The PW-LDA calculations were reported for 16- and 32-ion unit cells and E_s reported as 7.8 and 6.8 eV, respectively which is within the experimental range of 4–7 eV (Mackrodt, 1982). The plane wave basis set depends on the cell size and thus these calculations were subject to an error in the unit cell energy of several eV when comparing the 16- and 32-ion cells.

A systematic study of the convergence properties of this system has been performed within the LCAO-HF methodology (Dovesi and Orlando, 1994). As mentioned above, this technique is based on analytic integrals and thus the numerical precision of the calculations is very high; e.g. the energies of the ideal 16- and 32-ion cells agree to better than 0.01 meV. Cation and ion defect formation energies were reported for cells containing 16, 32, 64 and 128 ions. It was found that the convergence of these energies is rather slow and that for the 128-ion cell the convergence was not complete. Attempts at using the correction term in equation (1) to accelerate convergence were not successful. The use of the term with the experimental dielectric constant only slightly improved convergence. More rapid convergence was achieved by optimizing the choice of ε but this leads to very different effective dielectric properties for the cation and anion defects. It was concluded that the correction term is too simplistic. For instance, it implies that the defects interact via an undisturbed dielectric medium equivalent to the bulk crystal, whereas clearly the local environment has been strongly modified by the presence of the defect. The calculation of converged formation energies for charged defects thus requires very large supercells and further study using numerically precise techniques is required to examine methods for accelerating convergence.

3.4 Surface chemistry

The chemical reactions which occur at oxide surfaces are important in a wide variety of industrial, geological and environmental contexts. Nevertheless, the study of the atomic and electronic structure of these surfaces is poorly developed compared to that of metallic and semi-conducting systems. Most experimental work has been concerned with powder or polycrystalline samples. There are many technical difficulties to overcome in studying oxides. From the experimental point of view well-characterized samples are difficult to prepare, surfaces tend to fracture rather than cleave and the surface properties are unstable when probed with traditional surface science techniques which induce surface charging. Studies of surfaces well characterized both in terms of atomic structure and stoichiometry are rare. Until recently the vast majority of the theoretical studies have been based on empirical or semi-empirical models within both classical and quantum frameworks. While providing valuable insight into surface properties, the validation of these techniques is limited by the absence of detailed experimental data for the surface. The model parameters are determined from bulk properties and one must be cautious of predictions in environments that are significantly different from the bulk, although there is evidence that extrapolation of potentials from bulk to surfaces of ionic crystals works well, as discussed in Chapter 7. First principle QM methods can be used to study these systems reliably. Calculations of surface reconstructions, chemical reactions and the ground state electronic structure for comparison with surface spectroscopy are becoming a primary source of information in this field. The number of published investigations is currently limited only by the exceptional computational resources that are required. Recent algorithmic developments coupled with the use of massively parallel computers means that these limitations are slowly being removed.

For the most widely studied system, both experimentally and theoretically, we can turn again to MgO. The nature of the hydroxylation reaction at MgO surfaces has recently been studied carefully with a number of techniques. Scamehorn *et al.* (1994) used the LCAO-HF approach to examine the hydroxylation and water adsorption energies of the (100) surface and at a number of simple defects. The adsorption energies are displayed in Fig. 8.1. The results of this study may be analysed in simple terms. On the ideal bulk terminated surface, an isolated water molecule will not dissociate; the preference is for a physisorbed water molecule. At the four-coordinated (edge) sites the hydroxylation process is competitive with a stronger tendency for water adsorption. At three-coordinated (corner) sites hydroxylation is strongly favoured. Analysis of the ground state charge density indicated that the sites are differentiated by the ability of the exposed oxygen ion to form a chemical bond with hydrogen, the O–H bond length (and bond population) changing from 1.01 Å (0.20e) on the free surface to 0.96 Å (0.44e) at the three-coordinated defect. The electronic states of the defect sites are also calculated to shift by up to 5 eV relative to the ideal surface which provides a clear signal of defect sites in ultraviolet and X-ray photoelectron spectroscopy (UPS and XPS) measurements. These results confirm a simple picture of the surface reactivity of MgO and provide a conceptual framework within which the wide variety of disparate experimental data can be analysed.

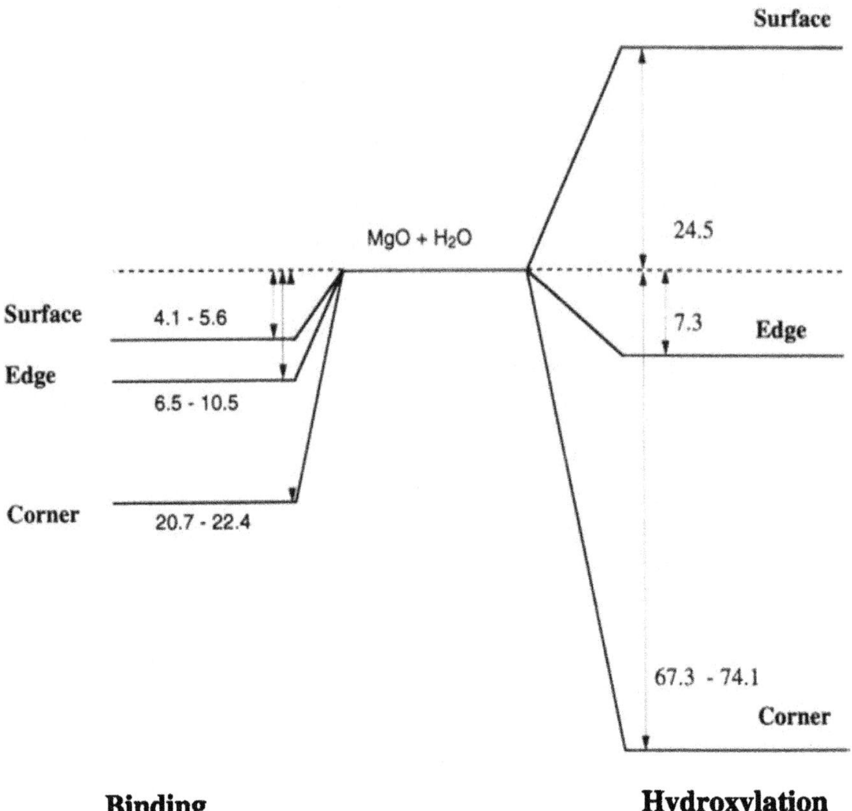

Figure 8.1 Binding and hydroxylation energies (kcal/mol H_2O) for water at threefold (corner), fourfold (edge) and fivefold (surface) coordinated defect sites on the MgO(100) surface (Scamehorn *et al.*, 1994).

This work is at the limit of what can be currently achieved with techniques using local basis sets. The main reason for this is the absence of accurate and efficient analytical forces to assist with structural optimization. The PW-LDA technique is unique in that analytic forces may be easily, and inexpensively, computed from the ground state wave function. There is therefore a need to develop soft and accurate pseudopotentials to allow the PW-LDA technique to be applied to these systems accurately and efficiently. This problem has been addressed by a number of groups in recent years, resulting in the application of this technique in a number of surface studies. Langel and Parrinello (1994) reported a simulated annealing simulation of the adsorption of water at the MgO(100) surface. Simulations on the ideal surface at low temperature (40 K) lead to physisorption of the water molecule and, at higher temperature (200 K), desorption to the gas phase. Defects were introduced by studying a small step model of (100)-like facets on a (110) surface. Surface hydroxylation is then achieved as the water molecule spontaneously dissociates at 25 K. As one would expect, the conclusions of the study are consistent with the LCAO-HF results. Refson *et al.* (1995) argued that the observed high coverage of OH groups and the roughening of MgO surfaces could not be explained by adsorption at defect sites alone. They demonstrated that it would be energetically favourable for the (100) surface to decompose into hydroxylated (111) facets rather like the $MgOH_2$ (brucite) structure and thus that in the naturally occurring mineral (periclase) the (111) hydroxyl surface will be predominant. The detailed picture of the MgO(100) + H_2O system which has emerged from these recent studies has yet to be firmly established by experimental observation and will, no doubt, be subject to further theoretical research.

QM theory has been applied to studies of several other oxide surfaces in recent years. The structure of the low index surfaces of Al_2O_3 has been computed within PW-LDA theory (Manassidis and Gillan, 1994), which predicts a very large relaxation of the basal plane surface (0001). Studies of this surface have been extended to the adsorption of niobium atoms (Kruse *et al.*, 1994) and of complex organic molecules (Frank *et al.*, 1995).

The interpretation of surface spectroscopy from oxide surfaces and the direct computation of optical properties is particularly difficult due to strong relaxation and final state effects. In a recent study the GW approximation has been used to correct the LDA eigenvalues for the bulk and surface states of MgO (Schonberger and Aryasetiawan, 1995). Reasonable agreement with the observed electron energy loss spectra was achieved.

4 COMPLEX ELECTRONIC STRUCTURE – TRANSITION METAL OXIDES

As has been documented above, the success of various independent electron approximations to the ground state solutions of the Schrödinger equation is remarkable. In one important class of materials this approximation appears to break down. In the Mott–Hubbard insulators the LDA solutions describe a ground state that is non-magnetic and metallic (or nearly so), whereas these materials are magnetically ordered wide band gap insulators (Mattheis, 1971; Terakura *et al.*, 1984; Zaanen and Sawatzky, 1990). While MgO is the paradigm for simple oxides, NiO is by far the most studied transition metal oxide and will suffice to illustrate the modern approach to these materials. This class of materials also includes many transition metal (also lanthanide bearing) compounds, and includes the superconducting cuprates; it is in this context that these problems have received a great deal of attention in recent years. The Hubbard model, which is based on an extreme tight binding or *atomistic* description of the electronic structure, provides an accurate description of the ground state, electronic phase transition and low lying excited states (e.g.

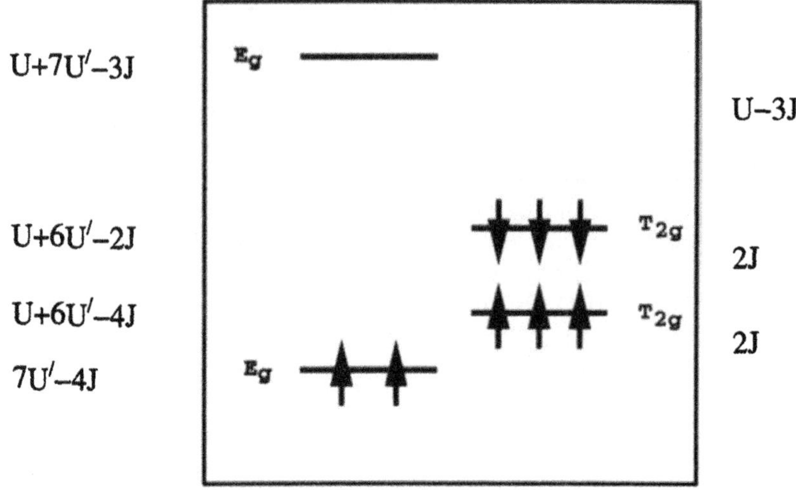

Figure 8.2 An extreme tight binding (atomistic) view of the electronic structure in NiO (see text for explanation).

Brandow, 1977, 1992). In the following, recent progress in this area will be described with regard to the accurate description of the *ground state* electronic structure and energetics. The *ab initio* description of the low lying excited states in these systems is an area fraught with difficulties which will not be discussed here (e.g. Manghi *et al.*, 1994; Aryasetiawan and Gunnarsson, 1995).

In Fig. 8.2 the electronic structure of NiO is characterized from an atomistic point of view (Brandow, 1977, 1992). The orbital energies of the Ni^{2+} (d^8) ion in an octahedral field are dominated by *intra-site* effects; the (averaged) diagonal (U) and off-diagonal (U') Coulomb interactions and exchange (J) determine the orbital energies shown in the figure. The orbitals are split into triple (T_{2g}) and double (E_g) degenerate sets. The unoccupied E_g orbital is split from the occupied states by a gap of order U ($\gg J$). In a simple band theory description of the electronic structure of NiO, the atomic d-level broadens to produce a partially filled d-band yielding a metallic state. If a spin polarized solution is adopted, the gap that is opened at the Fermi energy (E_f) will be generated by inter-site crystal field, covalency or magnetic effects of order the Stoner I parameter, not U (Anisimov *et al.*, 1991; Severin *et al.*, 1993). It is the atomistic viewpoint which more closely corresponds to the observed antiferromagnetically ordered insulating ground state of NiO. This view finds detailed and compelling confirmation in the observed low lying optical excitations from NiO and Ni-doped MgO (Reinen, 1965). As the concentration of Ni^{2+} ions is increased up to pure NiO there is hardly any change in the Frenkel exciton spectrum, the details of this spectrum being readily described within crystal field multiplet theory or, quantitatively, by cluster calculations (Janssen and Nieuwport, 1988). It is the correct treatment of the on-site Coulomb interactions that produces this simple picture of the electronic structure of NiO and underpins the Hubbard model. A theory based on the independent approximation will not produce a ground state consistent with this local description unless the energy of an orbital depends on the current occupancy of other orbitals (Fig. 8.2), the so-called orbital polarization. In the LDA the electronic mean field potential is a function of the local electron density and is thus rather insensitive to the occupancy of specific orbitals. It is for this reason that the LDA fails to correctly describe the ground state. In recent years this problem has been addressed by a number of researchers. Qualitatively correct ground states have been generated using a number of different corrections to the LDA. The Self-Interaction Correction (SIC-LDA) consists of transforming the Bloch orbitals to a localized representation and then computing the Coulomb interactions between them. It has been used to obtain the ground state charge density and band structure of a number of monoxides and cuprate materials (e.g. Svane and Gunnarsson, 1990; Svane, 1992; Szotek *et al.*, 1993). Alternatively, the on-site Coulomb interactions can be introduced as a parameter U to yield the LDA + U approximation which has also been used in studies of the band structure in the monoxides (Anisimov *et al.*, 1990) and cuprate materials (Anisimov *et al.*, 1992). The latter study is a rare example in which a full-potential LAPW approach was used to obtain an energy expression sufficiently accurate to enable structural relaxation and thus a study of the formation energy of the small polaron in $La_{2-x}Sr_xCuO_4$.

In this context the HF Hamiltonian is of particular interest. HF theory is distinguished from the LDA in that it attempts to compute the ground state wave function, whereas LDA computes the ground state density and energy. The HF approximation therefore contains explicit interactions between the orbitals and is automatically ‘self-interaction’ corrected. The on-site Coulomb interaction is therefore treated quite naturally and a qualitatively correct description of the ground states is obtained. The density of states of NiO projected onto the Ni E_g, T_{2g} and O orbitals computed in the HF approximation is shown in Fig. 8.3.

This is, of course, a band-theory calculation, but the energy ordering of the occupied orbitals is exactly as one would expect from the atomistic model (Fig. 8.2); the effect of the crystalline environment is to broaden the states slightly. Qualitatively similar results can be obtained within the LDA corrected theories described above. Having obtained a qualitatively correct description of the ground state, the essential test of the theory is whether or not the energy expression is consistent with the known properties of the material. In recent studies (Mackrodt *et al.*, 1993; Towler *et al.*, 1994) it has been shown that the HF description of the ground state energetics is on essentially the same footing as that in simple oxides, the lattice constant, elastic moduli and phonon frequencies following the trends expected for wide band gap ionic systems (bond lengths slightly too long and moduli somewhat too high). On reflection, this is exactly as one would expect; correlation effects are small in the presence

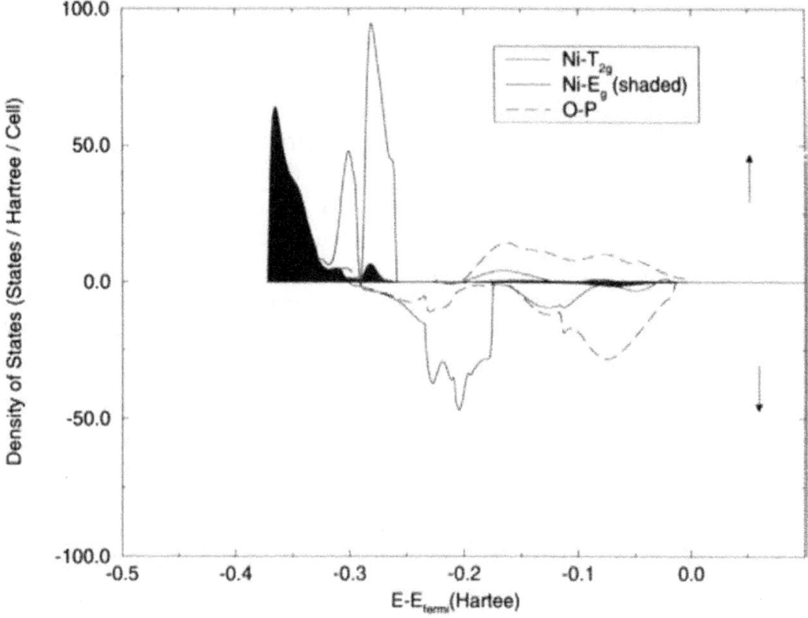

Figure 8.3 HF density of states of NiO projected onto the Ni E_g, T_{2g} and O orbitals.

of the large band gap and restricted to overestimation of the size of the negative ion.

The computation of the relative energetics of the ordered magnetic states observed in these oxides requires high numerical precision. Using analytic matrix elements, HF theory in a Gaussian basis set can be used to study these states. The observed antiferromagnetic ground state (AF2) is predicted to be the lowest in energy and the energy relative to the ferromagnetic state found to be consistent with the Néel temperature computed within a simple Ising model (Towler *et al.*, 1994). A rather delicate structural consequence of the magnetic ordering in the cubic monoxides is the exchange-striction induced rhombohedral distortion of the magnetic ground state. The HF energy expression quite naturally describes the inter-site exchange interactions that give rise to this effect and the distortion is quantitatively reproduced (Towler *et al.*, 1994). These studies have recently been extended to a variety of materials. Of particular note are studies of the cooperative Jahn Teller effect in $KCuF_3$ (Towler *et al.*, 1995) which require a high degree of numerical precision and probe the delicate balance between the Coulomb, exchange and kinetic energy contributions.

As noted above, the non-ground state properties of these materials (e.g. the optical band gap and electrical conductivity) are both difficult to compute reliably and of primary importance to our understanding of these systems. These properties remain an area of much current research.

5 CONCLUSIONS

The current state of the art in the use of *ab initio* quantum theory in materials science is limited by the length and time scales that need to be studied in order to examine realistic physical processes. Nevertheless, important information can be obtained about rather complicated systems, e.g. the protonation of zeolite frameworks, the formation of charged defects in ionic crystals and the energetics of simple chemical reactions at surfaces. In highly correlated systems, where traditionally the reliability of band theory methods have been in doubt, rapid progress is being made in the description of the ground state energetics, allowing us to study the transition metal oxides with almost as much confidence as we now study silicates and simple oxides.

REFERENCES

Allan, D.C. and Teter, M.P. (1987) *Phys. Rev. Lett.*, **59**, 1136.

Allan, D.C. and Teter, M.P. (1990) *J. Am. Ceram. Soc.*, **73**, 3274.

Anchell, J.J., White, J.C., Thompson, M.R. and Hess, A.C. (1994) *J. Phys. Chem.*, **98**, 4463.

Anisimov, V.I., Korotin, M.A. and Kurmaev, E.Z. (1990) *J. Phys. CM*, **2**, 3973.

Anisimov, V.I., Zaanen, J. and Andersen, O.K. (1991) *Phys. Rev. B*, **B44**, 943.

Anisimov, V.I., Korotin, M.A., Zaanen, J. and Andersen, O.K. (1992) *Phys. Rev. Lett.*, **68**, 345.

Aryasetiawan, F. and Gunnarsson, O. (1995) *Phys. Rev. Lett.*, **74**, 3221.

Blöchl, P.E. (1993) *Phys. Rev. Lett.*, **70**, 3263.

Brandow, B.H. (1977) *Adv. Phys.*, **24**, 651.

Brandow, B.H. (1992) *Journal of Alloys and Compounds*, **181**, 377.

Car, R. and Parrinello, M. (1985) *Phys. Rev. Lett.*, **55**, 2471.

Causà, M., Dovesi, R. and Roetti, C. (1991) *Phys. Rev. B*, **43**, 11937.

Ceperley, D.M. and Alder, B.J. (1980) *Phys. Rev. Lett.*, **45**, 566.

Colle, R. and Salvetti, O. (1975) *Theor. Chem. Acta*, **37**, 329.

Corà, F. and Pisani, C. (1994) *Mod. Simul. Mater. Sci. Engng*, **2**, 965.

Chelikowsky, J.R., King, H.E. Jr., Troullier, N., Martins, J.L. and Glinnemann, J. (1990) *Phys. Rev. Lett.*, **65**, 3309.

Chelikowsky, J.R., Troullier, N. and Martins, J.L. (1991) *Phys. Rev. B*, **44**, 489.

De Vita, A., Gillan, M.J., Lin, J.S., Payne, M.C., Stich, I. and Clarke, L.J. (1992) *Phys. Rev. B*, **46**, 12964; *Phys. Rev. Lett.*, **68**, 3319.

Dovesi, R. and Orlando, R. (1994) *Phase Transitions*, **52**, 151.

Dovesi, R., Roetti, C., Freyria-Fava, C., Apra, E., Saunders, V. R. and Harrison, N.M. (1992) *Phil. Trans. A*, **341**, 203.

Frank, I., Marx, D. and Parrinello, M. (1995) *J. Am. Chem. Soc.*, **117**, 8037.

Freyria-Fava, C., Dovesi, R., Saunders, V.R., Leslie, M. and Roetti, C. (1993) *J. Phys. CM*, **5**, 4793.

Gillan, M.J. (1991) In *Proc. NATA ASI on Computer Simulation in Material Science*, Vol. 257 (ed. M. Mayer and V. Pontikis), Kluwer, Dordrecht.

Goniakowski, J., Bouette-Russo S. and Noguera, C. (1993) *Surf. Sci.*, **284**, 315.

Greatbanks, S.P., Sherwood, P., Hillier, I.H., Hall, R.J., Burton, N.A. and Gould, I.R. (1995) *Chem. Phys. Lett.*, **234**, 367.

Henderson, B. (1980) *CRC Crit. Rev. Solid State Mater. Sci.*, **9**, 1.
Hess, A.C. and Saunders, V.R. (1992) *J. Phys. Chem.*, **96**, 4637.
Janssen, G.J.M. and Nieuwpoort, W.C. (1988) *Phys. Rev. B*, **38**, 3449.
Keskar, N.R., Troullier, N., Martins, J.L. and Chelikowsky, J.R. (1991) *Phys. Rev. B*, **44**, 4081.
Kohn, W. and Sham, L.J. (1965) *Phys. Rev.*, **140**, 1133.
Kruse, C., Finnis, M.W., Milman, V.Y., Payne, M.C., De Vita, A. and Gillan, M.J. (1994) *J. Am. Ceramic Soc.*, **77**, 431.
LaFemina, J.P. (1992) *Surf. Sci. Rep.*, **16**, 133.
Langel, W. and Parrinello, M. (1994) *Phys. Rev. Lett.*, **73**, 504.
Liu, A.Y., Singh, D.J. and Krakauer, H. (1994a) *Phys. Rev. B*, **49**, 17424 and references therein.
Liu, F., Garofalini, S.H., King-Smith, D. and Vanderbilt, D. (1994b) *Phys. Rev. B*, **49**, 12528.
Mackrodt, W.C. (1982) In *Computer Simulation of Solids* (eds C.R.A. Catlow and W.C. Mackrodt), Springer Verlag, Berlin.
Mackrodt, W.C. and Stewart, R.F. (1979) *J. Phys. C*, **12**, 5015.
Mackrodt, W.C., Harrison, N.M., Saunders, V.R., Allan, N.L., Towler, M.D., Aprà, E. and Dovesi, R. (1993) *Phil. Mag. A*, **68**, 653.
Manassidis, I. and Gillan, M.J. (1994) *J. Am. Ceramic Soc.*, **77**, 335.
Manghi, F., Calandra, C. and Ossicini, S. (1994) *Phys. Rev. Lett.*, **73**, 3129.
Maseras, F. and Morokuma, K. (1995) *J. Comp. Chem.*, **16**, 1170.
Mattheiss, L.F. (1971) *Phys. Rev. B*, **5**, 290.
McCarthy, M.I. and Harrison, N.M. (1994) *Phys. Rev. B*, **49**, 8574.
Nada, R., Catlow, C.R.A., Dovesi, R. and Pisani, C. (1990) *Phys. Chem. Mineral.*, **17**, 353.
Nicholas, J.B. and Hess, A.C. (1994) *J. Am. Phys. Soc.*, **116**, 5428.
Orlando, R., Dovesi, R., Roetti, C. and Saunders, V.R. (1994a) *Chem. Phys. Lett.*, **228**, 225.
Orlando, R., Dovesi, R., Azavant, P., Harrison, N.M. and Saunders, V.R. (1994b) *J. Phys. CM*, **6**, 8573.
Payne, M.C., Teter, M.P., Allan, D.C., Arias, T.A. and Joannopoulos, J.D. (1992) *Rev. Mod. Phys.*, **62**, 1045.
Perdew, J. (1991) In *Electronic Structure of Solids* (eds P. Ziesche and J. Eschrig), Akademie Verlag, Berlin.
Perdew, J. and Zunger, A. (1981) *Phys. Rev. B*, **23**, 5048.
Philipsen, P., Te Velde, G. and Baerends, E.J. (1994) *Chem. Phys. Lett.*, **226**, 583.
Pisani, C. (1993) *J. Mol. Catal.*, **85**, 229.
Pisani, C., Cora, F., Nada, R. and Orlando, R. (1994) *Comp. Phys. Commun.*, **82**, 139.
Refson, K., Wogelius, R.A., Fraser, D.G., Payne, M.C., Lee, M.H. and Milman, V. (1995) *Phys. Rev. B*, **52**, 10823.
Reinen, D. and Bunsenges, Ber. (1965) *Phys. Chem.*, **69**, 82.
Sarnthein, J., Pasquarello, A. and Car, R. (1995) *Phys. Rev. Lett.*, **74**, 4682.
Scamehorn, C.A., Harrison, N.M. and McCarthy, M.I. (1994) *J. Chem. Phys.*, **101**, 1547.
Schonberger, U. and Aryasetiawan, F. (1995) *Phys. Rev. B*, **52**, 8788.
Severin, L., Brooks, M.S.S. and Johansson, B. (1993) *Phys. Rev. Lett.*, **71**, 3214.
Sherman, D.M. (1993) *J. Geo. Res.*, **98**, 18795.
Sulimov, V.B., Pisani, C., Corà, F. and Sokolov, V.O. (1994) *Solid State Commun.*, **90**, 511.
Svane, A. (1992) *Phys. Rev. Lett.*, **68**, 1900.
Svane, A. and Gunnarsson, O. (1990) *Phys. Rev. Lett.*, **65**, 1148.
Szotek, Z., Temmerman W.M. and Winter, H. (1993) *Phys. Rev. B*, **47**, 4029.

Terakura, K., Oguchi, T., Williams, A.R. and Kübler, J. (1984) *Phys. Rev. B*, **30**, 4734.
Towler, M.D., Allan, N.L., Harrison, N.M., Saunders, V.R., Mackrodt, W.C. and Aprà, E. (1994) *Phys. Rev. B*, **50**, 5041.
Towler, M.D., Dovesi, R. and Saunders, V.R. (1995) *Phys. Rev. B*, **52**, 10150.
Troullier, N. and Martins, J.L. (1990) *Solid State Commun.*, **74**, 613.
Troullier, N. and Martins, J.L. (1991) *Phys. Rev. B*, **43**, 1883.
Vanderbilt, D. (1990) *Phys. Rev. B*, **41**, 7892.
Wang, Y., Stocks, G.M., Shelton, W.A., Nicholson, D.M.C., Szotek, Z. and Temmerman, W.M. (1995) *Phys. Rev. Lett.*, **75**, 2867.
White, J.C. and Hess, A.C. (1993) *J. Phys. Chem.*, **97**, 8703.
Zaanen, J. and Sawatzky, G. (1990) *Prog. Theor. Phys. Suppl.*, **101**, 231.
Zupan, A. and Causà, M. (1994a) *Int. J. Quantum Chem. S*, **28**, 633.
Zupan, A. and Causà, M. (1994b) *Chem. Phys. Lett.*, **220**, 145.
Zupan, A. and Causà, M. (1995) *Int. J. Qu. Chem.*, **56**, 337.

9

Silicates and Microporous Materials

R.G. Bell and G.D. Price

1 INTRODUCTION

The increasingly indispensable role of atomistic and quantum mechanical simulations in inorganic crystallography is perhaps no more strikingly illustrated than in the field of silicates and zeolitic materials. The two classes of material with which we shall be concerned in this chapter, namely microporous zeolites (including both aluminosilicate-based materials and their sister compounds, the aluminophosphates (ALPOs)), and the dense silicate materials and related oxides which constitute the bulk of the Earth's mantle, have in common the fact that their structural properties may be difficult to determine by conventional experimental means. Yet highly detailed and accurate structural information is critical in understanding the properties of both these important types of material.

Zeolites, characterized by their complex and beautiful crystal structures, continue to represent a vibrant field of research for scientists from a wide range of disciplines, not least because of their commercially important catalytic and sorptive properties. Much attention is devoted to structural studies, and it is frequently the case that subtle changes in structure can have a profound impact on the properties of these materials. Similarly, in the case of mantle minerals, the interpretation of seismic data, which is the main source of information about our planet's interior, relies heavily on detailed knowledge of the structures of these various minerals. In both cases computational modelling has proved an invaluable tool, and indeed the desire to achieve a fuller understanding of the properties of zeolites and silicate minerals has stimulated many innovations in computer simulation.

In this survey we review the most recent advances in the computer modelling of these materials, with particular regard to structure and structure-related properties.

COMPUTER MODELLING IN INORGANIC CRYSTALLOGRAPHY
ISBN 0-12-164135-X

2 ZEOLITIC MATERIALS

2.1 Background

Zeolites are microporous inorganic solids, i.e. their crystal structures contain large pores and voids which can vary from ca. 3 Å to over 12 Å in effective diameter. They have found wide application in the fields of ion-exchange, gas separation and, most importantly, in heterogeneous catalysis, for which zeolites provide the unique combination of a vast internal surface (typically 300–700 cm^2/g) accessible through channels of comparable size to organic molecules, and a well-defined pore size distribution. Molecules are therefore not only allowed to diffuse through the zeolite, but have additional size and shape-selective constraints imposed on them. Much literature is devoted to zeolite science, and for a more in-depth and recent overview of the field the reader may consult the proceedings of the latest International Zeolite Conference (Weitkamp *et al.*, 1994). General introductions to zeolite science are given by Newsam (1992) and Catlow (1992), with the latter emphasizing the role of computer modelling in the study of these materials.

The zeolite framework structures are built up from TO_4 tetrahedra, which are connected by corner sharing to form three-dimensional networks (see Fig. 9.1). Traditionally the term zeolite was applied to aluminosilicate framework materials, i.e. where T is mainly silicon or aluminium, but analogous ALPOs are now generally regarded as part of the zeolite canon, and the term microporous structure unambiguously embraces both types of compound. Since all oxygen atoms in the TO_4 tetrahedra are shared, it follows that the composition of a purely siliceous zeolite is SiO_2. The introduction of an aluminium, or other trivalent element, into a tetrahedral site in place of silicon results in the creation of a negatively charged framework. Similarly, in the case of the ALPOs (neutral composition $AlPO_4$), a divalent element such as Mg or Co^{II} may be substituted for Al, or silicon for phosphorus, with the same effect. It is this framework charge imbalance, and the ways in which it can be neutralized, which gives rise to most of the commercially important properties of zeolites.

One way in which the charge may be compensated is by the introduction of a hydroxyl proton onto a framework oxygen adjacent to the tetrahedral heteroatom, resulting in the formation of a Brønsted acid site. The ability of zeolites to catalyse a wide range of organic reactions derives mainly from this Brønsted acidity (although Lewis acidity and extra-framework species also play prominent roles). The acid strength of a particular zeolite depends, moreover, not only on the framework composition but also significantly on the local geometry around the acid site.

Negative framework charge may also be neutralized by extra-framework species, such as metal ions, which position themselves in the internal voids of the zeolite. Two important properties which arise from the presence of these ions

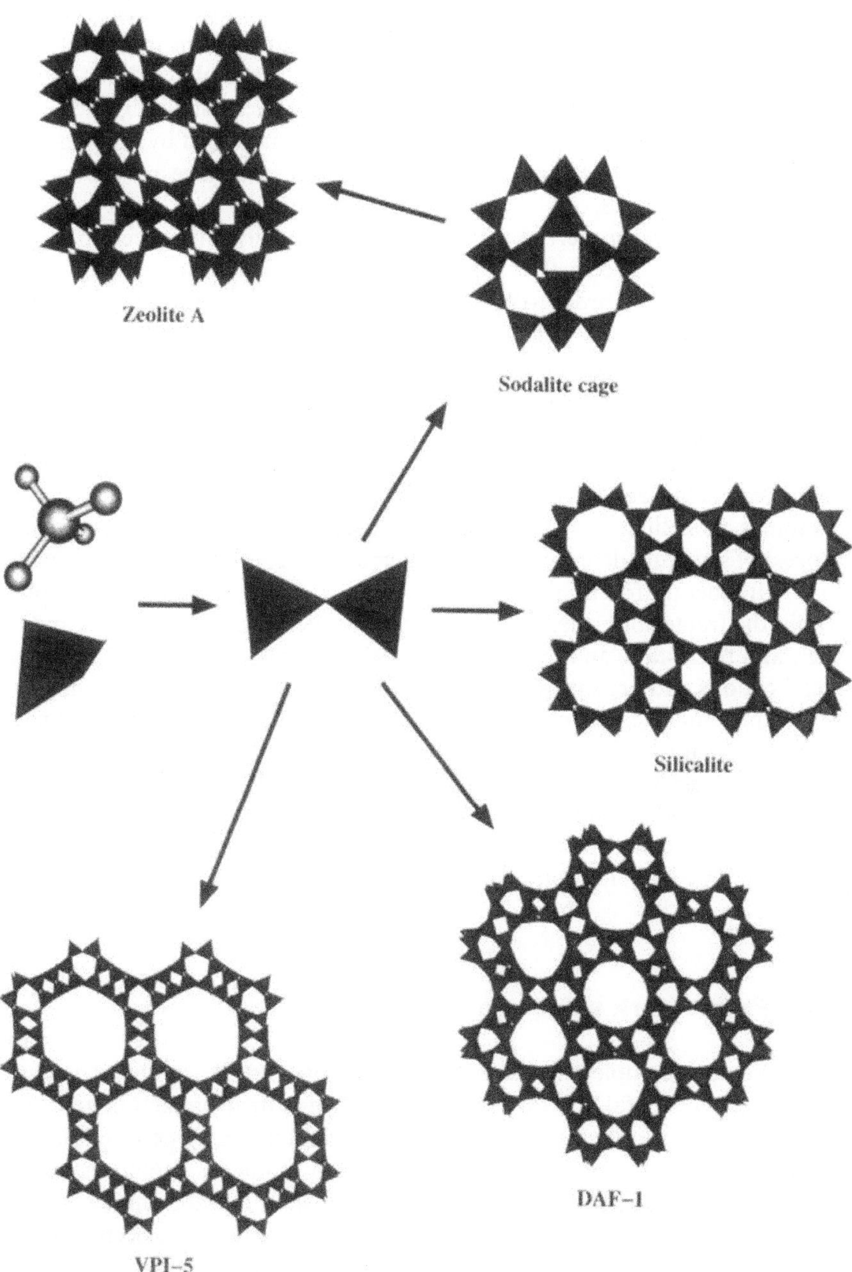

Figure 9.1 Diagram showing the way zeolite framework structures are built up from TO_4 units.

are ion-exchange and gas adsorption. An example of ion-exchange is found in the application of zeolites in water softening, where Na^+ ions within the zeolite are exchanged for alkaline earth metal ions. The ability of some zeolites to absorb selectively certain gases, which has led to applications in the field of air separation, is also highly dependent on the nature and location of extra-framework cations, both from the point of view of accessibility of adsorption sites and by the way in which the internal volume is modified by the presence of these ions. In addition, the positioning of the cations can have a profound effect on catalytic properties.

Since so many properties of zeolites are strongly influenced by crystal structure, it would seem that a great deal of detailed structural information would be required in order to gain any meaningful insight into these properties. Determining the crystal structures of zeolites is, however, rarely a straightforward matter. The great majority of microporous materials are only obtainable in powder form, rather than as single crystals; moreover, the structures are often quite complex, with large unit cells and significant numbers of asymmetric atoms. Although the advent of high resolution powder diffraction facilities based on synchrotron radiation and neutron sources has proved of great benefit, computer modelling is also increasingly regarded as a vital technique in the process of structure determination. Further challenges are posed by the disorder that is frequently found in zeolite structures, e.g. the presence of fault planes within the framework, the lack of long-range ordering of heteroatoms and acid sites, and both static and dynamic disorder of extra-framework species such as metal cations and organic absorbates. Here computer simulation again plays a very important role, and can be an essential adjunct to experimental techniques deployed to probe local structure and dynamics, such as X-ray adsorption spectroscopy and solid state NMR, which by themselves can only offer limited information. Indeed, obtaining unambiguous structural information about a zeolite will typically involve combining data from computer simulations with that from a variety of experimental methods.

2.2 Framework structure

The particular complexities of zeolite framework structures have spawned a number of theoretical approaches directed towards their prediction and refinement. In this section we shall concentrate principally on some recent applications of atomistic simulation methods to the study of zeolite framework structure. However, it is appropriate first to give a brief survey of other theoretical advances in this field, which have for the most part been concerned with developing an understanding of framework structure from an empirical standpoint. Fuller details of these topics may be found in the previous reviews of Freeman *et al.* (1992) and Price *et al.* (1992).

The systematic analysis of framework topology, i.e. the ways in which four-connected T-atoms may be linked together in three-dimensional networks (e.g. Smith, 1988) gave rise to a number of important concepts, such as the Secondary Building Unit (Meier and Olson, 1992 and references therein) and the Akporiaye–Price formalism (Akporiaye and Price, 1989), which considered zeolite frameworks as being constructed from two-dimensional sheets, using a variety of symmetry operators. Based on a geometrical analysis of the way the TO_4 tetrahedra are connected together in known zeolite structures, Deem and Newsam (1992) derived a series of empirical cost functions which they implemented in a Monte Carlo simulated annealing program, as discussed in Chapter 5. Using their method, an approximate structural model can be obtained from knowledge only of the unit cell dimensions and symmetry, and the framework density, thus paving the way for genuine structural prediction. Earlier, Meier and Villiger (1969) had developed a distance-least-squares (DLS) technique, also based on empirical geometric data, which enables the optimization of approximate framework coordinates and is now routinely used in zeolite structure determination. The problem of stacking disorder in zeolite structures has also been tackled by Treacy *et al.* (1991) with their DIFFaX program, which can quantitatively simulate diffraction patterns for materials with a given degree of two-dimensional faulting.

The most accurate simulations of zeolite crystal structure in recent years have been achieved by energy minimization methods using transferable potentials of the type discussed in Chapter 3. Lattice energy minimization lends itself most readily to the modelling of inorganic crystal structures, in that a minimum energy periodic structure is obtained from a given starting configuration using accurate potential functions, which have often been derived quantum mechanically, or fitted to a variety of structural and thermodynamic properties of a well-characterized structure, e.g. of a binary oxide. Recent applications of these techniques to microporous structures will now be discussed.

2.2.1 Silicalite

An early illustration of the accuracy of potential-based lattice energy minimization calculations was provided by the monoclinic distortion of silicalite, discussed in Chapter 1. It is known that silicalite, the siliceous analogue of the zeolite ZSM-5 (see Plate VII), undergoes a phase transition at around 300 K, with the low temperature phase having monoclinic symmetry and the high temperature phase orthorhombic. As observed by powder X-ray diffraction, the change in symmetry is very subtle, the lattice parameter α changing by a mere 0.64°. An important test of the model interatomic potentials for silicates is therefore whether this low temperature distortion can be reproduced, since we would expect lattice energy minimization to

Table 9.1 Calculated and experimental structural data for silicalite.

	Experimental (Olson *et al.*, 1981)	Experimental (van Koningsveld *et al.*, 1990)	Simulated (Bell *et al.*, 1990)
a (Å)	20.07	20.107	19.986
b (Å)	19.92	19.879	19.747
c (Å)	13.42	13.369	13.324
α (°)	90	90.67	90.803
β (°)	90	90	90
γ (°)	90	90	90
Cell volume (Å^3)	5365	5343	5298
Si–O bond lengths (Å)			
Range	1.52–1.67	1.582–1.607	1.595–1.608
Average	1.59	1.595	1.601
O–Si–O angles (°)			
Range	97.5–129.0	107.1–111.5	106.3–115.0
Si–O–Si angles (°)			
Range	142.6–175.0	141.3–169.0	143.2–157.9
Average	155.0	153.0	148.8

simulate the lowest temperature structure. This indeed is what was found by Bell *et al.* (1990): using the orthorhombic structure as the starting configuration, and employing a constant pressure minimization algorithm, the distortion was reproduced in the minimized structure. As can be seen from Table 9.1, the agreement between the calculated cell parameters and those obtained experimentally for monoclinic silicalite are in remarkable agreement. It is informative to note that the distortion was only simulated when the shell model (see Chapters 3 and 7) was employed to reproduce the polarizability of the oxygen ions. Using the rigid ion formalism, the structure remained orthorhombic.

2.2.2 NU-87

More recently lattice energy methods have been directly applied to the structure determination of novel zeolite framework types. Here we recall a highlight, outlined in Chapter 1. It concerns the case of NU-87, a high silica zeolite possessed of a multidimensional channel system based on 10- and 12-ring pores (see Plate VII). Shannon *et al.* (1991) were able to derive the correct framework topology (i.e. the connectivity of the tetrahedral units) from a combination of electron diffraction and powder synchrotron X-ray diffraction, assisted by DLS calculations. However, a number of weak reflections were observed in the synchrotron powder pattern which could not be accounted for within the predicted F-centred space group, and were initially supposed to be due to impurity phases. A lattice energy minimization using the shell model, in which no symmetry constraints were applied, revealed a minimum energy structure in the space group $P2_1/c$. As a result the anomalous reflections were able to be accounted for. The $P2_1/c$ asymmetric unit contained 56 independent atoms making it extremely unlikely that this result could have been achieved from the powder data without the aid of the minimization calculations.

2.2.3 MAPO-36

The previous two examples dealt with highly siliceous zeolites, in which the framework composition could be regarded as SiO_2 for the purposes of simulation. As we have seen, an analogous family of microporous materials based on aluminium phosphate also exists, known as ALPOs. One such material, again referred to briefly in Chapter 1, is MAPO-36 (or MeALPO-36 – the Me indicating the presence of a small amount of divalent metal substituting for framework Al). This material was the subject of a structural study by the Royal Institution group (Wright *et al.*, 1992). A combination of adsorption measurements and high resolution electron microscopy indicated that MAPO-36 possesses a 12-ring channel system similar to that of ALPO-5, and consideration of the unit cell dimensions and symmetry, derived from electron diffraction, suggested a model for the framework structure. Interestingly, the powder X-ray diffraction measurements showed a temperature dependence similar to that observed for silicalite, with the high temperature pattern (above 350°C) displaying higher symmetry than the low temperature. The ideal topology, refined by DLS, had a maximum possible symmetry of *Cmcm*, assuming complete disorder of Al and P over the framework T sites, and with this structure it was possible to match closely the experimental high temperature X-ray data, but not to explain the peak splittings observed at lower temperature (Fig. 9.2). Two lattice energy minimization simulations were then carried out in which no symmetry constraints were applied. In the first the T atoms were replaced by silicon, and in the second by aluminium and phosphorus in a strictly ordered configuration. While the 'SiO_2' simulation retained *Cmcm* symmetry, the ordered '$AlPO_4$' simulation, using a partial charge potential model, resulted in a reduction of space group symmetry to $C\bar{1}$ and a distortion of the unit cell. The simulated XRD pattern closely resembles that recorded at 150°C, although, as can be deduced from Fig. 9.2, the simulated unit cell (a = 13.46 Å, b=22.17 Å, c = 5.29 Å, α = 90.2°, β = 92.0°, γ = 90.0°) is about 2.5% larger than that observed experimentally (a = 13.16 Å, b=21.65 Å, c = 5.19 Å, α = 90.2°, β = 92.0°, γ = 90°). Much closer agreement can now be obtained by using the formal charge potential model of Gale and Henson (1994), which gives a = 13.11 Å, b=21.63 Å, c = 5.19 Å, α = 90.1°, β = 92.0°, γ = 90.3°. The framework structure of MAPO-36 is illustrated in Plate VII.

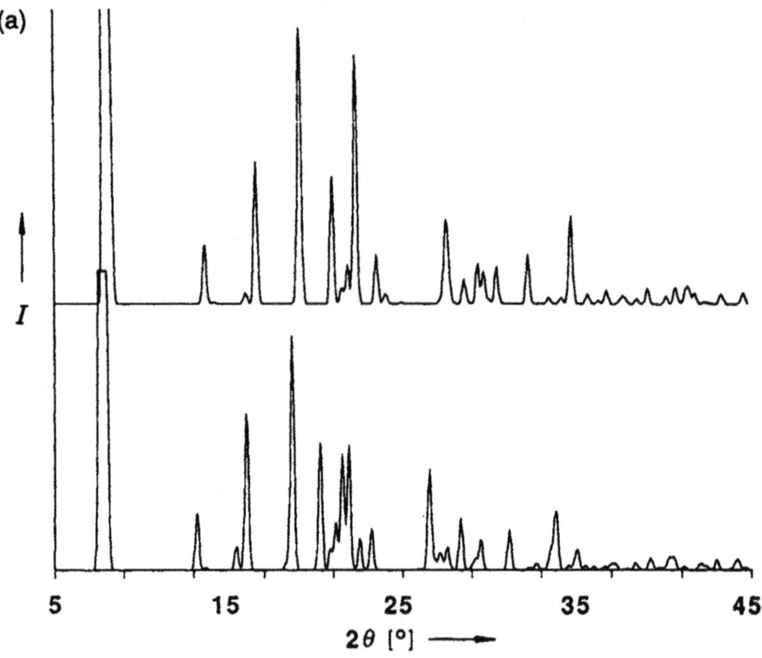
(a)
I
5
15
25
35
45
2θ [°]

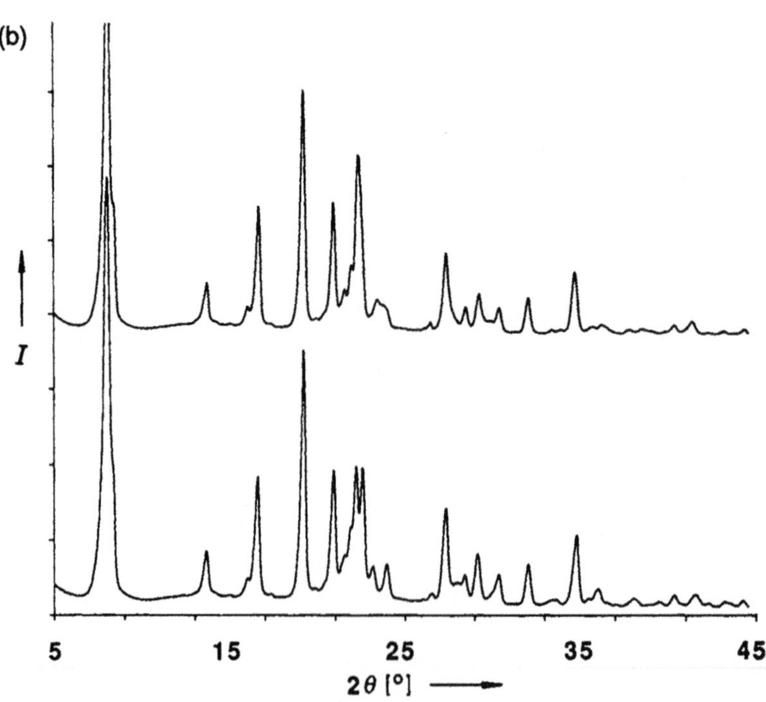
(b)
I
5
15
25
35
45
2θ [°]

2.3 Extra-framework species

The location of extra-framework species such as metal cations and adsorbed organic molecules presents a number of problems distinct from those associated with the framework structure itself. In the case of metal cations there is often considerable disorder, resulting in the partial occupancy of a number of unique sites. Measurements of these positions will involve the detailed analysis of accurate diffraction data, and the problem becomes more difficult when the loading of cations is low, as in the case of high silica zeolites. For adsorbates, such as catalytic reactants and organic template molecules which are used in zeolite syntheses, there are additional factors such as conformational disorder and, at finite temperatures, the dynamic processes of diffusion and rotation.

While energy minimization on its own can provide some insight into metal ion placement, e.g. in comparing cation sites which have already been determined experimentally (see the studies on mordenite described below), in most cases modelling the structure and properties of extra-framework species requires that the effects of temperature be taken into account at some stage, either explicitly to study dynamics, or as a means of overcoming local energy minima. Both molecular dynamics (MD) and Monte Carlo methodologies, often in combination with energy minimization, have been used successfully in this regard. In particular the automated Monte Carlo docking procedure of Freeman *et al.* (1991), which blends all three methods, has proved especially suitable for locating minimum energy sites in zeolite channel systems, and its application to locating extra-framework cations is described in Chapter 5. In the procedure, MD is first applied to the guest molecule, as though it were in the gas phase, in order to generate a library of conformations. Each of these conformations is then 'docked' into the zeolite framework using a Monte Carlo algorithm. The most favourable site/conformation combinations may subsequently be refined by energy minimization to yield the overall minimum energy configuration. Monte Carlo docking can be applied both to molecular sorbates, and to metal ions. In the latter case the guest MD calculations are obviously omitted, and the method then becomes a simple and efficient way of obtaining a Boltzmann-weighted cation distribution within the zeolite. Examples of recent studies of extra-framework cation location are now considered.

2.3.1 Metal cations in mordenite

The mordenite framework structure (see Plate VII and Fig. 9.3) has a complex topology, containing unidirectional 12-ring channels together with

---

Figure 9.2 (a) Simulated X-ray diffraction patterns of MAPO-36 using SiO_2 (above) and $AlPO_4$ (below) framework compositions. (b) Experimental X-ray powder diffraction patterns of MAPO-36 recorded at 375°C (above) and 150°C (below).

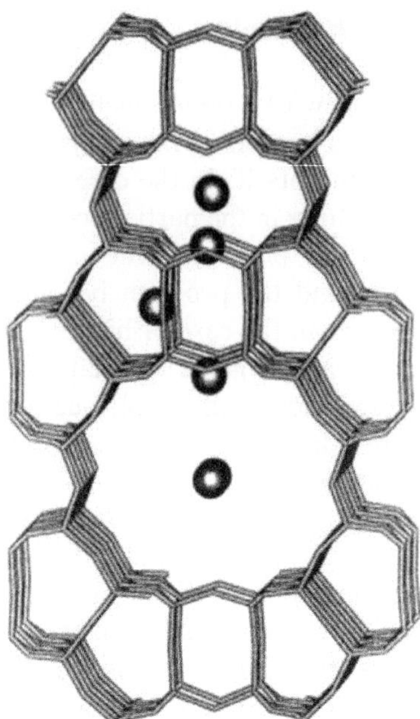

Figure 9.3 Section of framework structure of mordenite, showing the various extra-framework cation sites (represented as spheres).

smaller 8-ring channels and the so-called 8-ring 'side pocket'. Within these three environments, a variety of different cation positions have been determined experimentally. Two key factors which control the location of charged cations are (1) the distribution of aluminium within the framework, which strongly influences the electrostatic field at a given point and (2) the local topology of the framework, effectively the degree to which the metal may be coordinated to framework oxygen. Den Ouden *et al.* (1990) were the first to conduct lattice energy minimization calculations on this material specifically to study these effects. In the case of Ni^{2+}, they found that, although the most energetically favourable site was in the 8-ring side pocket, the dominant factor was the proximity of framework Al – placing two aluminium atoms on next-neighbour T positions would always favour the adjacent Ni^{2+} site over other extra-framework sites where the aluminiums were more widely separated. Subsequent simulations by Bell *et al.* (1993) confirmed the topological preference for the more highly coordinated side-pocket sites, both for Ni^{2+} and Na^+, but also found that, with increasing cation concentration, the Ni^{2+} cations were relatively less well-stabilized compared to the Na^+, suggesting that Al distribution has a less marked influence on the monovalent cation.

2.3.2 Structure of ETS-10

For *ab initio* modelling of extra-framework cations, when only the framework structure is known, it becomes convenient to use a Monte Carlo method, as described above. Recent calculations by Newsam *et al.* (1996) (discussed in greater detail in Chapter 5) have demonstrated the efficacy of this approach, by correctly reproducing the distribution of Na^+ in zeolite A, and also predicting the (as yet unknown) location of Li^+ in Li-ABW.

A similar method was used by Sankar *et al.* (1996) in a combined experimental and computational study of the novel microporous material ETS-10. In addition to four-coordinate silicon units, the framework contains chains of corner-sharing TiO_6 octahedra running in two orthogonal directions and imparting an overall negative change to the framework (see Fig. 9.4). The two-dimensional 12-ring channel system was known to contain a high concentration of alkali metal cations (Na^+ and K^+, as synthesized) but the exact cation locations, as well as the detailed structure of the titanium chains, could not be determined from diffraction experiments due to widespread faulting in the structure. Sankar *et al.* chose to probe the local environment of the titanium atom using X-ray adsorption spectroscopy. Having confirmed the six-fold coordination of Ti from X-ray Absorption Near Edge Structure (XANES) measurements, they were then able to use the results of computer simulations to fit a consistent structural model to Ti-Extended X-ray Absorption Fine Structure (EXAFS) data. The simulations involved symmetry-unconstrained lattice energy minimizations using several random distributions of the extra-framework cations. The minimizations suggested that far from being regular octahedra, the TiO_6 units were highly distorted, with long and short Ti–O bonds

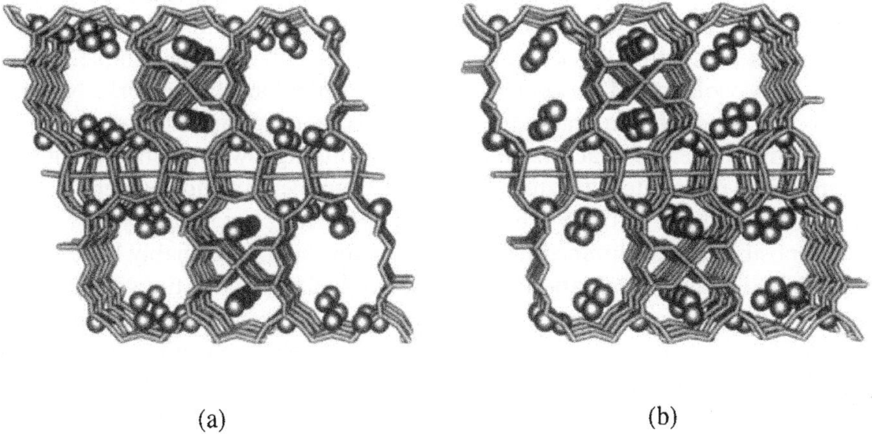

Figure 9.4 Minimum energy distributions of (a) Na^+ and (b) K^+ in ETS-10. In each case both the framework and cation position were optimized by lattice energy minimization.

along the Ti–O–Ti chain. Armed with this information, it proved possible to obtain an excellent fit to the EXAFS data using a multiple scattering approach, and also taking into account the Ti–cation distances derived from the simulations. In this instance computer modelling was again able to provide key information which could not have been obtained by experiment. The final structural model from the EXAFS is shown in Plate VIII. The cation positions shown in Fig. 9.4 are those obtained from the energy minimization.

2.3.3 *Para*-xylene adsorbed in ferrierite

Freeman *et al.* (1991) originally deployed their automated docking procedure to study the sorption of isomers of butene in silicalite. Other key applications have involved butene in DAF-1 (Bell *et al.*, 1994), butanols in ZSM-5 (Shubin *et al.*, 1994) and the modelling of template molecules in a range of zeolites (e.g. Lewis *et al.*, 1995; Stevens and Cox, 1995; Weigel *et al.*, 1996). Kaszkur and coworkers (1993) have also demonstrated how MD can be combined with powder X-ray diffraction, in a study of 1,4-dibromobutane in zeolite Y. In this work, MD trajectories were obtained for the guest molecules and used to interpret synchrotron powder diffraction data recorded at two different temperatures. The simulations revealed that, at low temperatures, the dibromobutane is sorbed at a position spanning two of the framework supercages, with each Br atom located close to an Na^+ cation. At higher temperatures, one terminal Br remains sorbed while the other exhibits considerable thermal motion.

Recent work by Kaszkur and colleagues (1996) used the Monte Carlo docking method to determine the position of *para*-xylene in ferrierite, a medium-pore zeolite which has found wide application as an acid catalyst. Rietveld refinement of high resolution powder diffraction data, both X-ray and neutron, was complicated by considerable displacements to the framework atoms, observed after the *p*-xylene had been loaded into the sample. This suggested a violation of the *Immm* symmetry of the bare framework. Nevertheless it was possible to obtain a fit within this space group, albeit with large anisotropic temperature factors, the possibility of using lower symmetry models having been discounted due to the excessive numbers of independent parameters. In the light of these problems with symmetry, the accuracy of the *p*-xylene position, as determined from Fourier maps, might justifiably have been regarded with some reservation. Monte Carlo docking calculations were then carried out to attempt to confirm the location of the sorbate. The simulations used a $1 \times 1 \times 3$ supercell of the ferrierite framework, in order to exclude unrealistic molecule–molecule interactions, and involved full energy minimization of both framework and guest molecule. The predicted minimum energy location of the molecule was in remarkable agreement with that obtained experimentally (see Plate IX) thus vindicating the approach taken in refining the diffraction data.

3 MANTLE-FORMING MINERALS

3.1 Background

The Earth's mantle is composed predominantly of magnesium–iron silicates (see Fig. 9.5). This is known from direct sampling of upper mantle material, seismic velocity–density systematics, and from inferences based upon solar abundances. The upper mantle largely contains forsteritic olivine ($(Mg,Fe)_2SiO_4$) and both orthorhombic and monoclinic pyroxene. The forsterite structure is characterized by having isolated SiO_4 tetrahedra while the pyroxene phases have linear chains of corner-sharing SiO_4 tetrahedra. At a depth of about 400 km the upper mantle minerals become unstable and transform to denser polymorphs that make up the mantle transition zone. Forsterite transforms first to wadsleyite (or the β-phase) and subsequently to ringwoodite (which has the spinel structure). Silicon remains in fourfold coordination in both these phases. In contrast, pyroxenes transform either to majorite (which has the garnet structure) or, less probably, to a polymorph with the ilmenite structure. Both of these phases have silicon in octahedral coordination. The 670 km seismic discontinuity marks the boundary of the lower mantle. It is believed that this discontinuity coincides with the breakdown reaction of ringwoodite to magnesium silicate perovskite and magnesiowüstite ($(Mg,Fe)O$ – rock-salt structure). Under about the same temperature–pressure conditions garnet also transforms to perovskite. In fact over 70% of the lower mantle and over 40% of the entire planet is composed of magnesium silicate perovskite.

Perovskite is a simple structure based upon the corner sharing of octahedra, enclosing a large 12-fold coordinated site. Magnesium silicate perovskite appears to be orthorhombically distorted at lower pressures, with silicon in the octahedral sites and the magnesium ion in an 8-fold distorted environment. In order to gain deeper insight into the physics and chemistry of the Earth's interior, and in particular to be able to interpret new seismic tomographic data, it is essential to determine the high pressure ($>$ 25 GPa) and high temperature ($>$ 2000 K) behaviours of these mantle minerals. However, such pressures and temperatures are currently very difficult to achieve (certainly simultaneously) in the laboratory, and so one of the major goals of computational mineral physics has been to describe fully the physical and defect behaviour of Earth-forming phases under the conditions found in the Earth's interior.

3.2 Methodologies – overview

Simulating the properties of materials at extremes of temperature and pressure places high demands on many of the traditional atomistic modelling techniques, not least because of the need to derive robust interatomic potentials

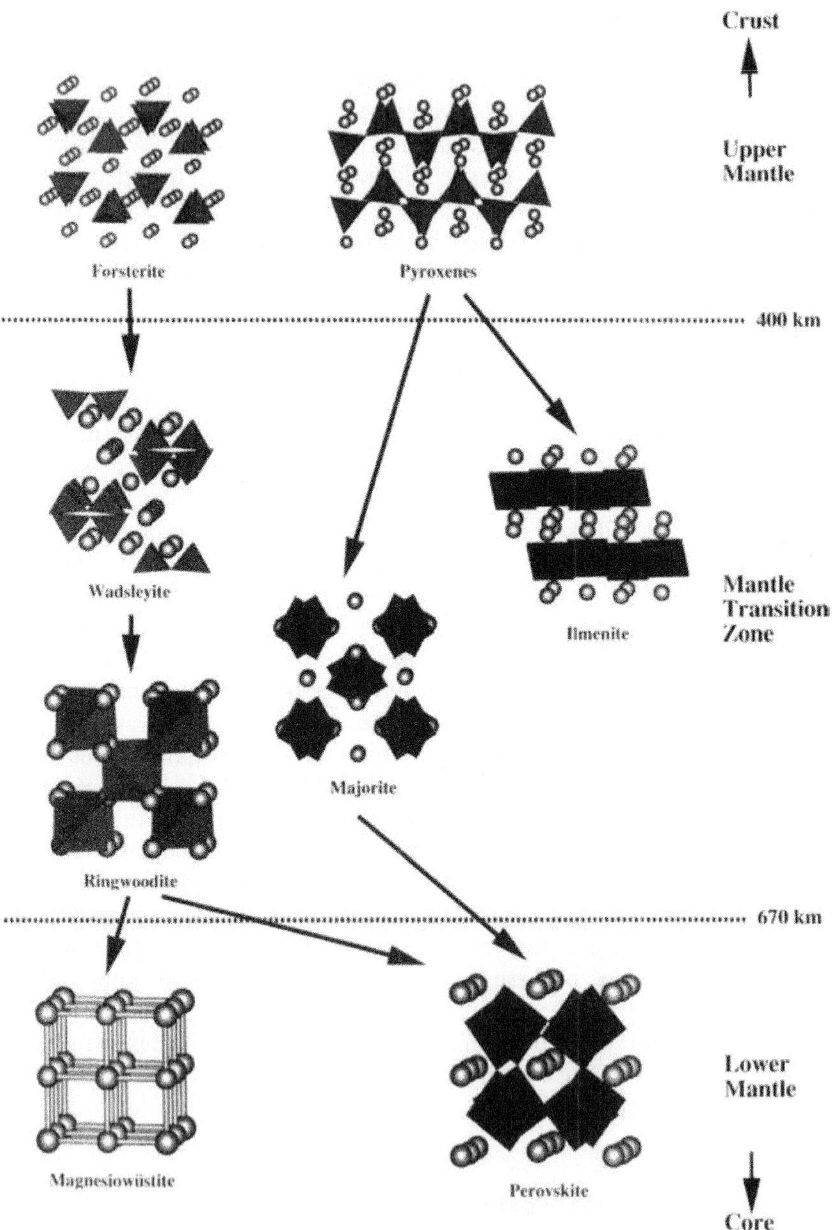

Figure 9.5 Diagram showing molecular structures of various mantle minerals, the regions within the Earth's mantle in which they are thought to be found and the phase transitions that are believed to occur. For each mineral, the polyhedra shown represent the coordination sphere of silicon.

which will be valid under conditions for which no experimental data are available, and over a wide range of interatomic separations. The two most obviously suited classical methods are free energy minimization based on lattice dynamics (described fully in Chapter 3), which permits the effect of temperature to be calculated via the evaluation of lattice vibrational frequencies, and molecular dynamics (see Chapter 4). Lattice dynamical models are based on the Quasi-Harmonic Approximation (QHA), which assumes that lattice vibrational modes are independent of one another; at elevated temperatures, above the Debye temperature, θ_D, vibrational amplitudes become larger and therefore the QHA breaks down as the displacements interact with one another and phonon–phonon scattering becomes prevalent. In contrast, MD describes these anharmonic effects implicitly; the problem with this technique lies rather in the low temperature regime, where quantum effects, not accounted for in classical MD, play an important role in determining thermodynamic properties such as heat capacity. In order to compensate for this shortcoming, Matsui (1989) investigated the possibility of applying a quantum correction to the structural and thermodynamic properties of MgO, as calculated from MD simulations at zero pressure and 200–2000 K. Using a Wigner–Kirkwood expansion of the free energy in powers of Planck's constant up to h^2, he found that the quantum contribution to the incompressibility was insignificant. For the thermal expansion coefficient and heat capacity, however, the applied correction became increasingly important as the temperature lowered. Down to around 500 K the corrected MD results showed close agreement with experimental data for all structural and thermodynamic properties. Below 500 K agreement was not as good, suggesting the need to include higher order terms in h for accurate predictions of thermal expansion coefficients and heat capacity.

In both molecular dynamics and lattice dynamics, the effect of pressure, essential if one is to obtain accurate predictions of phenomena such as phase transitions and anisotropic compression, can be modelled by allowing constant stress, variable geometry cells.

In order to investigate the effect of pressure on the validity of the QHA, by comparison to molecular dynamics methods, Matsui *et al.* (1994) undertook a parallel set of lattice dynamical and MD calculations on $MgSiO_3$ perovskite. They found that at zero pressure and 500 K the two techniques were in good agreement, but that as the temperature increased (remaining at 0 Pa), the molar volume of the perovskite, as calculated by the lattice dynamical method, was increasingly overestimated. However, at the temperatures (2000–3000 K) and pressures (<100 Gpa) found in the lower mantle, the molar volumes and incompressibilities predicted by the two techniques (Fig. 9.6) again became comparable, but the thermal expansion coefficients differed significantly due to the increased anharmonicity in this high temperature region.

It can thus be seen that to model the behaviour of mantle materials accurately

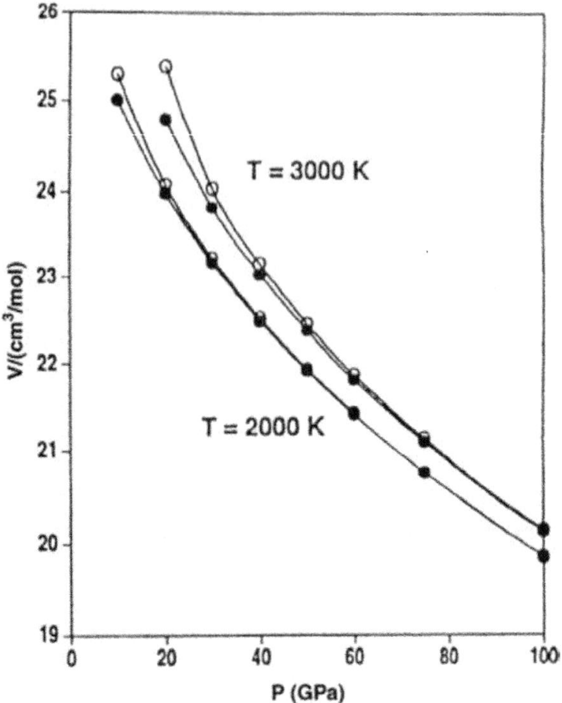

Figure 9.6 Simulated lattice dynamics and molecular dynamics molar volumes of $MgSiO_3$ perovskite at 2000 and 3000 K as a function of pressure up to 100 GPa (after Matsui *et al.*, 1994); ●, MD, ○, LD.

using lattice dynamics requires a more rigorous treatment of the effects of anharmonicity than has currently been implemented. Much recent work, of which key examples are described below (see also Vočadlo *et al.*, 1995), has therefore been directed towards developing and applying accurate MD techniques to the study of these minerals.

Latterly, increasing use has also been made of Quantum Molecular Dynamics (QMD), based on the pioneering work of Car and Parrinello (1985) (see Chapter 8). The Car–Parrinello method makes use of Density Functional Theory to calculate explicitly the energy of a system and hence the interatomic forces, which are then used to determine the atomic trajectories and related dynamic properties, in the manner of classical MD. As an *ab initio* technique, QMD has the advantage over classical simulation methods that it is not reliant on interatomic potentials, and should in principle lead to far more accurate results. The disadvantage is that it demands far greater computing resources, and its application has thus far been limited to relatively simple systems.

3.3 Phase behaviour and polymorphism of $MgSiO_3$ – molecular dynamics simulations

An early application of molecular dynamics to the study of phase transitions in mantle minerals concerned the behaviour of $MgSiO_3$ perovskite under the conditions of temperature and pressure found in the lower mantle. Matsui and Price (1991), using a rigid ion two-body potential to describe the interatomic interactions, carried out a series of constant temperature–constant pressure MD runs between 0 and 150 GPa and with the temperature gradually increased towards the melting temperature. At pressures above 10 GPa the calculations predicted a phase transition from orthorhombic to cubic just below the melting temperature, accompanied by melting of the oxygen sublattice. For example, at 30 GPa the diffusion coefficient of oxide ions at 5000 K was calculated to be 1×10^{-5} cm^2/s. Solid electrolyte behaviour at lower mantle temperatures and pressures had previously been inferred from laboratory measurements, and has important implications in attempting to understand this key region of the Earth's interior. Although the actual temperatures involved would be expected to be somewhat lower than those found from the simulation, this can be explained by the absence of any consideration of defect or surface effects in the simulations as well as the use of the rigid ion, rather than the polarizable shell, model – factors which one would expect to lead to an overestimate of melting temperature. Interestingly a parallel simulation by Kapusta and Guillopé (1993), using a different potential model, also simulated the solid electrolyte transition, but predicted an orthorhombic to tetragonal, rather than cubic, phase transition.

A comprehensive MD study of various $MgSiO_3$ polymorphs thought to be present in the mantle was subsequently carried out by Matsui and Price (1992). For each of six phases, the simulations predicted both structural and thermodynamic properties to within a few percent of the experimental values (see Table 9.2). Furthermore, these authors were able to predict the existence of two pyroxene phases, whose presence had previously only been inferred from experiment. One of these was a high temperature post-protoenstatite phase and the other a high pressure high clinoenstatite polymorph (space group *C2/c*). The predicted structural parameters of the *C2/c* phase were afterwards used by Angel *et al.* (1992) to solve the single crystal X-ray structure of that phase when it was later synthesized.

3.4 Quantum molecular dynamics studies

The recent development of QMD techniques offers the possibility of simulating the properties of minerals at finite temperatures and pressures to a much higher degree of accuracy than hitherto achieved although, as we have already noted, current computing technology permits the study only of relatively simple systems. In the approach of Wentzcovitch and Martins (1991), the electronic

Table 9.2 Observed and simulated structural and physical properties of the six $MgSiO_3$ polymorphs ($P = 0$ GPa) (after Matsui and Price, 1992).

Phase		Ortho	Clino	Proto	Garnet	Ilmenite	Perovskite
Space group		*Pbca*	*P2₁/c*	*Pbcn*	*I4₁/a*	*R3̄*	*Pbnm*
Cell lengths, cell angle and molar volume							
a (Å)	Obs	18.227	9.605	9.306	11.501	4.728	4.775
	Calc	18.146	9.600	9.378	11.516	4.740	4.772
b (Å)	Obs	8.819	8.813	8.892	11.501	4.728	4.929
	Calc	8.727	8.672	8.820	11.516	4.740	4.925
c (Å)	Obs	5.179	5.166	5.349	11.480	13.56	6.897
	Calc	5.262	5.244	5.458	11.523	13.33	6.942
β or γ (°)	Obs	90.0	108.5	90.0	90.0	120.0	90.0
	Calc	90.0	108.6	90.0	90.0	120.0	90.0
V (cm^3/mol)	Obs	31.33	31.22	31.32	28.58	26.35	24.44
	Calc	31.36	31.14	33.98	28.76	26.03	24.56
Bulk modulus (K_0), volume thermal expansivity (α) and enthalpy (H)							
K_0 (GPa)	Obs	108	—	112	154	212	247
	Calc	84	92	88	137	224	250
α (10^{-5}/K)	Obs	2.5	2.5	4.0	2.2	2.4	3.2
	Calc	3.8	4.1	6.0	2.2	2.8	2.9
H (kJ mol)	Calc	−7187.7	−7187.7	−7186.5	−7151.0	−7157.0	−7150.0

structure and molecular dynamics aspects of the simulation are kept separate, thus permitting relatively long MD time steps. These authors validated their technique in a study of liquid lithium, which gives excellent agreement with experimental properties. Their method can additionally incorporate a variable number of basis functions, thus making it well suited to variable cell shape simulations. Thus, combined with a novel constant pressure algorithm (Wentzcovitch, 1991), a study was made of the behaviour of $MgSiO_3$ perovskite at zero temperature and pressures up to 150 GPa (Wentzcovitch *et al.*, 1993, 1995), conditions well beyond the extrapolation range for empirical potential models. The low pressure (< 11 GPa) results were in excellent agreement with single crystal X-ray measurements for orthorhombic perovskite under compression (see Table 9.3), but the calculated relative compressibilities of the orthorhombic axes were in conflict with those inferred from Brillouin scattering. Calculations on the relative enthalpies of the orthorhombic (*Pbnm*) and cubic (*Pm3m*) structures indicated a significant and increasing difference in enthalpy with pressure between the two structures, and suggested that the orthorhombic phase would always be more stable than the cubic phase throughout the lower mantle, at variance with the predictions of classical MD simulation described in the previous section. Since the QMD simulations were carried out at zero temperatures, this conclusion was based on estimates of the vibrational entropies of the phases; we still await QMD simulations at finite pressure *and* temperature – now becoming feasible with the latest computer technology – to reinforce this

Table 9.3 Experimental and theoretical parameters of zero pressure *Pbnm* phase of $MgSiO_3$ (after Wentzcovitch *et al.*, 1993).

	Calc.(*Pbnm*)	Exp.(*Pbnm*)	Calc.(*Pm3m*)
a	4.711	4.7787(4)	4.909
b	4.880	4.9313(4)	4.909
c	6.851	6.9083(8)	6.942
Mg_x	0.5174	0.5141(1)	0.500
Mg_y	0.5614	0.5560(1)	0.500
O^1_x	0.1128	0.1028(2)	0.000
O^1_y	0.4608	0.4660(2)	0.500
O^2_x	0.1928	0.1961(1)	0.250
O^2_y	0.1995	0.2014(2)	0.250
O^2_z	0.5582	0.5531(1)	0.500

finding. In the later, more detailed, study (Wentzcovitch *et al.*, 1995) it was additionally predicted that $CaSiO_3$ would be preferentially found as a cubic perovskite phase at 150 GPa. Elastic constants and their pressure derivatives were derived for both the Mg- and Ca-containing minerals. The similarity of the bulk moduli strongly suggests that any enrichment of the lower mantle by calcium would have little effect on the compressibility of that region.

4 SUMMARY

Silicate systems provide some of the greatest challenges and rewards for computer modelling techniques. Their use in the study of both microporous and mantle materials described in this chapter will continue to grow.

REFERENCES

Akporiaye, D.E. and Price, G.D. (1989) *Zeolites*, **9**, 23.

Angel, R.J., Chopelas, A. and Ross, N.L. (1992) *Nature*, **358**, 322.

Bell, R.G., Jackson, R.A. and Catlow, C.R.A. (1990) *J. Chem. Soc. Chem. Commun.*, 782.

Bell, R.G., Jackson, R.A. and Catlow, C.R.A. (1993) In *Proc. 9th International Zeolite Conference*, Vol. 1 (eds R. Von Ballmoos, J.B. Higgins and M.M.J. Treacy), Butterworth-Heinemann, Boston, pp. 703.

Bell, R.G., Lewis, D.W., Voigt, P., Freeman, C.M., Thomas, J.M. and Catlow, C.R.A. (1994) *Stud. Surf. Sci. Catal.*, **84**, 2075.

Car, R. and Parrinello, M. (1985) *Phys. Rev. Lett.* **55**, 2471.

Catlow, C.R.A. (1992) In *Modelling of Structure and Reactivity in Zeolites* (ed. C.R.A. Catlow), Academic Press, London.

Deem, M.W. and Newsam, J.M. (1992) *J. Am. Chem. Soc.*, **114**, 7189.
den Ouden, C.J.J., Jackson, R.A., Catlow, C.R.A. and Post, M.F.M. (1990) *J. Phys. Chem.*, **94**, 5286.
Freeman, C.M., Catlow, C.R.A., Thomas, J.M. and Brode, S. (1991) *Chem. Phys. Lett.*, **186**, 137.
Freeman, C.M., Levine, S.M., Newsam, J.M., Sauer, J., Tomlinson, S.M., Brickmann, J. and Bell, R.G. (1992) In *Modelling of Structure and Reactivity in Zeolites* (ed. C.R.A. Catlow), Academic Press, London.
Gale, J.D. and Henson, N.J. (1994) *J. Chem. Soc. Faraday Trans.*, **90**, 3175.
Kapusta, B. and Guillopé, M. (1993) *Phys. Earth. Planet. Int.*, **75**, 205.
Kaszkur, Z.A., Jones, R.H., Waller, D., Catlow, C.R.A. and Thomas, J.M. (1993) *J. Phys. Chem.*, **97**, 426.
Kaszkur, Z.A., Jones, R.H., Bell, R.G., Catlow, C.R.A. and Thomas, J.M. (1996) *Mol. Phys.* (in press).
Lewis, D.W., Freeman, C.M. and Catlow, C.R.A. (1995) *J. Phys. Chem.*, **99**, 11194.
Matsui, M. (1989) *J. Chem. Phys.*, **91**, 489.
Matsui, M. and Price, G.D. (1991) *Nature*, **351**, 735.
Matsui, M. and Price, G.D. (1992) *Phys. Chem. Min.*, **18**, 365.
Matsui, M., Price, G.D. and Patel, A. (1994) *Geophys. Res. Lett.*, **21**, 1659.
Meier, W.M. and Olson, D.H. (1992) *Atlas of Zeolite Structure Types*, 3rd edn, Butterworth-Heinemann, London.
Meier, W.M. and Villiger, H. (1969) *Z. Kristallogr.*, **129**, 411.
Newsam, J.M. (1992) In *Solid State Chemistry: Compounds* (eds A.K. Cheetham and P. Day), Oxford University Press, Oxford.
Newsam, J.M., Freeman, C.M., Gorman, A.M. and Vessal, B. (1996) *Chem. Commun.*, **1945**.
Olson, D.H., Kokotailo, G.T., Lawton, S.L. and Meier, W.M. (1981) *J. Phys. Chem.*, **85**, 2238.
Price, G.D., Wood, I.E. and Akporiaye, D.E. (1992) In *Modelling of Structure and Reactivity in Zeolites* (ed. Catlow, C.R.A.), Academic Press, London.
Sankar, G., Bell, R.G., Thomas, J.M., Anderson, M.W., Wright, P.A. and Rocha, J. (1996) *J. Phys. Chem.*, **100**, 449.
Shannon, M.D., Casci, J.L., Cox, P.A. and Andrews, A. (1991) *Nature*, **353**, 417.
Shubin, A.A., Catlow, C.R.A., Thomas, J.M. and Zamaraev, K.I. (1994) *Proc. Roy. Soc. Lond. A*, **446**, 411.
Smith, J.V. (1988) *Chem. Rev.*, **88**, 149.
Stevens, A.P. and Cox, P.A. (1995) *J. Chem. Soc. Chem. Commun.*, 343.
Treacy, M.M.J., Newsam, J.M. and Deem, M.W. (1991), *Proc. R. Soc. Lond. A*, **433**, 499.
van Koningsveld, H., Jansen, J.C. and van Bekkum, H. (1990) *Zeolites*, **10**, 235.
Vočadlo, L., Patel, A. and Price, G.D. (1995) *Min. Mag.*, **59**, 597.
Weigel, S.J., Gabriel, J.C., Puebla, E.G., Bravo, A.M., Henson, N.J., Bull, L.M. and Cheetham, A.K. (1996) *J. Am. Chem. Soc.*, **118**, 2427.
Weitkamp, J., Karge, H.G., Pfeifer, H. and Hölderich, W. (eds) (1994) *Zeolites and Related Microporous Materials: State of the Art 1994*, Elsevier, Amsterdam. (*Stud. Surf. Sci. Catal.*, **84**.)
Wentzcovitch, R.M. (1991) *Phys. Rev. B*, **44**, 2358.
Wentzcovitch, R.M. and Martins, J.L. (1991) *Solid State Comm.*, **78**, 831.
Wentzcovitch, R.M., Martins, J.L. and Price, G.D. (1993) *Phys. Rev. Lett.*, **78**, 3947.
Wentzcovitch, R.M., Ross, N.L. and Price, G.D. (1995) *Phys. Earth. Planet Int.*, **90**, 101.
Wright, P.A., Natarajan, S., Thomas, J.M., Bell, R.G., Gai-Boyes, P.L., Jones, R.H. and Chen, J. (1992) *Angew. Chem. Int. Ed. Engl.*, **31**, 1472.

10

High-T_C Superconductors

N.L. Allan and W.C. Mackrodt

1 INTRODUCTION

Since the historic discovery of high-temperature superconductivity at $\sim$30 K in La_2CuO_4 (Bednorz and Müller, 1986) and subsequently in other oxides (for a review, see Singh and Edwards, 1993) and even doped C_{60} (Hebard *et al.*, 1991), there has been intense speculation as to possible coupling mechanisms. Rather less theoretical attention has been paid to aspects of these materials related to their solid-state chemistry, including structure, defect characteristics and redox behaviour. The purpose of this chapter is to show how computer modelling, and in particular atomistic simulation techniques of the type described earlier in this volume, which have been widely used to study conventional oxide ceramics, can be used to investigate normal-state properties of these high-T_C oxides that are clearly relevant to their superconducting behaviour. This new area has provided a unique opportunity to enlarge the application of these simulation techniques at a time when for some years there has been unprecedented industrial and technological interest in these materials.

The occurrence of high-temperature superconductivity seems to be restricted to a small number of systems and here we concentrate on cuprates, which encompass the vast majority of high-T_C materials. Even within the cuprates, high-T_C behaviour appears to occur only in a (small) subset of these compounds with the observation that minor changes in structure, composition, pressure or dopant content can alter the superconducting properties dramatically. This chapter describes a small number of examples to show how atomistic lattice simulations, working within the relatively simple framework of an ionic model, can nevertheless be used to rationalize many of the puzzling features of these materials. It is worth emphasizing here that superconductivity in these materials is observed only for a small range of compositions and the changes in the total energy at the superconducting transition are so small that it is reasonable to assume that many of the normal-state properties must be essentially unchanged in the superconducting state.

Defects play a crucial role in all the known high-T_C materials. The charge carriers are extrinsically controlled, either by doping or by oxidation/reduction

COMPUTER MODELLING IN INORGANIC CRYSTALLOGRAPHY
ISBN 0-12-164135-X

so that these carriers, at least as unpaired entities, can be considered as point defects and treated computationally as for many other oxides for which defect energies can often be estimated to an accuracy comparable with experiment. Important defect properties include the nature, location (and migration) of oxygen vacancies and interstitial oxygen, the nature of the electronic defects, i.e. the charge carriers, and their stability with respect to lattice defects. These defects are a key to understanding why La_2CuO_4 gives rise to a p-type superconductor when doped with BaO or treated with O_2 at high pressure, whereas Nd_2CuO_4 does not, but forms an n-type superconductor on doping with CeO_2.

An exhaustive survey of the computer modelling of high-temperature superconductors is not possible in this chapter, so only a small number of specific examples are presented here. These involve well-characterized materials to illustrate the diversity of information that can be obtained using these techniques, with some emphasis, as is suitable for this volume, on structural aspects. The possible consequences of these results for high-temperature superconductivity are discussed. We have concentrated on static simulations of perfect and defective lattices, and have not been able to include a discussion of lattice dynamics, molecular dynamics or Monte Carlo studies of aspects of these oxides. Attention is focused on the La–Cu–O, Nd–Cu–O, Y–Ba–Cu–O and Sr–Cu–O systems, for, again, brevity has not permitted a discussion of modelling studies of the families of Bi, Tl or Hg cuprate superconductors (e.g. Islam and Winch, 1995) or the high-T_C bismuthates which do not contain Cu (e.g. Zhang and Catlow, 1991).

The format of this chapter is as follows. First we compare the structures and defect chemistry of La_2CuO_4 and Nd_2CuO_4. Oxygen incorporation in $YBa_2Cu_3O_6$ is then considered in detail, concentrating on anion disorder and the consequences of having two types of Cu atom in very different crystallographic environments. Calculations of the cation distribution in $RE_2MCu_2O_6$ (RE = La,Nd; M = Sr,Ca) are then surveyed and related to why superconductivity is apparently found only in doped systems with RE = La and M = Ca. We conclude with a discussion of Sr–Cu–O superconductors, including the incorporation of fluorine, and the possible reasons why $SrCuO_2$ appears to be unique in giving rise to both p- and n-type superconducting materials.

2 THEORETICAL METHODS

The simulations reported here of perfect (bulk) and defective lattices are described in some detail in Chapters 3 and 7 of this volume. The methods have been applied to an ever increasing range of ceramic materials over the past 15 years (e.g. Catlow and Mackrodt, 1982; Cormack 1993). The same ionic model is used throughout wherein charges are assigned to ions according to the conventional valence rules, i.e. 2+ for copper, calcium and strontium,

2− for lattice oxygen and 3+ for lanthanum and neodymium. A consistent set of two-body electron-gas potentials (e.g. Allan and Mackrodt, 1994a) are used to specify the non-Coulombic interactions and the Dick–Overhauser shell model is included, in the usual way, to allow for the effects of electronic polarization. By the term 'consistent set' we mean a set of potentials for the *binary oxides*, La_2O_3, Nd_2O_3, CuO, CaO, BaO, etc., all of which use the *same* electron-gas potential for O^{2-}/O^{2-}. In each case, the electron-gas M/O^{2-} potential, V, was modified, by introducing a constant, R_0, such that the new potential, $\tilde{V}$, takes the form

$$\tilde{V}(R) = V(R - R_0) \tag{1}$$

R_0 is determined by fitting to the experimental lattice parameter. Shell charges and spring constants were derived for the binary oxides by fitting to dielectric and elastic constants. All the binary potentials and shell parameters were transferred *unchanged* for the ternary and quaternary cuprates.

Despite its simplicity, one advantage of the ionic model is the formal identification of valence band holes as Cu^{3+} or O^- and defect electrons as Cu^+, both of which play a crucial role in p- and n-type superconductivity, respectively. Short-range potentials are assumed to be independent of ion charge for cations, while for O^- (oxygen holes) electron-gas potentials were generated using an O^- electron density. For cations the ionization energy/electron affinity terms which contribute to the formation energies of these defects are taken as the free ion values, and the ionization potential of O^{2-} is taken from previous work (see Allan and Mackrodt, 1988; Mackrodt and Stewart, 1979), with the same value assumed for all the compounds considered here. Lattice contributions to the formation energy of electronic defects were calculated as described below for point lattice defects.

Static simulations of perfect lattices give the crystal structure and the corresponding lattice energy of the low-temperature phase. In the athermal limit (i.e. at 0 K and in the absence of lattice vibrations) the lattice structure is determined by the condition that it is in mechanical equilibrium, i.e.

$$\partial E/\partial \mathbf{X}_i = 0 \tag{2}$$

where E is the (static) internal energy and the $\{\mathbf{X}_i\}$ the variables that define the structure. These are the three lattice vectors, the positions of the basis atoms in the unit cell and, for calculations based on the shell model, the corresponding shell displacements, which represent the polarization of ions not at a centre of inversion symmetry.

The treatment of the defective lattice follows the customary two-region approach (Catlow and Mackrodt, 1982; this volume Chapter 7) in which the total energy of the defective system is minimized by variation of the nuclear positions (and shell displacements) around the defect. The crystal is partitioned into an inner region, immediately surrounding the defect where the relaxation is assumed to be greatest, and a less perturbed outer region. In the inner region the elastic equations for the force are solved explicitly to determine the relaxations; in the outer region these are estimated using the Mott–Littleton approximation (Mott and Littleton, 1938).

3 CRYSTAL STRUCTURES

Elements of the perovskite structure, which provide the putative superconducting two-dimensional CuO_2 planes, appear to be essential for high-temperature superconductivity in cuprate systems. Copper ions in these planes can be six-coordinate (La_2CuO_4, $Bi_2Sr_2CuO_6$), five-coordinate ($YBa_2Cu_3O_7$) or four-coordinate (Nd_2CuO_4). The crystal structures are shown in Fig. 10.1. The highest T_C's appear to be associated with a double layer of CuO_5 pyramids in which the Cu is five-coordinate, a structural unit found in $YBa_2Cu_3O_{7-\delta}$ (and, e.g. in $Tl_2Ba_2Ca_2Cu_3O_{10}$, $Bi_2Sr_2CaCu_2O_8$ and $Bi_2Sr_2Ca_2Cu_3O_{10}$) but absent in La_2CuO_4 and Nd_2CuO_4.

Accurate computed structures are a prerequisite for any theoretical approach to high-T_C oxides. We have tested our potentials for a wide range of cuprates, including examples such as Al_2CuO_4 which adopts a spinel structure (with tetrahedrally coordinated Cu) rather than a perovskite-related structure, and does not lead to a superconducting material. For La_2CuO_4 and Nd_2CuO_4 the following structures are obtained in order of increasing lattice energy:

La_2CuO_4: *orthorhombic K_2NiF_4* < tetragonal K_2NiF_4 < Nd_2CuO_4 < inverse spinel < spinel

Nd_2CuO_4: *Nd_2CuO_4* < tetragonal K_2NiF_4 < inverse spinel < spinel

The crystal structure in italics is that observed at low temperature. Clearly the size of the trivalent ion is crucial in determining the preferred structure ($r(La^{3+}) > r(Nd^{3+})$). Calculated lattice constants for these and other cuprates are compared with experiment in Table 10.1; overall, the agreement with experiment is very satisfactory. Even for $Y_2Cu_2O_5$, which has relatively low symmetry with distorted CuO_2 chains and Cu with an irregular five-coordination to oxygen, all three lattice constants are within 2.7% of the experimental values.

For La_2CuO_4, a comment on the difference between the equatorial (CuO_2 plane) and axial (c-axis) Cu–O bond lengths is worth adding. Pickett (1989) has emphasized that differences of this magnitude can result *simply* from the layered type of structure, as well as from Jahn–Teller distortion (see also Allan and Mackrodt, 1990a). There can be no Jahn–Teller effect in La_2NiO_4, an important material for comparison with La_2CuO_4. This has the same structure as tetragonal La_2CuO_4 with axial Ni–O and equatorial Ni–O bond lengths equal to 2.24 Å and 1.93 Å, respectively (Müller-Buschbaum and Lehmann, 1978).

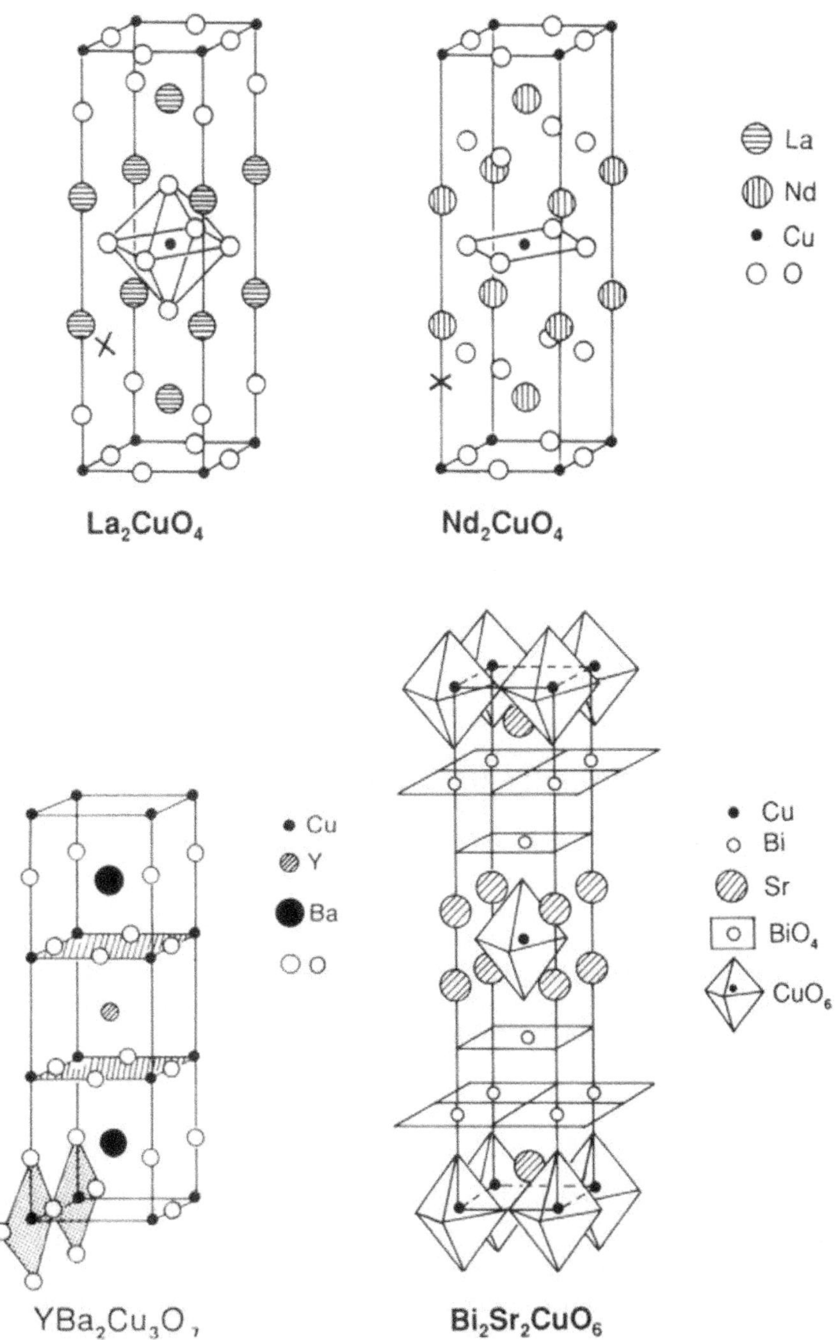

Figure 10.1 Crystal structures of tetragonal La_2CuO_4, Nd_2CuO_4, $YBa_2Cu_3O_7$ and $Bi_2Sr_2CuO_6$.

Table 10.1 Calculated and experimental lattice parameters (Å) of a range of ternary and quaternary cuprates. Experimental values are in parentheses.

Oxide	Structural type	Lattice constants (Å)		
		a_0	b_0	c_0
La_2CuO_4	Orthorhombic K_2NiF_4	5.427	5.352	12.994
		(5.406)	(5.370)	(13.150)[a]
La_2CuO_4	Tetragonal K_2NiF_4	3.783	–	13.050
		(3.778)	–	(13.093)[b]
Nd_2CuO_4	Nd_2CuO_4	3.898	–	11.985
		(3.939)	–	(12.147)[c]
Y_2CuO_4	Nd_2CuO_4	3.830	–	11.57
		(3.864)	–	(11.70)[d]
Al_2CuO_4	Spinel	8.025	–	–
		(8.086)	–	–
$Bi_2Sr_2CuO_6$		3.724	–	25.00
		(3.79)	–	(24.62)[e]
$Y_2Cu_2O_5$		10.51	3.450	12.56
		(10.805)	(3.496)	(12.461)[f]
$YBa_2Cu_3O_6$		3.900		11.82
		(3.854)		(11.82)[g]

[a] Grande *et al.* (1977), [b] Longo and Racah (1973), [c] Gopalkrishnan *et al.* (1989), [d] Okada *et al.* (1990), [e] Torardi *et al.* (1988), [f] Aride and Flandrois (1989), [g] Cava *et al.* (1990a).

4 DEFECT CHEMISTRY OF La_2CuO_4 AND Nd_2CuO_4

La_2CuO_4 doped with Ca, Sr or Ba, and the oxygen-rich material La_2CuO_{4+X} are p-type superconductors. Ce-doped Nd_2CuO_4 exhibits n-type superconductivity. La_2CuO_4 does not appear to give rise to n-type superconductors, neither has p-type superconductivity been observed in Nd_2CuO_4. Besides the implications for possible high-T_C coupling mechanisms, this striking difference in behaviour between two materials with closely related structures provides an important test for simulation methods.

4.1 Lattice defects

We turn first to lattice defects. For both systems the anion Frenkel energy (1.09 eV and 1.46 eV for tetragonal La_2CuO_4 and Nd_2CuO_4, respectively) is lower than the Schottky energy (per defect). These magnitudes suggest only a small degree of intrinsic disorder in the absence of oxygen. In La_2CuO_4 the lowest energy interstitial site is calculated (Zhang *et al.*, 1990; Allan and Mackrodt, 1994a) to be (0,0.5,~0.25) (see Fig. 10.1) in which the interstitial is eightfold coordinated as suggested by the neutron diffraction studies of Chaillout *et al.*

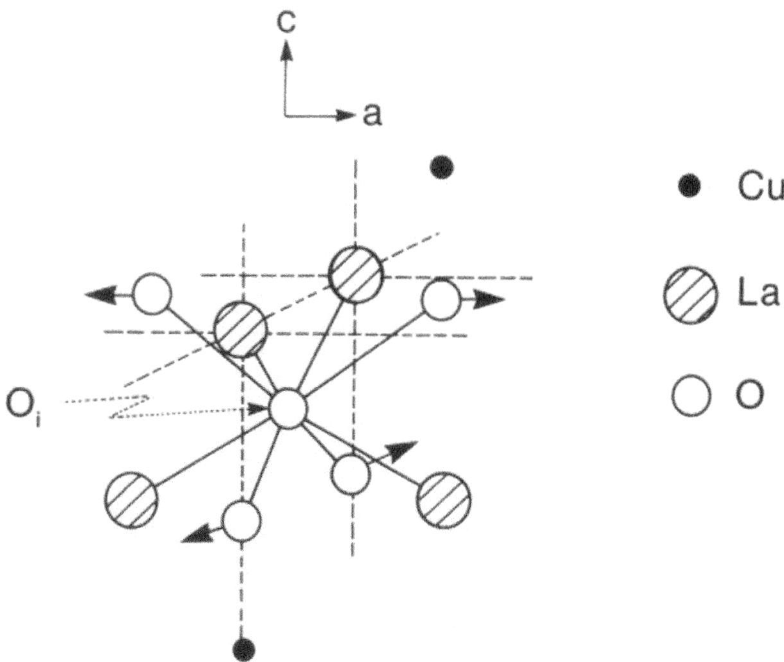

Figure 10.2 Calculated local environment surrounding a doubly charged interstitial in La_2CuO_4 . The calculated interionic separations are La–O ≈ 2.3 Å, O–O ≈ 2.7 Å. (Reproduced from Allan and Mackrodt (1994a) with permission of Gordon and Breach Publishers.)

(1989), and this site is also that occupied by excess oxygen in La_2NiO_{4+X} (Jorgensen *et al.*, 1989). Calculations indicate that in La_2CuO_4 the four nearest oxygens are displaced about 0.5 Å from their normal lattice sites away from the doubly charged interstitial (Fig. 10.2), a local relaxation close to that reported for La_2NiO_{4+X} (Jorgensen *et al.*, 1989). However, the neutron diffraction studies of La_2CuO_{4+X} (Chaillout *et al.*, 1989) indicate one short interstitial (O)–lattice (O) distance of <1.8 Å for which we are unable to account.

In Nd_2CuO_{4+X} the lowest energy interstitial position (Fig. 10.1) is (0,0,∼0.25) (confirmed in the EXAFS study of Ghigna *et al.*, 1995) so that the favoured interstitial sites in La_2CuO_4 and Nd_2CuO_4 are those normally occupied by the oxide ion in the other structure. The formation of doubly charged oxygen interstitials compensated by holes according to*

$$La_2CuO_4 + (x/2)O_2 \rightarrow La_2CuO_{4+X}$$

$$\tfrac{1}{2}O_2(g) \longrightarrow O_i'' + 2h^{\bullet}$$

*As is common when dealing with ceramics, we use the Kröger–Vink notation throughout this chapter to describe lattice and electronic defects. A summary of this notation is given in the appendix.

is consistent with the occurrence of superconductivity in La_2CuO_{4+X} since holes are thought to be a necessary prerequisite for high-temperature superconductivity in the La–Cu–O system. In addition, the maximum value of T_C occurs for $X \approx 0.15$–0.2 in $La_{2-X}Sr_XCuO_4$ and for $X \approx 0.08$–0.1 in La_2CuO_{4+X}, which is entirely consistent with the proposed mode of oxygen incorporation since oxygen excess leads to twice the hole concentration of an equivalent doping with Sr. Nd_2CuO_{4+X} is not superconducting, in keeping with this type of structure exhibiting only n-type superconductivity.

Oxygen vacancies have been shown to play a major role in determining T_C, since they introduce local scattering potentials, perturb the phonon modes and may disrupt the superconducting CuO_2 planes. In La_2CuO_4, oxygen vacancies at equatorial sites in the CuO_2 planes have lower formation energies than those at axial sites in the LaO planes, although the equatorial Cu–O bond is shorter, and hence stronger, than the axial bond. This is due entirely to differences in lattice relaxation at the two sites: the energy needed to form an unrelaxed axial oxygen vacancy is actually slightly smaller than that for an equatorial vacancy. At low temperatures, oxygen vacancies will be confined to equatorial sites in agreement with experiment, which is a possible explanation for the sensitivity of T_C to the presence of oxygen vacancies (Tarascon *et al.*, 1987).

4.2 Electronic defects and redox energies

The nature and energetics of electronic defects are major factors which control the properties of high-temperature superconductors. Atomistic simulation techniques allow an estimation of the formation energies of these defects. Within the framework of an ionic model, valence band holes are described in localized terms as Cu^{3+} (h_{Cu}) or O^- (h_O) and defect electrons as Cu^+ (e_{Cu}). The formation energy of these defects involves a contribution from the ionization potential/electron affinity of the appropriate ion in addition to a lattice energy term (including the effects of relaxation) and a band contribution, in the case of delocalized carriers (large polarons).

Defect energies in which both cores and shells are relaxed are often referred to as 'thermal' energies in that they correspond to lattices in thermal equilibrium. The corresponding energies in which only the shells are allowed to relax are referred to as 'optical' energies. In the present context of electronic defects in oxides, the difference between 'optical' and 'thermal' energies is important because it permits a quantitative distinction to be made between descriptions of free carriers (either holes or electrons) in the large ($^lh_{Cu}$, ^lh_O, $^le_{Cu}$) and small ($^sh_{Cu}$, ^sh_O, $^se_{Cu}$) polaron limits (Catlow *et al.*, 1977). Full details of all the relevant terms are given in our earlier work (Allan and Mackrodt, 1988, 1990b; Allan *et al.*, 1989). The concept of the polaron involves the coupling of the free carrier to the oscillating electric field of a vibrating polar lattice and so is essentially dynamic in nature. The justification of static calculations is that a

Table 10.2 Calculated electronic defect properties of tetragonal La_2CuO_4 and Nd_2CuO_4. ${}^{l}h_{Cu}$, for example, denotes a *l*arge polaron *h*ole on *Cu*.

	Energy			
	La_2CuO_4		Nd_2CuO_4	
Formation energy of valence band holes	${}^{l}h_{Cu}$	1.5	${}^{l}h_{O}$	3.1
	${}^{s}h_{Cu}$	1.7	${}^{s}h_{O}$	4.1
	${}^{l}h_{O}$	3.0	${}^{s}h_{Cu}$	4.2
Difference in energy between lowest energy holes	0.2		1.0	
Defect electron formation energy	${}^{s}e_{Cu}$	2.7	${}^{s}e_{Cu}$	−0.9

static description (the zero-frequency limit of the dynamic case) contains all the essential physics of the case (Anderson, 1975; Chakraverty, 1981).

Table 10.2 collects together the calculated hole and defect electron formation energies for tetragonal La_2CuO_4 and Nd_2CuO_4. Our estimates of hole energies, and from these the essential character of the lowest energy holes, indicate a strong dependence on crystal structure and/or Cu coordination. Our ionic model predicts Cu 3*d* holes in La_2CuO_4 (six-coordinate Cu) and O 2*p* holes in Nd_2CuO_4, with the latter considerably higher in energy than the former. Defect electrons have much smaller formation energies in Nd_2CuO_4 than La_2CuO_4. These differences have a marked effect on the redox properties of these two systems. It is important to emphasize that the relatively crude estimates of hole (and defect) electron energies which are obtained from atomistic simulations cannot serve the role of electronic structure calculations for these materials. The differences in the calculated energies of Cu and O holes are small, bearing in mind the approximations made in the calculations. Nevertheless, there is a clear variation with Cu coordination, which suggests an increase in the O character of holes with decreasing Cu coordination. Figure 10.3 collects together estimates of hole and defect electron energies for the wide range of cuprates we have considered in previous work, which serves to emphasize further the importance of the copper coordination to oxygen.

Superconductivity is intimately connected with the energetics of the redox reactions which control the non-stoichiometry in the presence of oxygen and the equilibrium between lattice and electronic defects. Using the Kröger–Vink notation, the most important of these are:

$$\tfrac{1}{2}O_2(g) \longrightarrow O_i'' + 2h^{\bullet} \tag{3}$$

$$2M_M^x + Cu_{Cu}^x + 2O_2(g) \longrightarrow 2V_M''' + V_{Cu}'' + 8h^{\bullet} + M_2CuO_4 \tag{4}$$

$$O_O^X \longrightarrow \tfrac{1}{2}O_2(g) + V_O^{\bullet\bullet} + 2e' \tag{5}$$

$$V_O^{\bullet\bullet} + \tfrac{1}{2}O_2(g) \longrightarrow O_O^x + 2h^{\bullet} \tag{6}$$

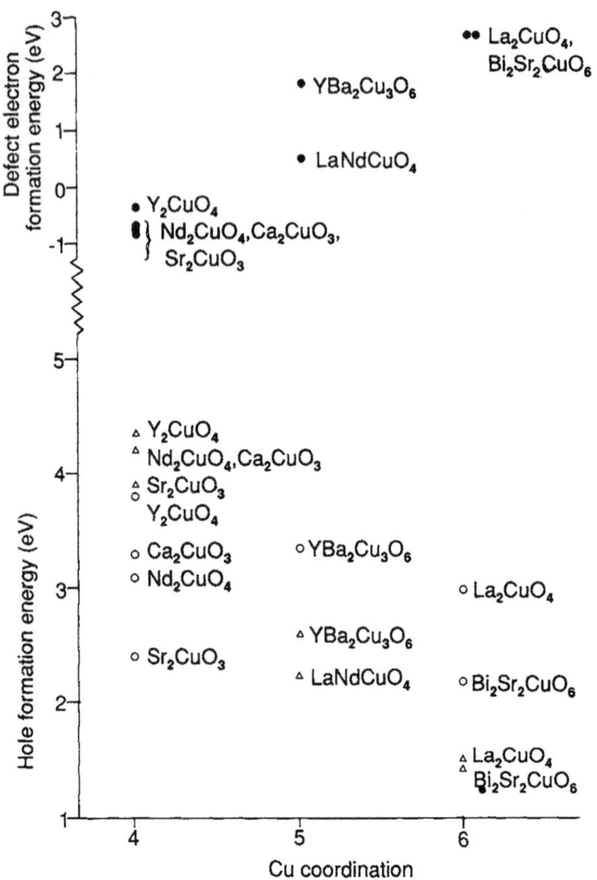

Figure 10.3 Estimated Cu hole (△), O hole (○) and defect electron formation energies (eV) (●) for a range of cuprates containing CuO_2 planes and/or CuO_2 chains versus the Cu coordination to oxygen. (Reproduced from Allan *et al.* (1992) with permission from Gordon and Breach Publishers.

$$O_i'' \longrightarrow \tfrac{1}{2}O_2(g) + 2e' \tag{7}$$

$$2V_M''' + V_{Cu}'' + 4O_O^x \longrightarrow 2O_2(g) + 8e' \tag{8}$$

$$O_i'' \longrightarrow O_i' + e' \tag{9}$$

with energies E_1–E_7, respectively. These are listed in Table 10.3. The figures in brackets are estimates of E_4, E_5 and E_6 at 1000 K, which is approximately the synthesis temperature of these materials. These have been calculated allowing for the significant contribution to the total free energies of these reactions from the entropic contribution of oxygen gas, and assuming all the relevant defect energies are independent of temperature. These values predict that La_2CuO_4 and Nd_2CuO_4 will be oxygen rich rather than oxygen or cation deficient. Since

Table 10.3 Calculated redox energies (eV) for tetragonal La_2CuO_4 and Nd_2CuO_4. Details of reactions 1–7 are given in the text. Energies all refer to the athermal limit except for those in brackets, which refer to 1000 K.

	E_1	E_2	E_3	E_4	E_5	E_6	E_7
La_2CuO_4	0.7	4.5	9.9	−1.5(−0.3)	7.7 (6.5)	29.1 (24.3)	5.8
Nd_2CuO_4	3.1	18.6	4.2	0.2 (1.4)	1.1(−0.1)	−1.0(−5.8)	1.4

interstitial ionization (reaction (9)) is endothermic, doubly charged interstitials in both materials are predicted, together with a $p(O_2)$ power dependence of 1/6 for $[h^\bullet]$, which is in agreement with the high-temperature conductivity measurements of Su *et al.* (1990).

We now return to the question for which, to date, no fundamental explanation in terms of electronic structure considerations appears to have been forthcoming, namely why La_2CuO_4 apparently leads only to p-type 'hole' superconductors and Nd_2CuO_4 only to n-type 'electron' superconductors, despite the similarity in their crystal structures. A tentative explanation lies in their redox equilibria. Substitution of RE^{3+} (RE = La,Nd) by divalent ions (M^{2+}) leads in principle to charge compensation by *either* oxygen vacancies (forming $RE_{2-X}M_XCuO_{4-X}$), *or* holes ($RE_{2-X}M_XCuO_4$), and substitution by 4+ ions (N^{4+}) to cation vacancies, oxygen interstitials ($RE_{2-X}N_XCuO_{4+(X/2)}$) or defect electrons ($RE_{2-X}N_XCu^{2+}_{2-X}Cu^{+}_{X}O_4$). The criteria for electronic compensation by either holes or defect electrons are determined by the values of E_4–E_6 and so are the relative formation energies of lattice and electronic defects.

It is clear from Table 10.3 that oxygen vacancies in La_2CuO_4 are unstable with respect to the formation of holes ($E_3 < 0$) in oxygen-deficient La_2CuO_4, whereas cation vacancies are calculated to be stable with respect to defect electrons in the corresponding cation deficient system. From this it can be shown that the solution of CaO, SrO and BaO in La_2CuO_4 leads to systems in which the impurity defects M'_{La} are compensated by holes. On the other hand, holes are unstable with respect to oxygen vacancies in Nd_2CuO_4 and electrons stable with respect to cation vacancies, so that divalent impurities are compensated by oxygen vacancies and four-valent dopants (or substitution of O^{2-} by F^-) by defect electrons above 1000 K. In summary, La_2CuO_4 is readily oxidized, Nd_2CuO_4 readily reduced. Thus, purely on stability grounds La_2CuO_4 would be expected to lead to hole superconductivity and Nd_2CuO_4 to high-T_C electron superconductivity. Allan *et al.* (1992) have discussed the calculated redox characteristics of a wider range of cuprates, including examples such as $LaNdCuO_4$ containing five-coordinate Cu in which, as with La_2CuO_4, holes are stable with respect to oxygen vacancies.

4.3 Solution energies

The doping, and in particular, the mode of solution, of dopant ions are of crucial importance in these materials due to the extrinsic control of the charge carriers, which has already been stressed. As an example, we consider +1 and +2 cation dopants in La_2CuO_4. Table 10.4 contains calculated solution energies for the two possible substitution reactions, bearing in mind that holes are the preferred compensating defect:

$$XO + La_{La}^{X} + \tfrac{1}{4}O_2(g) \longrightarrow X'_{La} + h^{\bullet} + \tfrac{3}{2}O_O^{X} \qquad E_{La}$$

$$\tfrac{1}{2}X_2O + La_{La}^{X} + \tfrac{1}{2}O_2(g) \longrightarrow X''_{La} + 2h^{\bullet} + La_{La}^{X} + \tfrac{3}{2}O_O^{X} \qquad E_{La}$$

or

$$XO + Cu_{Cu}^{X} \longrightarrow X_{Cu} + Cu_{Cu}^{X} + O_O^{X} \qquad E_{Cu}$$

$$\tfrac{1}{2}X_2O + Cu_{Cu}^{X} + \tfrac{1}{4}O_2(g) \longrightarrow X'_{Cu} + h^{\bullet} + Cu_{Cu}^{X} + O_O^{X} \qquad E_{Cu}$$

The site preference of a given dopant is governed by the relative values of E_{Cu} and E_{La}. The results show clearly that Ca, Sr and Ba will substitute for La^{3+}. The holes formed couple in some manner to form superconducting Cooper pairs. Mg, on the other hand, is predicted to replace Cu, thereby disrupting the superconducting CuO_2 planes *without* the formation of holes, so that magnesium doping would not be expected to give rise to a superconducting material, in agreement with experiment. It is interesting that, although many individual energy terms contribute to the overall solubility, the site preference is consistent with very simple arguments based on the ionic radii of La^{3+}, Cu^{2+} and the dopant ion. The values in Table 10.4 also show that addition of Na_2O (and by inference K_2O) to La_2CuO_4 is predicted to lead to the substitution of La by Na (K) with the formation of holes and thus to high-T_C behaviour. Li, however, is predicted to substitute for Cu and so have a deleterious effect on the superconductivity. Since the calculation of these solution energies (Allan and Mackrodt, 1988), the predictions have been confirmed experimentally on several occasions (see Cheong *et al.* (1989) for a summary of experimental results).

Turning briefly to anion substitution, it is straightforward within our model to rationalize why fluoridation of Nd_2CuO_4 leads to F^- substitution of O^{2-} ($F_O^{\bullet}$) and the formation of defect electrons, giving rise to n-type superconductivity

Table 10.4 Calculated solution energies (eV) in the athermal limit of Group 1 and Group 2 oxides in tetragonal La_2CuO_4. E_{La} denotes the mode of solution involving substitution for La and E_{Cu} substitution for Cu.

	MgO	CaO	SrO	BaO	Li_2O	Na_2O
E_{La}	2.4	0.3	−0.3	−0.1	0.94	−0.7
E_{Cu}	0.8	4.0	7.3	11.0	0.85	3.7

(James *et al.*, 1989). In contrast, calculations (Allan and Mackrodt, 1994a) indicate that direct fluorination of La_2CuO_4 leads to the incorporation of interstitial fluoride (F_i') (cf. excess oxygen), to the formation of holes and to p-type superconductivity, as is found experimentally (e.g. Chevalier *et al.*, 1990).

5 Y–Ba–Cu–O SYSTEMS: $YBa_2Cu_3O_{7-x}$

The Y^{3+} ion is considerably smaller than La^{3+}, Pr^{3+} or Nd^{3+} and the compounds formed in the Y–Ba–Cu–O system differ from those formed by La–Cu–O and Nd–Cu–O. For instance, Y_2CuO_4, which has the Nd_2CuO_4 structure, is only prepared at high pressure (Okada *et al.*, 1990). In this section we consider the calculated defect chemistry of $YBa_2Cu_3O_6$, which can be regarded as the 'parent' oxide of $YBa_2Cu_3O_{7-\delta}$, the first system found to possess a T_C (≈ 90 K) at liquid nitrogen temperatures (Wu *et al.*, 1987). Formally, as is clear from Fig. 10.4, $YBa_2Cu_3O_7$ is related to semiconducting $YBa_2Cu_3O_6$ by the introduction of oxygen into the vacant O(4) site. $YBa_2Cu_3O_6$ and $YBa_2Cu_3O_7$ contain the two-dimensional CuO_2 planes present

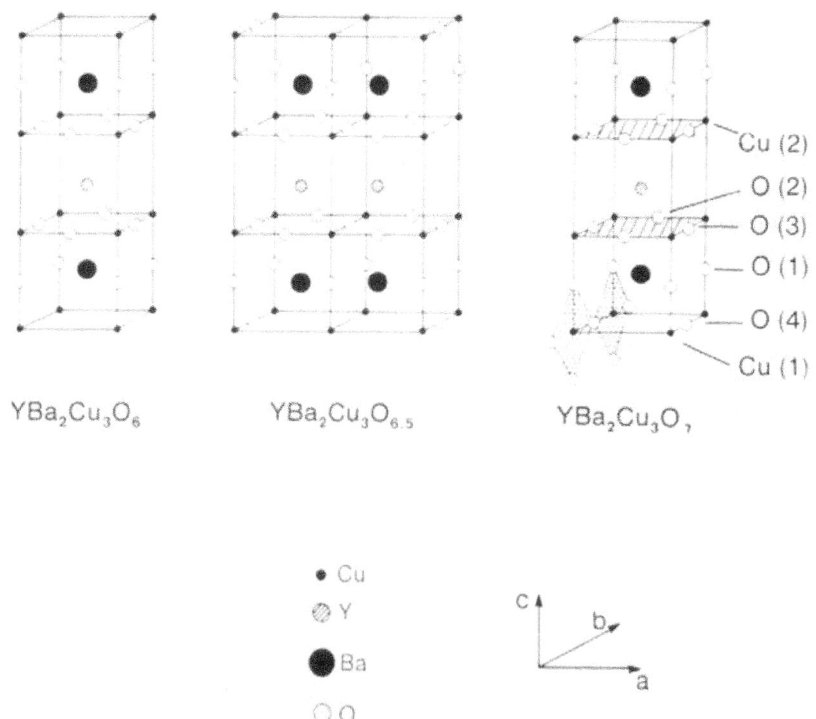

Figure 10.4 Crystal structures of $YBa_2Cu_3O_6$, $YBa_2Cu_3O_{6.5}$ and $YBa_2Cu_3O_7$ showing the atom labelling used here. In $YBa_2Cu_3O_7$ the two-dimensional CuO_2 planes are shown hatched and the CuO_2 chains along the *b*-axis are shaded.

in La_2CuO_4 and Nd_2CuO_4. The Cu atoms (Cu(2)) in these planes are fivefold coordinate to O and there is a double layer of CuO_5 pyramids, the unit present in those cuprates with the highest T_C's. $YBa_2Cu_3O_7$ also contains CuO_2 chains (Fig. 10.4). In $YBa_2Cu_3O_6$, since the O(4) site is vacant, Cu atoms at the Cu(1) position in the so-called *basal plane* are only twofold linearly coordinated to oxygen.

The simulation of $YBa_2Cu_3O_6$ is less straightforward than for La_2CuO_4 and Nd_2CuO_4 since at first sight it is not clear how to treat the two types of Cu which have very different environments. The band structure calculations of, for example, Temmerman *et al.* (1988) suggest charges of +1 and +2 for copper at Cu(1) and Cu(2), respectively, i.e. that $YBa_2Cu_3O_6$ is formally a mixed-valence compound. Using the same set of potentials as that used previously for the ternary cuprates discussed earlier, the lattice energy of $YBa_2Cu_3O_6$, with charges of +1 and +2 for Cu(1) and Cu(2), respectively, is more than 6 eV lower than the value obtained for an equal assignment of charge to all the copper ions, which is consistent with the *ab initio* studies. The calculated value of a_0 is within 1.2% of experiment and that of c_0 2% smaller than that observed.

In view of the different charges and coordinations of Cu(1) and Cu(2) it is not unreasonable that the Cu(1)–O short-range potential should be somewhat different from the Cu(2)–O potential, suggesting a small modification of the Cu(1)–O potential similar to those used in the original generation of a consistent set of potentials for these mixed oxides (cf. equation (1)):

$$V_{Cu(1)-O}(R) = V_{Cu(2)-O}(R - R_0) \tag{10}$$

Using a small positive value of R_0, which makes the Cu(1)–O non-Coulombic interaction more repulsive than Cu(2)–O, it is possible to reproduce the experimental value of c_0, while keeping the value of a_0 unchanged (Table 10.1). Nevertheless, the qualitative conclusions concerning the defect chemistry given below are unchanged even if this small alteration is not made.

Charge carriers in $YBa_2Cu_3O_{6+X}$ are formed *not* by aliovalent substitution of Y^{3+} (cf. $La_{2-X}Sr_XCuO_4$ and $Nd_{2-X}Ce_XCuO_4$) but by oxygen incorporation (cf. La_2CuO_{4+X}). The variation of T_C with oxygen content X (Cava *et al.*, 1990a; Namgung and Irvine, 1990) shown in Fig. 10.5 is particularly intriguing. There is considerable variation between samples prepared by different techniques and, in particular, between samples prepared by quenching from high temperature and those prepared by low-temperature reduction of $YBa_2Cu_3O_7$. However, despite these differences it is generally accepted that high-T_C behaviour does not occur for $X < \sim 0.3$. In contrast, superconductivity has been observed at ~50 K in $Y_{1-Z}Ca_ZBa_2Cu_3O_6$ (McCarron *et al.*, 1989; Legros-Glédel *et al.*, 1991), in which the charge carriers are introduced by the substitution of Y^{3+} by Ca^{2+}, as in $La_{2-X}Ca_XCuO_4$. Figure 10.5 also shows the variation in T_C with X for $Y_{1-Z}Ca_ZBa_2Cu_3O_{6+X}$. The superconducting properties of $Yb_{1-Z}Ca_Z(Ba_{0.8}Sr_{0.2})_2Cu_3O_{6+X}$, reported by Wada *et al.* (1991), are very similar to those of $Y_{1-Z}Ca_ZBa_2Cu_3O_{6+X}$, in that these systems are superconducting even for very small X.

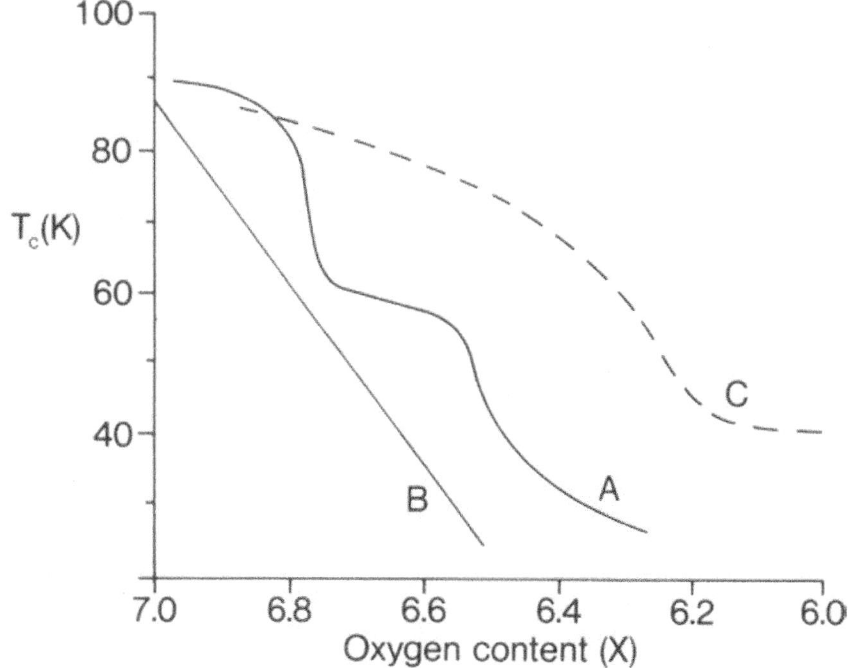

Figure 10.5 Schematic variation of T_c with oxygen content X for $YBa_2Cu_3O_{6+X}$ samples prepared by different techniques. (A: Cava *et al.*, 1990a, B: Namgung *et al.*, 1990) and for $Y_{0.7}Ca_{0.3}Ba_2Cu_3O_{6+X}$ (C: Legros-Glédel *et al.*, 1991). Reproduced from Allan *et al.* (1992) with permission of Gordon and Breach Publishers.

Calculated hole formation energies in $YBa_2Cu_3O_6$ (Fig. 10.3, see also Allan and Mackrodt (1993)) show that the lowest energy *free* carrier is a Cu hole in the (super)conducting CuO_2 planes. Oxygen holes in the CuO_2 planes are predicted to have similar energies, as are holes at the O(1) apical position ($\sim$2.8 eV), allowing for the uncertainties associated with terms that contribute to the total energy of formation. As with La_2CuO_4, calculated redox energies indicate that holes are stable with respect to the formation of oxygen vacancies and defect electrons are unstable with respect to cation vacancies and oxygen interstitials.

The calculated energy (-0.5 eV) for the incorporation of oxygen at the O(4) site:

$$YBa_2Cu_3O_6 + (x/2)O_2 \rightarrow YBa_2Cu_3O_{6+X}$$

$$\tfrac{1}{2}O_2(g) \longrightarrow O_i'' + 2h^{\bullet}$$

is lower than the corresponding value for La_2CuO_4 and compares with the experimental value of ~ -1.0 eV (Parks *et al.*, 1989; Verweij *et al.*, 1990). O(4) is the lowest energy interstitial position in agreement with experiment. This interstitial is *strongly* bound to two neighbouring Cu(1) holes (i.e. two Cu^{2+}

ions), and allowance for this binding has been made in the estimate of the energy of O_2 incorporation. All other configurations of the oxygen interstitial and the two holes are much greater in energy. This strong interstitial–hole association is not found in La_2CuO_4 (Allan and Mackrodt, 1994a) where, as we have seen, the interstitial oxygen is in a different location away from the CuO_2 planes (at (0,0.5,~0.25) and is stabilized by four nearest-neighbour La^{3+} cations.

Our calculations suggest that the absence of superconductivity in the $YBa_2Cu_3O_{6+X}$ system for $X < \sim 0.3$ can be accounted for by the predicted formation of holes which are trapped by negatively charged oxygen interstitials at the O(4) position in the basal plane, rather than free holes in the CuO_2 planes. This is in broad agreement with Goodenough's (1990) view. In $Y_{1-Z}Ca_ZBa_2Cu_3O_6$, on the other hand, the basal plane is devoid of both lattice and interstitial oxygen, so that holes introduced by Ca will be located, at least in part, in the CuO_2 planes leading to superconductivity. This would appear to be a plausible explanation for high-T_C behaviour in Ca-doped materials with oxygen content less than ~6.3. As in La_2CuO_4 (Allan and Mackrodt, 1988), calculated association energies between the Ca dopant and CuO_2 plane holes are very small, which contrasts markedly with the strong association of basal plane holes to added oxygen. The occurrence of two types of Cu atom clearly has very important implications for the superconducting behaviour of Y–Ba–Cu–O ceramics.

For completeness, it is worth noting briefly that it is now fairly well-established that the ordering of the oxygen atoms in the basal plane controls the concentration of holes in the CuO_2 planes and, as such, is the key to the variation in T_C with oxygen content (Goodenough, 1990; Goodenough and Manthiram, 1990). In the range $0 < X < 0.25$ half of the Cu^+ at the Cu(1) sites is oxidized to Cu^{2+}: every insertion of an oxygen atom results in the oxidation of its two neighbouring Cu^+ ions, which are now threefold coordinate, to Cu^{2+}. No holes are generated in, or transferred to, the Cu(2)–O (super)conducting planes. When X exceeds 0.25 the optimum ordering of the oxygen atoms in the basal plane is such that only *alternate* chains of *b*-axis sites are occupied (see Figs. 10.4 and 10.6 for $X = 0.5$ in which alternate chains are complete). At this stage, insertion of oxygen into the partially complete chains generates holes in the CuO_2 planes. Half the ions at Cu(1) sites remain as Cu^+, and fourfold-coordinate Cu atoms at Cu(1) sites are not oxidized beyond Cu^{2+}. T_C rises sharply. In $YBa_2Cu_3O_{6.5}$ this results in a formal hole density of 0.25 holes per CuO_2 unit in the CuO_2 planes, half that for $YBa_2Cu_3O_7$, and, consistent with this, T_C for $YBa_2Cu_3O_{6.5}$ is half that of $YBa_2Cu_3O_7$. When $0.5 < X < 0.75$ the empty chain sites are filled with oxygen and the remaining Cu^+ ions are oxidized to Cu^{2+}. The oxygens in the new *b*-axis chain order such that each Cu(1) is threefold coordinate; again no holes are generated in the CuO_2 planes. For $X > 0.75$, all the Cu(1) sites are Cu^{2+} and additional oxidation creates holes in the CuO_2 planes with a resultant increase in T_C since, as before, the Cu at the Cu(1) sites are only fourfold coordinate to oxygen and are not oxidized

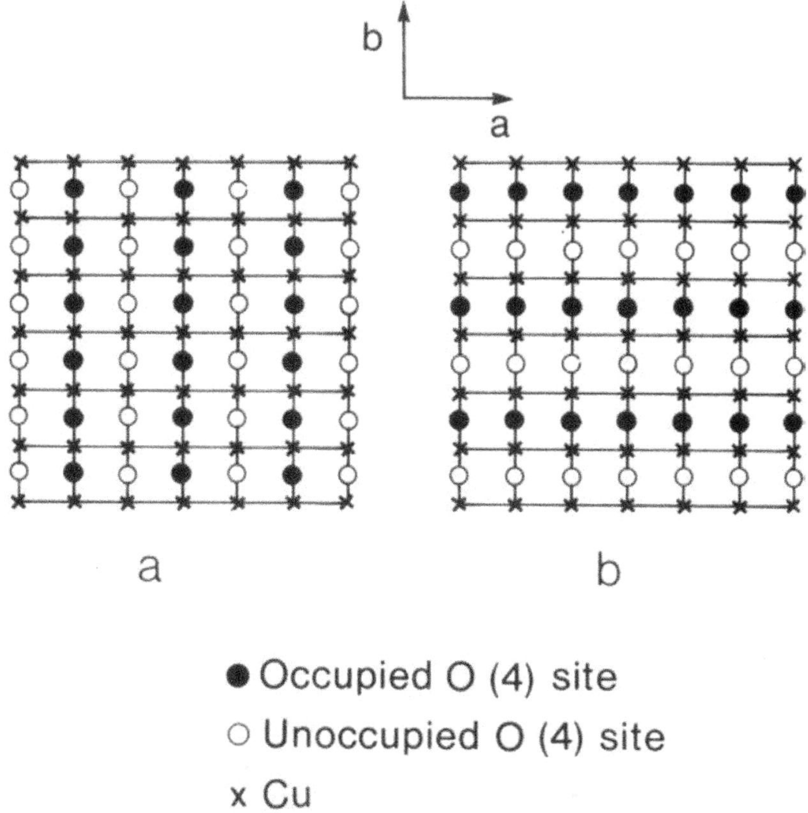

Figure 10.6 Possible basal plane oxygen orderings in $YBa_2Cu_3O_{6.5}$. Filled and open circles are occupied and unoccupied oxygen O(4) sites, respectively. Crosses denote Cu sites. (a) Alternate *b*-axis chains empty and so the Cu atoms are either four- or twofold coordinate to oxygen; (b) all basal Cu sites equivalent and all Cu atoms threefold coordinate. (Reproduced from Allan and Mackrodt (1993) with permission from JAI Press Ltd.)

beyond Cu^{2+}. A different oxygen ordering, arising from a variation in sample treatment, will obviously lead to a change in the distribution of holes between the (inactive) basal and the (super)conducting CuO_2 planes and so show a different variation in T_C with oxygen content.

Atomistic simulations of the type described here are strictly applicable only to insulators and semiconductors since the effects of long-range screening in the metallic state are not incorporated in the model. Nevertheless, useful studies of $YBa_2Cu_3O_{6.5}$, $YBa_2Cu_3O_7$ and other metallic compositions such as $YBa_2Cu_4O_8$ based on an ionic model have been reported (e.g. Islam 1990; Zhang and Catlow 1992a; Allan and Mackrodt, 1993; Islam 1994), as well as Monte Carlo studies of oxygen ordering in the basal plane (e.g. Poulsen *et al.*, 1991) and molecular dynamics studies of oxygen diffusion (e.g. Zhang and Catlow, 1992b).

6 $La_2CaCu_2O_6$, $La_2SrCu_2O_6$ AND RELATED COMPOUNDS

Systems with the highest T_C's ($\geqslant 80$ K), such as $YBa_2Cu_3O_7$ possess CuO_2 planes which are also part of a double layer of CuO_5 pyramids. Figure 10.7 shows that this structural unit is also present in $RE_2MCu_2O_6$ (RE = La, Nd, M = Ca,Sr). The distinct cation sites are also shown in this figure, in which A_1 cations are sandwiched between the CuO_2 planes, while the A_2 cations lie

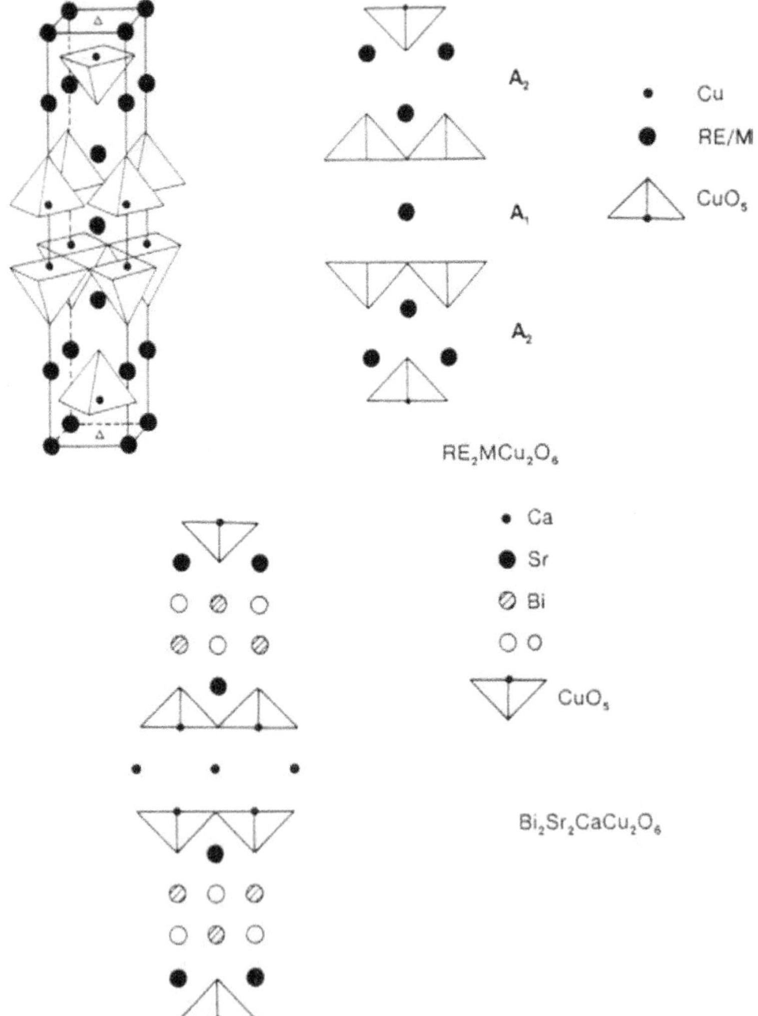

Figure 10.7 Crystal structures of $RE_2MCu_2O_6$ (RE = La,Nd: M = Ca,Sr) and $Bi_2Sr_2CaCu_2O_8$. Labelling of the distinct cation sites in the $RE_2MCu_2O_6$ structure is also shown. (Reproduced from Allan and Mackrodt (1993) with permission of JAI Press Ltd.)

between the double layers of CuO_5 pyramids. RE^{3+}/M^{2+} ions occupy both the A1 and A2 sites in $La_2SrCu_2O_6$. Compare the structure of $Bi_2Sr_2CaCu_2O_8$, also shown in Fig. 10.7, where Ca^{2+} and Sr^{2+} ions are located at A1 and A2 positions, respectively.

Many non-stoichiometric materials of the type $RE_{2-X}M_{1+X}Cu_2O_6$ and $RE_{2-X}M'XMCu_2O_6$ (cf. $La_{2-X}MXCuO_4$ have been reported and superconductivity observed in $La_{2+X}Ca_{1+X}Cu_2O_6$ (Ishii *et al.*, 1991) ($T_C \approx 40$ K, which is appreciably lower than for $YBa_2Cu_3O_7$, $Bi_2Sr_2CaCu_2O_8$ and Tl_2Ba_2Ca-Cu_2O_8) and $La_{1.6}Sr_{0.4}CaCu_2O_6$ ($T_C \sim 55$ K) (Cava *et al.*, 1990b,c). In both these compounds charge carriers are introduced by aliovalent substitution for RE^{3+} as in $La_{2-X}M_XCuO_4$ (e.g. M = Ca,Sr,Ba,Na). There appear to be no reports of high-T_C behaviour in the analogous Sr compound, $La_{2-X}M_XSrCu_2O_6$ (M = Ca,Sr) (e.g. Lightfoot *et al.*, 1990). Nd analogues such as $Nd_{1.9}Sr_{1.1}Cu_2O_6$ (Heyri and Larese, 1990) and $Nd_{1.7}Bi_{0.1}Sr_{0.9}Ca_{0.3}Cu_2O_6$ (Huang and Sleight, 1990) are apparently not superconducting materials. This behaviour, which shows an intriguing difference between the calcium compound and its strontium analogue, suggests that a study of the family of compounds $RE_2MCu_2O_6$ would be worthwhile (for a fuller account see Allan *et al.*, 1994).

With reference to Fig. 10.7, $RE_2MCu_2O_6$ compounds can possess three types of tetragonal structure depending on the distribution of RE and M. They can be written as: A_1 [1.0 Re; 0.0 M] A_2 [0.5 RE; 0.5 M] (3 structures); A_1 [0.5 RE; 0.5 M] A_2 [0.75 RE; 0.25 M] (2 structures); A_1 [0.0 RE; 1.0 M] A_2 [2.0 RE; 0.0 M] (1 structure). The possible cation stacking sequences in these structures, which we designate X, Y and Z respectively, are;

	X1	X2	X3	Y1	Y2	Z
A_2	RE	RE	M	RE	RE	RE
A_2	RE	M	RE	RE	RE	RE
- - - -	- - - -	- - - -	- - - -	- - - -	- - - -	- - - -
A_1	RE	RE	RE	M	M	M
- - - -	- - - -	- - - -	- - - -	- - - -	- - - -	- - - -
A_2	M	RE	RE	M	RE	RE
A_2	M	M	M	RE	M	RE
- - - -	- - - -	- - - -	- - - -	- - - -	- - - -	- - - -
A_1	RE	RE	RE	RE	RE	M
- - - -	- - - -	- - - -	- - - -	- - - -	- - - -	- - - -

where the horizontal lines (- - - -) denote the CuO_2 planes.

Atomistic simulation techniques provide a convenient means of examining the relative energetics of the different possible stackings. Table 10.5 collects together the calculated a_0 and c_0 parameters and lattice energies of the six structures for $La_2CaCu_2O_6$ and $La_2SrCu_2O_6$. For $La_2SrCu_2O_6$ the lowest energy structure is predicted to be the X1 structure; X2 is less stable by

Table 10.5 Calculated lattice parameters (Å) and lattice energies, E_L (eV), of $La_2CaCu_2O_6$ and $La_2SrCu_2O_6$. Experimental lattice parameters for $La_2SrCu_2O_6$ are also given. For each compound, ΔE_L is the difference in lattice energy (eV) between the structure listed and the structure lowest in energy.

		Lattice parameter (Å)		Energy (eV)	
Compound	Structure	a_0	c_0	E_L	ΔE_L
$La_2CaCu_2O_6$	X1	3.879	19.45	−252.06	0.07
	X2	3.883	19.21	−252.13	0.00
	X3	3.883	19.15	−251.98	0.15
	Y1	3.863	19.37	−252.51	0.62
	Y2	3.866	19.28	−252.09	0.04
	Z	3.853	19.22	−252.09	0.04
$La_2SrCu_2O_6$	X1	3.899	19.83	−250.47	0.00
	X2	3.900	19.67	−250.44	0.03
	X3	3.902	19.54	−250.11	0.36
	Y1	3.904	19.77	−249.75	0.72
	Y2	3.903	19.71	−250.20	0.27
	Z	3.906	19.70	−249.99	0.48
Expt.: Caignaert *et al.* (1990)		3.864	19.94		
Nguyen *et al.* (1980)		3.865	19.89		

0.03 eV. The calculated lattice parameters of both X1 and X2 structures are within ~1% of the measured values (Nguyen *et al.*, 1980; Caignaert *et al.*, 1990). In contrast, for $La_2CaCu_2O_6$ the X2 structure is lowest in energy, but the Y2 and Z structures are only 0.04 eV higher in energy, and all the structures bar Y1 are separated by only 0.15 eV. For comparison, experimental lattice constants of 3.825 Å (a_0) and 19.420 Å (c_0) have been reported for $La_{1.9}Ca_{1.1}Cu_2O_6$ (Fuertes *et al.*, 1990).

These results predict that in $La_2SrCu_2O_6$ the A_1 sites will be occupied only by La^{3+} cations, whereas in $La_2CaCu_2O_6$ both Ca^{2+} and La^{3+} ions will be found at these sites. It is worth noting that the relative magnitudes of the ionic radii of La^{3+}, Cu^{2+} and Sr^{2+} are

$$r(Sr^{2+}) > r(La^{3+}) > r(Ca^{2+})$$

with the larger difference in size being that between Sr^{2+} and La^{3+}. The diffraction studies of $La_2SrCu_2O_6$ by Caignaert *et al.* (1990) (which correct the earlier study of Nguyen *et al.* (1980)) and of $La_{1.9}Ca_{1.1}Cu_2O_6$ by Izumi *et al.* (1989) and Cava *et al.* (1990b) broadly confirm these conclusions in that in $La_{1.9}Ca_{1.1}Cu_2O_6$ the A_1 site is almost 90% occupied by Ca^{2+}.

An obvious extension of these studies is to the relative energies of possible structures of $RE_2MCu_2O_6$ where RE = La,Nd,Y and M = Ca,Sr,Ba. The X1 structure in which the RE^{3+} ions occupy A_1 sites is lowest in energy in every

Table 10.6 Calculated lattice energies, E_L (eV), of $La_2BaCu_2O_6$, $Nd_2CaCu_2O_6$, $Nd_2SrCu_2O_6$ and $Nd_2BaCu_2O_6$. ΔE_L is the energy difference (eV) between the given structure and the lowest energy structure for the given compound.

Structure	$La_2BaCu_2O_6$		$Nd_2CaCu_2O_6$		$Nd_2SrCu_2O_6$		$Nd_2BaCu_2O_6$	
	E_L	ΔE_L	E_L	ΔE_L	E_L	ΔE_L	E_L	ΔE_L
X1	−248.51	0.00	−258.27	0.00	−256.57	0.00	−254.49	0.00
X2	−248.23	0.28	−258.25	0.02	−256.39	0.18	−254.12	0.37
X3	−247.70	0.81	−257.92	0.35	−255.79	0.78	−253.40	1.09
Y1	−247.26	1.25	−257.28	0.99	−255.45	1.10	−252.56	1.84
Y2	−247.97	0.54	−258.01	0.26	−255.99	0.58	−253.66	0.83
Z	−247.72	0.79	−257.76	0.51	−255.59	0.98	−253.23	1.26

case except $La_2CaCu_2O_6$. From this, two trends are apparent: first, for a given RE^{3+} ion, the difference in energy between the X1 structure and the other structures decreases as $r(M^{2+})$ decreases; and second, for a given M^{2+} ion, the difference in energy between the X1 structure and the others increases as $r(RE^{3+})$ decreases. The results in Table 10.6, which list the calculated lattice energies of the possible structures for $La_2BaCu_2O_6$, $Nd_2SrCu_2O_6$ and $Nd_2CaCu_2O_6$, and those in Table 10.5 illustrate these conclusions. The energy differences are such that X, Y and Z structures are all accessible only at typical synthesis temperatures for $La_2CaCu_2O_6$.

Neutron diffraction results for $LaNdSrCu_2O_6$ have been reported by Grasmeder and Weller (1990), so that this compound provides a further test of our approach. The calculated lowest energy structure for this compound is found to be of X1 type with the stacking sequence:

A_1	A_2	A_2	A_1	A_2	A_2
Nd	Sr	Sr	Nd	La	La

and lattice parameters $a_0 = 3.848$ Å, $c_0 = 19.63$ Å. These are within 0.8% of the experimental values of 3.854 Å and 19.77 Å, respectively. Experiment also indicates that the A_1 site is occupied exclusively by Nd^{3+}, which is the smaller of the two rare-earth ions. The lowest energy structure without Nd^{3+} at both A_1 sites is calculated to lie 0.25 eV higher in energy than the structure above, which suggests that A_1 sites are occupied only by Nd^{3+} ions. Thus overall agreement with experiment is very satisfactory.

So far our discussion has been confined to stoichiometric $RE_2MCu_2O_6$, rather than the potentially superconducting doped materials in which there is some substitution of M^{2+} for RE^{3+}. We have also considered the different RE/M′/M cation orderings in the systems $RE_{1.5}M'_{0.5}MCu_2O_6$ (unit cell $RE_3M'M_2Cu_4O_{12}$), with RE = La,Nd and M,M′ = Ca,Ba,Sr (M′ may be the same as M). There are 5 cation orderings for M′ = M and 16 otherwise, which again fall into X-, Y- and Z-type structures. A further complication is that allowance must be made for the holes which are the compensating defects for the additional divalent cation, thereby incorporating either one Cu^{3+} or one O^- per formula unit ($RE_3M'M_2Cu_4O_{12}$). The lowest energy structure for $La_{1.5}Ca_{1.5}Cu_2O_6$ is Z type assuming a Cu hole, and Y type assuming an O hole, so that compared with $La_2CaCu_2O_6$, the explicit incorporation of holes makes more favourable structures in which the A_1 site is occupied by the divalent cation. For $La_{1.5}Sr_{0.5}CaCu_2O_6$ the lowest energy structures are Y type, irrespective of the nature of the holes. Once again structures in which Ca^{2+} ions occupy A_1 sites are relatively lower in energy than in $La_2CaCu_2O_6$. Sr^{2+} ions are predicted to occupy only A_2 sites, in agreement with the results of Cava *et al.* (1990b), but quite contrary to the conclusions of Adachi *et al.* (1990) for a sample with the same overall composition but which was not superconducting.

A further instructive comparison is between the occupancy of cation sites in these systems and that in superconducting Bi–Sr–Ca–Cu–O, Y–Ba–Cu–O and La–Ba–Cu–O materials. In the Bi–Sr–Ca–Cu–O family, divalent alkaline-earth ions occupy the A_1 sites, whereas in $YBa_2Cu_3O_7$ and in $LaBa_2Cu_3O_7$, which has the same structure as $YBa_2Cu_3O_7$ and a critical temperature of ~72 K, the trivalent cation is found at these sites. It cannot, therefore, just be the occupancy of the A_1 site that is responsible for the absence of superconductivity. Several possibilities have been discussed by Allan *et al.* (1994). These include rumpling of the CuO_2 planes, impurity solubility, oxygen interstitials between the CuO_5 pyramids, oxygen vacancies in the CuO_2 planes and additional cation disorder in the *a*- and *b*-directions, which for $RE_2MCu_2O_6$ is possible for the X1 but not the Z type of structure, and cannot occur in $REBa_2Cu_3O_{7-\delta}$ simply because of the different RE:M ratio.

7 Sr–Ca–Cu–O SUPERCONDUCTORS

It is clear from the examples presented so far that atomistic simulation can reveal important contrasts within and between families of superconducting cuprates. The final family of compounds we briefly examine in this chapter are those in the range Sr–Ca–Cu–O, beginning with $SrCuO_2$, the defect chemistry of which remains highly controversial despite the apparent simplicity of its structure, and concluding with the recently synthesized oxyfluorides.

7.1 $SrCuO_2$

Sigriest *et al.* (1988) have synthesized the 'infinite layer' tetragonal 'parent' compound $Ca_{0.86}Sr_{0.14}CuO_2$ in which there are CuO_2 planes separated only by Ca and Sr cations, as shown in Fig. 10.8. This infinite-layer form of $SrCuO_2$ is prepared at high pressure, for under normal conditions it adopts an orthorhombic structure without CuO_2 planes, which in the athermal limit is

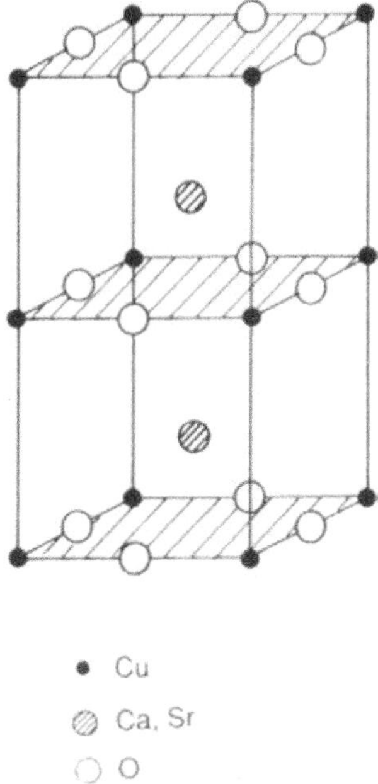

Figure 10.8 Crystal structure of tetragonal infinite layer form of $MCuO_2$.

calculated to be 0.7 eV per formula unit lower in energy than the tetragonal form (Allan *et al.*, 1994). The structure shown in Fig. 10.8 is idealized for $Ca_{0.86}Sr_{0.14}CuO_2$ insofar as the cation disorder leads to local rumplings of the CuO_2 planes due to oxygen displacements along the *c*-axis (Billinge *et al.*, 1991; Allan *et al.*, 1994). To date this compound appears not to show high-T_C behaviour.

n-Type superconductivity ($T_C \sim 40$ K) has been reported in $Sr_{1-x}RE_xCuO_2$ (RE = La (Er *et al.*, 1991), RE = Nd (Smith *et al.*, 1991), and p-type superconductivity ($T_C \approx 110$ K) in $(Sr_{1-x}Ca_x)_{1-y}CuO_2$ (Azuma *et al.*, 1992). It is particularly intriguing that these infinite layer compounds appear to be the only known systems that give rise to both 'hole' and 'electron' superconductivity. Calculated hole and defect electron energies and calculated redox energies (Allan *et al.*, 1994) are all broadly similar to those for Nd_2CuO_4 and other systems containing four-coordinate Cu, in that holes are predicted to be unstable with respect to oxygen vacancies and defect electrons stable with respect to cation vacancies. Thus it is clear that, as in Nd_2CuO_4, $CaCuO_2$ and $SrCuO_2$ are potential precursors for electron superconductivity, provided that aliovalent dopant is sufficiently soluble. The existence of hole superconductivity is more problematic. One possibility is the generation of charge carriers (holes) not by aliovalent substitution but by cation vacancies (Azuma *et al.*, 1992). The calculated energy for this is high (Allan *et al.*, 1994). In contrast Zhang *et al.* (1994) have proposed that the planar defects observed by Azuma *et al.* could be accounted for by the substitution of an SrO layer for a CuO_2 layer, with the incorporation of extra apical oxygen atoms in the neighbouring Sr layers. Our preliminary results (N.L. Allan, P. Akkaraju, J.P. Hill and W.C. Mackrodt, in preparation) indicate that the incorporation of such an extended defect appears to be relatively facile, as suggested by the estimated heat of formation of the Sr-rich material, $Sr_{28}Cu_{20}O_{48}$, from $SrCuO_2$ and SrO, or the energy of decomposition of $SrCuO_2$ to $Sr_{28}Cu_{20}O_{48}$ and CuO. Since the incorporation of apical oxygens leads to five-coordinate Cu, this model seems in principle to be able to account for both p-type superconductivity (following high-pressure O_2 treatment), n-type superconductivity (high-pressure reduction) or lack of superconductivity. Further studies of these and related systems such as Sr_2CuO_{3+X} (Hiroi *et al.*, 1993), which is of considerable interest since a recent neutron diffraction study (Shimakawa *et al.*, 1994) indicated oxygen vacancies in CuO_2 planes even in the superconducting material, would clearly be worthwhile.

7.2 $Sr_2CuO_2F_2$

In this final section we consider fluorine insertion into Sr_2CuO_3, which at ambient pressure yields $Sr_2CuO_2F_{2+\delta}$($T_C \approx 46$ K) (Al-Mamouri *et al.*, 1994). Whereas Sr_2CuO_3 contains CuO_2 chains rather than CuO_2 planes, $Sr_2CuO_2F_{2+\delta}$ has the La_2CuO_4 structure (Fig. 10.1), but neither X-ray nor neutron diffraction can differentiate between oxygen and fluorine ions. The locations of the oxygens and fluorines are clearly of considerable importance for theories of high-T_C behaviour (are there fluorines in the CuO_2 planes?) and here atomistic simulation can be a powerful technique.

In a preliminary study of such oxyfluorides (Hill *et al.*, 1996) using the same short-range potentials for F^- as O^-, $Sr_2CuO_2F_2$ is predicted to adopt the La_2CuO_4 structure in preference to that of Nd_2CuO_4, with an energy difference of ~1.5 eV per formula unit between the two. Furthermore, the energy difference per formula unit between the structure with both fluorines axial and that with two equatorial fluorines is ~1.1 eV. Accordingly, F^- ions are predicted to occupy only *axial* positions and so do not lie in the CuO_2 planes. The lowest interstitial position for additional fluorine in $Sr_2CuO_2F_{2+\delta}$ is the same as that for extra oxygen in La_2CuO_{4+x} and, analogously, the incorporation of interstitial fluorine generates holes in the CuO_2 planes leading to high-T_C behaviour:

$$Sr_2CuO_2F_2 + (\delta/2)F_2 \rightarrow Sr_2CuO_2F_{2+\delta}$$

$$\tfrac{1}{2}F_2(g) \longrightarrow F_i' + h^\bullet$$

Very recently the corresponding Ca compound, $Ca_2CuO_2F_{2+\delta}$, has been prepared (Al-Mamouri *et al.*, 1995): unlike $Sr_2CuO_2F_{2+\delta}$ it adopts the Nd_2CuO_4 structure. It is non-superconducting, which is consistent with the absence of hole superconductivity in this type of structure. As with $Sr_2CuO_2F_2$, lattice simulations of $Ca_2CuO_2F_2$ indicate a strong preference of F^- for sites not in the CuO_2 planes. The calculated lattice constants of $Ca_2CuO_2F_2$ are 3.852 Å and 11.86 Å for a_0 and c_0 which are within 0.1% and 0.2% of the experimental values of 3.850 Å and 11.84 Å, respectively, for $Ca_2CuO_2F_{2.1}$. Calculated static lattice energies, which of course refer to the athermal limit, indicate that the La_2CuO_4 structure is also favoured for $Ca_2CuO_2F_2$ by 0.5 eV per formula unit. While this suggests a possible failure of our potential model, it is interesting to note that the structural refinements of $Ca_2CuO_2F_{2.1}$ by Al-Mamouri *et al.* (1995) suggest the existence of regions with the La_2CuO_4 structure.

8 CONCLUSIONS

In this chapter we have selected a number of case studies to show how atomistic lattice simulations can be used to investigate the structural and defect chemistry of a wide range of high-T_C cuprates. Space limitations have necessarily restricted the number of examples to a small range of compounds, chiefly from the La–Cu–O, Nd–Cu–O, Y–Ba–Cu–O and Sr–Cu–O systems.

Calculations of the type reported here are relatively simple and inexpensive. There are clear limitations in that they are based on an ionic model which cannot serve the role of appropriate *ab initio* electronic structure studies, particularly for electronic defects. Neither can aspects related to the magnetism be addressed. Nevertheless, the approach is highly specific and versatile. It is clear from the results presented here that its particular strength lies in the insight that can be gained from comparative studies of interrelated families of materials with related structures and other similarities in their chemical or physical behaviour. It seems likely that simulation techniques will continue to provide valuable information on these complex materials and in particular on the new superconducting phases prepared by novel synthetic strategies.

APPENDIX: KRÖGER–VINK NOTATION

Here we describe briefly the Kröger–Vink notation (Kröger and Vink, 1956), used throughout the study of ceramics to denote defect type and *effective* charge. Tilley's (1987) book contains many examples. Defects are denoted by symbols comprised of three parts:

(i) A major symbol – either the chemical symbol for the atom type or a 'V' to denote a vacancy.
(ii) A subscript denoting the lattice site occupied, or an '*i*' to denote an interstitial site.

(iii) A superscript denoting the charge on the site *relative* to the normal charge on the site; ′ denotes an effective negative charge, x a neutral defect and • an effective positive charge.

A free hole is represented by $h^{\bullet}$ and a defect electron by e'. For example, an oxygen vacancy is denoted $V_O^{\bullet\bullet}$, an oxygen interstitial O_i'', and a calcium on a lanthanum site, Ca_{La}'.

In writing equations for defect reactions using this notation, there must be conservation of charge, mass and ratio of lattice sites.

REFERENCES

Adachi, S., Adachi, H., Setsune, K. and Wasa, K. (1990) *Physica C*, **169**, 377.
Allan, N.L. and Mackrodt, W.C. (1988) *Phil. Mag A.*, **58**, 555.
Allan, N.L. and Mackrodt, W.C. (1990a) *J. Chem. Soc. Faraday Trans.*, **86**, 1227.
Allan, N.L. and Mackrodt, W.C. (1990b) *J. Am. Ceram. Soc.*, **73**, 3175.
Allan, N.L. and Mackrodt, W.C. (1993) *Adv. Solid State Chem.*, **3**, 221.
Allan, N.L. and Mackrodt, W.C. (1994a) *Mol. Simul.*, **12**, 89.
Allan, N.L. and Mackrodt, W.C. (1994b) *Phil. Mag. B.*, **69**, 871.
Allan, N.L., Lawton, J.M. and Mackrodt, W.C. (1989) *Phil. Mag. B.*, **59**, 191.
Allan, N.L., Baram, P.S., Mackrodt, W.C. and Turner, M.J. (1992) *Mol. Simul.*, **9**, 115.
Allan, N.L., Baram, P.S., Gormezano, A. and Mackrodt, W.C. (1994) *J. Mater. Chem.*, **4**, 817.
Al-Mamouri, M., Edwards, P.P., Greaves, C. and Slaski, M. (1994) *Nature*, **369**, 382.
Al-Mamouri, M., Edwards, P.P., Greaves, C., Slater, P.R. and Slaski, M. (1995) *J. Mat. Chem.*, **5**, 913.
Anderson, P.W. (1975) *Phys. Rev. Lett.*, **34**, 953.
Aride, J. and Flandrois, S. (1989) *Solid State Commun.*, **72**, 459.
Azuma, M., Hiroi, Z., Takano, M., Bando, Y. and Takeda, Y. (1992) *Nature*, **356**, 775.
Bednorz, J.G. and Müller, K.A. (1986) *Z. Phys. B: Condens. Matter*, **64**, 189.
Billinge, S.J.L., Davies, P.K., Egami, T. and Catlow, C.R.A. (1991) *Phys. Rev. B*, **43**, 10340.
Caignaert, V., Nguyen, N. and Raveau, B. (1990) *Mater. Res. Bull.*, **25**, 199.
Catlow, C.R.A., Mackrodt, W.C., Norgett, M.J. and Stoneham, A.M. (1977) *Phil. Mag. B*, **35**, 177.
Catlow, C.R.A. and Mackrodt, W.C. (1982) In *Computer Simulation of Solids* (eds C.R.A. Catlow and W.C. Mackrodt), Springer-Verlag, Berlin, p. 3.
Cava, R.J., Hewat, A.W., Hewat, E.A., Batlogg, B., Marezio, M., Rabe, K.M., Krajewski, J.J., Peck, W.F. Jr and Rupp, L.W. Jr (1990a) *Physica C*, **165**, 419.
Cava, R.J., Santoro, A., Krajewski, J.J., Fleming, R.M., Waszczak, J.V., Peck, W.R. Jr and Marsh, P. (1990b) *Physica C*, **172**, 138.
Cava, R.J., Batlogg, B., van Dover, R.B., Krajewski, J.J., Waszczak, J.V., Fleming, R.M., Peck, W.F. Jr, Rupp, L.W. Jr, Marsh, P., James, A.C.W.P. and Scheemeyer, L.F. (1990c) *Nature*, **345**, 602.
Chaillout, C., Cheong, S-W., Fisk, Z., Lehmann, M.S., Marezio, M., Morosin, B. and Schirber, J.E. (1989) *Physica C*, **158**, 183; erratum, *Ibid.*, **159**, 892.
Chakraverty, B.K. (1981) *J. Phys. (Paris)*, **42**, 1351.
Cheong, S-W, Thompson, J.D. and Fisk, Z. (1989) *Physica C*, **158**, 109.

Chevalier, B., Tressaud, A., Lepine, B., Amine, K., Dance, J.M., Lozano, L., Hickey, E. and Etourneau, J. (1990) *Physica C*, **167**, 97.
Cormack, A. (1993) *Adv. Solid State Chem.*, **3**, 63.
Er, G., Miyamoto, Y., Kanamaru, F. and Kikkawa, S. (1991) *Physica C*, **181**, 206.
Fuertes, A., Obradors, X., Navarro, J.M., Gomez-Romero, P., Casañ-Pastor, N., Pérez, F., Fontcuberta, J., Miravitlles, C., Rodriguez-Carvajal, J. and Martinez, B. (1990) *Physica C*, **170**, 153.
Ghigna, P., Spinolo, G., Filipponi, A., Chadwick, A.V. and Hamner, P. (1995) *Physica C*, **253**, 147.
Goodenough, J.B. (1990) *Supercond. Sci. Technol.*, **3**, 26.
Goodenough, J.B. and Manthiram, A. (1990) *J. Solid State Chem.*, **88**, 115.
Gopalkrishnan, J., Subramanian, M.A., Torardi, C.C., Attfield, J.P. and Sleight, A.W. (1989) *Mater. Res. Bull.*, **24**, 321.
Grande, B., Müller-Buschbaum, Hk. and Schweizer, Z. (1977) *Z. Anorg. Allg. Chem.*, **428**, 120.
Grasmeder, J.R. and Weller, M.T. (1990) *J. Solid State Chem.*, **85**, 88.
Hebard, A.F., Rosseinsky, M.J., Haddon, R.C., Murphy, D.W., Glarum, S.H., Palstra, T.T.M., Ramirez, A.P. and Kortan, A.R. (1991) *Nature*, **350**, 600.
Heyri, E.A. and Larese, J.Z. (1990) *Physica C*, **170**, 239.
Hill, J.P., Allan, N.L. and Mackrodt, W.C. (1996) *J. Chem. Soc., Chem. Commun.* (in press).
Hiroi, Z., Takano, M., Asuma, M. and Takeda, Y. (1993) *Nature*, **364**, 315.
Huang, J. and Sleight, A.W. (1990) *Physica C*, **169**, 169.
Ishii, T., Watanabe, T., Kinoshita, K. and Matsuda, A. (1991) *Physica C*, **179**, 39.
Islam, M.S. (1990) *Supercond. Sci. Technol.*, **3**, 531.
Islam, M.S. (1994) *Mol. Simul.*, **12**, 101.
Islam, M.S. and Winch, L.J. (1995) *Phys. Rev. B*, **52**, 10510.
Izumi, F., Takayama-Muromachi, E., Nakai, Y. and Asano, H. (1989) *Physica C*, **157**, 89.
James, A.C.W.P., Zahurak, S.M. and Murphy, D.W. (1989) *Nature*, **338**, 240.
Jorgensen, J.D., Dabrowski, B., Pei, S., Richards, D.R. and Hinks, D.G. (1989) *Phys. Rev. B*, **40**, 2187.
Kröger, F.A. and Vink, H. (1956) *J. Solid State Phys.*, **3**, 307.
Legros-Glédel, C., Marucco, J-F., Vincent, E., Favrot, D., Poumellec, B., Touzelin, B., Gupta, M. and Alloul, H. (1991) *Physica C*, **175**, 279.
Lightfoot, P., Pei, S., Jorgensen, J.D., Tang, X.-X., Manthiram, A. and Goodenough, J.B. (1990) *Physica C*, **169**, 464.
Longo, J.M. and Racah, P.M. (1973) *J. Solid State Chem.*, **6**, 526.
Mackrodt, W.C. and Stewart, R.F. (1979) *J. Phys. C*, **12**, 431.
McCarron, E.M. III, Crawford, M.K. and Parise, J.B. (1989) *J. Solid State Chem.*, **78**, 192.
Mott, N.F. and Littleton, M.T. (1938) *Trans. Faraday Soc.*, **34**, 485.
Müller-Buschbaum, Hk. and Lehmann, U. (1978) *Z. Anorg. Allg. Chem.*, **447**, 47.
Namgung, C. and Irvine, J.T.S. (1990) *Physica C*, **168**, 346.
Nguyen, N., Er-Rakho, L., Michel, C., Choisnet, J. and Raveau, B. (1980) *Mater. Res. Bull.*, **15**, 891.
Okada, H., Takano, M. and Takeda, Y. (1990) *Physica C*, **166**, 111.
Parks, M.E., Navrotsky, A., Mocala, K., Takayama-Muromachi, E., Jacobson, A. and Davies, P.K. (1989) *J. Solid State Chem.*, **79**, 53.
Pickett, W.E. (1989) *Rev. Mod. Phys.*, **61**, 433.
Poulsen, H.F., Andersen, N.H., Andersen, J.V., Bohr, H. and Mouritsen, O.G. (1991) *Nature*, **349**, 594.

Shimakawa, Y., Jorgensen, J.D., Mitchell, J., Hunter, B.A., Shaked, H., Hinks, D.G., Hitterman, R.L., Hiroi, Z. and Takano, M. (1994) *Physica C*, **228**, 73.

Siegrist, T., Zahurak, S.M., Murphy, D.W. and Roth, R.S. (1988) *Nature*, **334**, 231.

Singh, K.K. and Edwards, P.P. (1993) *Adv. Solid State Chem.*, **3**, 99.

Smith, M.G., Manthiram, A., Zhou, J., Goodenough, J.B. and Markert, J.T. (1991) *Nature*, **351**, 549.

Su, M-Y., Cooper, E.A., Elsbernd, C.E. and Mason, T.A. (1990) *J. Am. Ceram. Soc.*, **73**, 453.

Tarascon, J.M., Greene, L.H., McKinnon, W.R., Hull, G.W. and Geballe, T.H. (1987) *Science*, **235**, 1373.

Temmerman, W.M., Szotek, Z. and Guo, G.Y. (1988) *J. Phys. C*, **21**, L867.

Tilley, R.J.D. (1987) *Defect Crystal Chemistry and its Applications*, Blackie, New York.

Torardi, C.C., Subramanian, M.A., Calabrese, J.C., Gopalakrishnan, J., McCarron, E.M., Morrissey, K.J., Askew, T.R., Flippen, R.B., Chowdhry, U. and Sleight, A.W. (1988) *Phys. Rev. B*, **38**, 225.

Verweij, H., Bruggink, W.H.M., Steeman, R.A., Frikkee, E. and Helmholdt, R.B. (1990) *Physica C*, **166**, 372.

Wada, T., Yaegashi, Y., Ichinose, A., Yamauchi, H. and Tanaka, S. (1991) *Phys. Rev. B*, **44**, 2341.

Wu, M.K., Ashburn, J.R., Torng, C.J., Hor, P.H., Meng, R.L., Gao, L., Huang, Z.J., Wang, Q. and Chu, C.W. (1987) *Phys. Rev. Lett.*, **58**, 908.

Zhang, X. and Catlow, C.R.A. (1991) *Physica C*, **173**, 25.

Zhang, X. and Catlow, C.R.A. (1992a) *Physica C*, **193**, 221.

Zhang, X. and Catlow, C.R.A. (1992b) *Phys. Rev. B*, **46**, 457.

Zhang, X., Catlow, C.R.A. and Zhou, W. (1990) *Physica C*, **168**, 417.

Zhang, H., Wang, Y.Y., Zhang H., Dravid, V.P., Marks, L.D., Han, P.D., Payne, D.A., Radaell, P.G. and Jorgensen, J.D. (1994) *Nature*, **370**, 352.

11

Molecular Crystals

S.L. Price

1 THE CONTRAST BETWEEN MOLECULAR AND IONIC SOLIDS

The major differences between crystals composed of organic molecules, such as benzene, and those composed of ions, such as sodium chloride, has led to different approaches to their modelling. These contrasting methods need, however, to be brought together for modelling many inorganic systems which include both covalent and ionic bonds. A topical example is provided by bio-inorganic interfaces where covalently bonded organic entities interact with ionic solids.

Structures of ionic solids may be understood in terms of the packing of spheres, whereas molecular crystals result from the packing of diverse shapes. Only a few simple polyatomic molecules are even approximately spherical; more molecules are plate like, but the majority of organic molecules have very irregular shapes. These shapes crystallize into closely packed structures, with the ratio of the molecular volume to the unit cell volume being 0.7 ± 0.05 for the overwhelming majority of organic crystal structures (Pertsin and Kitaigorodsky, 1987), which approaches the packing efficiency of hard spheres (0.74). This close packing principle can almost define the crystal structure for many irregularly shaped organic compounds. Indeed, the founding principle of molecular crystallography, put forward by Kitaigorodsky (1973), is that 'the protrusions of one molecule are inserted into the hollows of others'. He made considerable progress in rationalizing crystal structures using this dovetail principle and a mechanical structure seeker which used space-filling models. The equivalent for ionic solids, the radius ratio rules, are notoriously unsuccessful even for simple binary compounds.

Differences in shape lead to considerable differences in the ranges of structures adopted, and these modify the questions that computer modelling is seeking to address. Many molecular crystal structures have already been determined; the Cambridge Structural Database (Allen *et al.*, 1991; Allen and Kennard, 1993) contains 140 268 crystal structures of organic and organometallic compounds (April 1995 release). Various statistical analyses suggest that over 90% of these structures contain only one independent molecule in the

COMPUTER MODELLING IN INORGANIC CRYSTALLOGRAPHY
ISBN 0-12-164135-X

unit cell (Belsky and Zorkii, 1977) and that 75% of organic structures belong to only five space groups (Mighell *et al.*, 1983). Thus, structures with more than eight molecules in the unit cell are rare, and most molecules crystallize with one, two or four molecules in the unit cell. Hence, it is generally trivial to find a molecular packing consistent with a given set of cell dimensions. The aim of molecular crystal structure modelling is to predict the unit cell dimensions and packing of a molecule prior to its synthesis.

The prediction of molecular crystal structures is a fundamental problem in physical chemistry, which tests both the model for the forces between molecules and the model for the kinetic and thermodynamic effects that control crystallization. It is of practical importance in the design of new materials. For example, a molecule with a high non-linear optical coefficient will only maintain this desirable property in the crystal if it packs in a non-centrosymmetric space group. Many organic molecules can crystallize in more than one form, and the different polymorphs have different physical properties such as solubility or ease of tableting. Polymorphism can take many forms; it can result from different conformations of the molecule in the crystal, or the incorporations of different solvent molecules in the crystal structure, as well as simply different packings of identical molecules whose relative stability depends on the conditions of temperature and pressure. Polymorphism is poorly understood, as it often seems that the number of polymorphs known is mainly dependent on the experimental effort expended. Indeed, it often takes so much effort to persuade a new compound to form well-defined crystals, that few organic chemists have the motivation to experiment with further crystallization conditions to find new structures. Convincing computer predictions of novel polymorphs may catalyse their discovery. Emerging patent law on different crystal forms of drugs, such as ranitidine, will also encourage the systematic search for polymorphic crystal structures.

Before theoreticians and experimentalists can understand polymorphism, we need to be able to predict known molecular crystal structures. As in the study of polar materials described in Chapter 9 and 10, there are two components to such studies: first we need a realistic model for the forces between the molecules and second a simulation method that will find all the possible crystal structures. Progress is being made on both fronts. Model potentials that are sufficiently realistic, in that a minimum in the lattice energy lies acceptably close to the experimental structure, are now being developed for polar organic molecules. These potentials are theoretically based, giving greater confidence in the lattice energy estimates for hypothetical structures. Similarly, various systematic methods of generating hypothetical crystal structures as starting points for minimizations have recently been proposed. The extent to which these methods extend to different types of molecule is uncertain, nor have there been detailed comparisons with simulated annealing techniques, but the combination of these methods with realistic potentials should enable us to predict molecular crystal structures by the turn of the century.

2 THERMAL EFFECTS AND IMPLICATIONS FOR TARGET ACCURACY

In the past decade, the number of molecules for which there was an intermolecular potential that would predict a minimum in the static lattice energy for a structure acceptably close to experiment was very limited (Pertsin and Kitaigorodsky, 1987). It is worth asking what constitutes an 'acceptably close' prediction of a crystal structure, given that most crystal structure determinations are at room temperature and most organic molecular crystal structure modelling methods use energy minimization, which does not explicitly include temperature effects. However, since empirical potentials generally have to be fitted to available (usually room temperature) crystal structures, thermal effects will be partly absorbed into the potential parameterization, and so their neglect in structure prediction (albeit at a crude level) is consistent. Nevertheless, thermal expansion effects are unlikely to be reliably reproduced by this method, as they are far from uniform in molecular crystals. From the small sample of low and room temperature crystal structures in Table 11.1, it is clear that most cell edge lengths will expand by at least 1% between 0 K and room temperature, but this is very variable and some cell edges may even contract with temperature. Williams and co-workers aimed for an accuracy of 1% in lattice constants, 1° in cell angles, 2° in molecular rotations and 0.1 Å in molecular translations in their derivations of potentials for perchlorohydrocarbons (Hsu and Williams, 1980), oxohydrocarbons (Cox *et al.*, 1981) and azahydrocarbons (Williams and Cox, 1984), though many of their published structural predictions were admitted to be significantly outside these limits. Such accuracy may be unobtainable for many of the softer crystals using energy minimization methods, even with more sophisticated effective potential forms. Nevertheless, unless the molecule has anomalously anisotropic thermal expansion, an accuracy of a few percent in the cell edges should be obtainable with energy minimization methods and good model potentials. Conversely, although thermal expansion is appreciable, its neglect is not the major factor limiting our ability to predict more organic crystal structures.

3 MOLECULAR STRUCTURES WITHIN CRYSTALS

Central to the modelling of molecular crystal structures is the assumption that the forces between the molecules are very much weaker than the forces acting within the molecule. Thus the intermolecular forces do not change the molecular structure significantly. Comparisons of isolated molecule structures with those derived by crystallography show that, apart from the apparent shortening of bonds to hydrogen in X-ray studies, there are negligible

Table 11.1 Illustration of the variable effects of temperature on molecular crystal structures.

Crystal structure	Refcodes			Temperatures		Contraction on cooling $T_1 \rightarrow T_2$			Vol. coeff.
	Refcodes			T_1(K)	T_2(K)	$\Delta a(\%)$	$\Delta b(\%)$	$\Delta c(\%)$	$\Delta V(\text{Å}^3/\text{K})$
Benzene	BENZEN	02	01	270	138	0.94	2.48	3.13	0.062
Naphthalene	NAPHTD		01	295	12	2.06	0.45	0.24	0.030
Anthracene	ANTCEN	14	09	295	94	1.62	0.42	0.69	0.042
Formamide	FORMAM		02	295	90	2.33	1.51	−1.80	0.008
Urea	UREAXX	09	12	295	12	1.42	1.42	0.43	0.009
Benzamide	BZAMID	01	02	295	15	1.39	0.26	3.22	0.027
Pyrimidine	PRMDIN		01	271	107	1.22	0.34	2.97	0.029
p-Benzoquinone	BNZQUI		02	295	113	4.14	0.88	0.97	0.040
Alloxan	ALOXAN		01	295	42	0.76	0.76	1.75	0.016
Cyanuric acid	CYURAC	10	12	295	100	1.91	−0.06	0.33	0.014
Imidazol	IMAZOL	13	06	295	103	2.11	1.69	−0.06	0.025

These molecules are chosen fairly randomly from the set of small organic, essentially rigid molecules containing C, H, N and O only, where an approximately room temperature (T_1) and low temperature (T_2) crystal structure is available. The crystal structures used are denoted by their refcodes in the Cambridge Structural Database. The contractions are defined by:

$$\Delta a(\%) = 100\,\frac{a(T_1) - a(T_2)}{a(T_1)}$$

The volume contraction is defined by:

$$\Delta V = \frac{V(T_1) - V(T_2)}{T_1 - T_2}$$

differences in bond lengths or angles on crystallization (Pertsin and Kitaigorodsky, 1987). It is only the intramolecular degrees of freedom associated with very shallow potential energy surfaces, such as rotation around some single bonds, that will change from a dynamic equilibrium in the gas phase to a constrained conformation in the crystal structure. Thus, the torsion angles observed in the solid state may not correspond to the global minimum energy structure of the isolated molecule, but will correspond to a low energy conformation.

Many methods of predicting the structures of isolated molecules have been developed in recent years which are also capable of identifying which aspects of the molecular structure (usually torsion angles) can change significantly within a population of gas phase molecules at finite temperature or on crystallization. *Ab initio* methods can determine the structure of most small polyatomics extremely accurately, and are quite reliable for most moderately sized molecules. For many families of organic molecules, force fields have been developed which can quite accurately predict structures by energy minimization, e.g. MM3 for small molecules (Allinger and Yan, 1993), AMBER for biomolecules (Weiner *et al.*, 1986) and many others. The choice of force field should be governed by the published predictions for molecules closely related to the unknown structure. Thus there is unlikely to be any problem in the prediction of the molecular structure for a molecule that has yet to be synthesized and, indeed, a conformational structure prediction and energy analysis is a sensible precursor to the synthesis.

The more usual course is to take the molecular structure from an existing crystal structure. If it is an X-ray structure, then the published hydrogen atom positions correspond to the centre of the hydrogen electron density, not the nuclear position, which may not be determined very accurately. Such positions are usually corrected to give a standard bond length of 1.08 Å for C–H, for example, by the scheme given by Gavezzotti (1989). (However, some potential modelling schemes, such as that of Williams and Cox (1984), do not use non-nuclear positions for the hydrogen atom sites.)

Most crystal structure modelling of organic molecules has deliberately avoided the complication of possible conformational change on crystallization, by using molecules that can safely be assumed to be rigid. This has not been particularly limiting for developing methods or potentials, as conjugated systems are so prevalent in organic chemistry. Thus, the rest of this chapter will assume that the molecules are being treated as rigid. However, the need to include torsional flexibility will arise for modelling the crystal structure of many interesting molecules. The experience in treating conformational freedom in proteins and biomolecules, and the parameterization available in their force fields, suggests that the inclusion of these effects could be a relatively straightforward development, once we are able to predict the crystal structures of the rigid fragments of which the conformationally flexible molecules are composed.

4 DEVELOPMENTS IN MODEL INTERMOLECULAR POTENTIALS

4.1 Current potentials - the isotropic atom-atom potential method

The type of model potential that is sufficient to predict a molecular crystal structure is very dependent on the shape of the molecule. If the shape has many well-defined protrusions and cavities, so that there is only one way that it can pack densely with a tight fitting of the protrusions of one molecule into the cavities of its neighbours, then any potential which represents this shape and has an attractive component will be acceptable. However, many shapes, such as discs or cylinders, can close pack in many ways, each generating a range of structures with differing tilt angles, etc. In these cases, the chemical nature of the atoms and resultant intermolecular forces determine which of the reasonably close-packed structures is adopted.

The simplest model, which implicitly includes a contribution from each atom to the shape and a long-range attractive component, is the isotropic *6-exp* atom–atom intermolecular potential:

$$U = \sum_{ik} U_{ik} = \sum_{ik} A_{\iota\kappa} \exp(-B_{\iota\kappa} R_{ik}) - C_{\iota\kappa}/R_{ik}^{6} \tag{1}$$

where atom i in one molecule and atom k in the other, are of types ι and κ, respectively. (We recall that potentials of this form have been experimentally used in modelling ionic crystals as discussed in Chapters 3 and 4, although a slightly different convention is adopted with the exponential terms.) An alternative, equivalent form is a function of the minimum energy separation, $R^{0}_{\iota\kappa}$, and well depth, $\varepsilon_{\iota\kappa}$, of each individual atom–atom interaction:

$$U_{ik} = \frac{\varepsilon_{\iota\kappa}}{(\lambda_{\iota\kappa} - 6)} \left[6 \exp(\lambda_{\iota\kappa}) \exp\left(-\lambda_{\iota\kappa} \frac{R_{ik}}{R^{0}_{\iota\kappa}}\right) - \lambda_{\iota\kappa} \left(\frac{R^{0}_{\iota\kappa}}{R_{ik}}\right)^{6} \right] \tag{2}$$

where $\lambda_{\iota\kappa}$ is a 'steepness' parameter.

This atom–atom potential is essentially modelling the interaction between molecules as the sum of the interactions between spherical atomic charge distributions. The *6-exp* form is a simple realistic model for the repulsion and dispersion forces, which can give a qualitatively reasonable description of the properties of argon, but lacks the flexibility to represent accurately the argon pair potential (Maitland *et al.*, 1981).

The ability of this isotropic *6-exp* atom–atom potential to represent a wide range of crystal structures has recently been investigated by Filippini and Gavezzotti (1993). They derived model potentials of this form by empirical fitting to the structures of 217 crystals and their heats of sublimation (available for 122 compounds), including hydrocarbons, oxahydrocarbons, azahydrocarbons, chlorohydrocarbons, sulphohydrocarbons, sulphones, sulphoxides and nitrocompounds, avoiding any hydrogen bonded compounds. Guidance on the $R^0_{ικ}$ values was taken from a survey of the closest contacts in the crystal structures, and the well depths were empirically fitted without using combining rules for heteroatom interactions. The results from this large dataset of non-polar compounds were remarkably good for the heats of sublimation, with, on average, an agreement better than 10 kJ/mol. However, the quality of the structure predictions was variable, many being acceptable, but with cell edges frequently being in error by 5%. The scheme was also extended to hydrogen bonded substances (Gavezzotti and Filippini, 1994), where again the heats of sublimation were well predicted, but some structures were well represented while others were not.

This simple *6-exp* potential does not explicitly include the electrostatic contribution to the intermolecular forces. Indeed, these effects have been partially absorbed into the empirical potential parameters which is most apparent in the very deep potential wells for interactions involving polar hydrogen atoms, but also contributes to the major deviations from the traditional combining rules for the more polar atoms. This absorption of the electrostatic term is computationally advantageous, as the long-range nature of the electrostatic term requires special consideration in calculating the lattice energy. However, it is far from theoretically rigorous and so one can have little confidence that the model potential would extrapolate reliably to hypothetical polymorphs, even when the observed structure is well predicted.

The electrostatic contribution appears to play an important role in determining the minimum lattice energy structure for many molecules where the packing is not determined simply by the shape. Even for hydrocarbons, Williams and Starr (1977) showed that the introduction of atomic charges produced a significant improvement in the predicted crystal structures for the aromatic compounds. The empirically fitted charge separation in the C–H bond ($q_H = -q_c = 0.153e$) provides an electrostatic model that reproduces the total quadrupole moments of the considered aromatic hydrocarbons remarkably well (Price 1985), emphasizing the link between realism in the model potential and good crystal structure predictions. Empirically fitted charge-separation electrostatic models have also been derived for predicting the crystal structures of perchlorohydrocarbons (Hsu and Williams, 1980), amides (Hagler *et al.*, 1974) and carboxylic acids (Lifson *et al.*, 1979), but for most molecules the constraint implicit to such models of assuming certain functional groups are neutral is clearly unrealistic. The best procedure is to derive the electrostatic model from the *ab initio* charge distribution of the molecule. If the repulsion-dispersion parameters are empirically fitted to observed crystal structures, these parameters may absorb some of the errors in the electrostatic model.

An electrostatic model is most readily appended to the isotropic atom–atom potential method by assigning point charges at each nuclear site. There are many ways of calculating atomic charges, ranging from Mulliken analysis to 'potential derived' charges. In the latter method, the atomic charges are determined by fitting to the electrostatic potential calculated directly from the *ab initio* wave function, at a large grid of points outside the molecule. The resulting charges can be sensitive to the choice of grid, but clearly this method gives the best charges for modelling the electrostatic interactions in the region of the grid. There is an extensive literature on methods of determining atomic charges for molecular modelling, but as the careful comparison of many methods by Wiberg and Rablen (1993) demonstrates, no atomic charge model is capable of reproducing both the dipole moments and electrostatic potentials of a representative set of molecules in a satisfactory manner.

Many attempts have been made to extend the success of the isotropic atom–atom potential for hydrocarbons to include heteroatoms using various atomic point charge models. These have been reviewed by Pertsin and Kitaigorodsky (1987). A lack of transferability and sensitivity to the choice of model molecules and the electrostatic model is evident, even though many of the studies have only one type of heteroatom and avoid highly polar or hydrogen bonded molecules. For example, Cox *et al.* (1981) fitted a non-bonded O···O potential to β-oxygen, carbon dioxide, trioxane, tetraoxocane and succinic anhydride using potential derived charges, which transferred reasonably well to the structures of pentoxecane, 1,4-cyclohexanedione and diglycolic anhydride (r.m.s. error of 0.27 Å in cell lengths), but not to furan or *p*-benzoquinone. The extension of this approach to azahydrocarbons by Williams and Weller (1983) is particularly significant, in that fitting atomic charges to the electrostatic potential not only give a poor fit, but models with these charges were unable to predict many of the crystal structures. However, when additional lone pair charge sites were introduced, not only was there a dramatic improvement in the fit to the electrostatic potential outside the molecule, but also in the predicted crystal structures. Thus, assuming that the azabenzene molecules are superpositions of spherical atomic charge densities is inadequate for describing their crystal packing.

Introducing lone pair directionality has also been proposed as a possible means of improving the predicted crystal structures of carboxylic acids and amides, given the tendency of various isotropic atom–atom potentials to overestimate the H···O=C intermolecular hydrogen bond angle (Hagler *et al.*, 1979). This comparison of alternative force fields yielded an r.m.s. error in the cell vector lengths of 0.307 Å for 14 carboxylic acids and 0.235 Å for 12 amide crystals for the best of the potentials investigated.

There are many other indications that the electrostatic effects of non-spherical features of the charge distribution, such as lone pairs and π electrons, can be important in determining molecular crystal structures. At the extreme of homonuclear diatomics (X_2), the electrostatic potential outside the molecule arises from the non-spherical distribution of the valence electrons. Just as there are considerable variations in the bonding orbitals in the diatomics, there are also considerable variations in the lowest temperature ordered crystal structure.

Thus we find a cubic *Pa3* structure for *ortho*-hydrogen and nitrogen, a monoclinic structure (*C2/c*) for fluorine, and an orthorhombic layer structure *Cmca* for chlorine, bromine and iodine. β-Oxygen has stacked rafts of upright molecules in the $R\bar{3}m$ structure, with the *C2/m* α-O_2 structure being stabilized by the magnetic dipoles. Shape alone (as quantified by the ratio of the isotropic atom–atom zero-potential separation to bond length) only predicts stacked rafts of upright ($P\bar{6}m2$ or $R\bar{3}m$) or tilted (*C2/c*) molecules (English and Venables, 1974). Addition of a central quadrupole moment can stabilize the *Pa3* structure. However, whether the same total quadrupole moment is described by atomic dipole moments (approximating axial lone pair density) or atomic quadrupole moments (approximating equatorial lone pairs) can change whether the lowest lattice energy is *Pa3* or *Cmca*, for the same repulsion-dispersion potential (Price, 1987). Thus, for these highly symmetrical polyatomics, the adopted crystal symmetry is sensitive to the detailed description of the electrostatic forces.

This effect is not confined to diatomics and azabenzenes. Ritchie *et al.* (1993) also found the predicted structure of molecular crystals of explosive materials to be very sensitive to the electrostatic model used. They compared the use of Mulliken charges, potential derived charges and an accurate atomic multipole series description of the same wave function, in conjunction with the same repulsion–dispersion potential. Different electrostatic models gave significantly different calculated densities, with the Mulliken charges yielding the poorest results with most crystals having one unit-cell edge in error by over 10%. The potential derived charges often gave significantly different cell parameters from the atomic multipole model, but neither could account for all the crystal structures satisfactorily.

The opposing view, that the electrostatic term is not important in determining the crystal structure, has been put forward by Perlstein (1994). This was based on his success in predicting the one-dimensional chain motifs within crystal structures randomly selected from the Cambridge Structural Database and the low ratio of the electrostatic energy to van der Waals energy estimated using Gasteiger charges. The resolution of this conflict probably lies in the shapes of the molecules being studied and their ability to pack with translational symmetry into a tightly interlocking structure. Crystal structures where the molecules can slide and tilt relative to each other and remain reasonably close packed will be sensitive to the attractive forces, particularly the electrostatic forces which can vary dramatically from attractive to repulsive with orientation. The mixed success of Gavezzotti's *6-exp* potentials in predicting the structures of molecules with the same functional groups (Filippini and Gavezzotti, 1993; Gavezzotti and Filippini, 1994) supports this argument. The crystal packing of some molecules can be predicted by a range of simple repulsion-dispersion models, whereas other structures cannot be adequately described by any isotropic atom–atom potential, even when good atomic charges are included. The sensitivity of the minimum energy crystal structure to the model intermolecular potential cannot be predicted simply from the functional groups involved. In particular, it is not possible to know *a priori* which crystal structures are sufficiently sterically controlled to be predicted by a simple *6-exp* potential excluding explicit electrostatics without knowledge of the crystal structure. Therefore better, theoretically based potentials are required for all molecules if we are to predict unknown crystal structures with any confidence.

4.2 Distributed multipole electrostatic models

The first improvement that can be made to the isotropic atom–atom potential approach is to incorporate a realistic electrostatic model which includes the electrostatic effects of non-spherical features in the atomic charge density, such as lone pair and π electron density. This can be achieved by representing the charge associated with each atom by a multipole series, with a point charge, dipole, quadrupole, etc., on each atom. The anisotropy of the higher multipoles represents the non-spherical features of the charge distribution. The series can be taken to sufficient terms to ensure an accurate representation of the molecular electrostatic potential, excluding the effects of penetration within the molecular charge density. The electrostatic contribution to the intermolecular potential is most readily derived from the *ab initio* charge distribution of the isolated molecule and so is the only term where a theoretically rigorous model can be readily quantified for organic molecules at the moment (Price, 1995). However, apart from the feasibility, there is also considerable evidence that the electrostatic term dominates the orientation dependence of many intermolecular interactions, such as hydrogen bonding and π–π interactions. Buckingham and Fowler (1985) showed that optimizing the electrostatic energy, within accessible orientations, could predict the structures of over two dozen, mainly hydrogen bonded, van der Waals complexes of polar molecules. An accurate electrostatic model, where the atomic dipoles, quadrupoles, etc., represent the lone pair and π electron density sought out by the hydrogen bond donor is essential in these systems. Indeed the introduction of accurate electrostatic models transformed our ability to predict the structures of van der Waals complexes (Dykstra, 1993), which, like molecular crystals, depend on the interactions between molecules in van der Waals contact, which prompted us to investigate the introduction of distributed multipole models into crystal structure prediction.

4.2.1 Derivation of distributed multipole models

There are many ways of dividing up the molecular charge distribution so that it can be represented by a multipole expansion at several sites within the molecule, the most common choice being an expansion site at each nucleus, so that each multipole series represents an atomic charge distribution. Methods can be based on the physical division of the molecule into atoms by the zero flux surfaces in the *ab initio* charge density (Laidig, 1993) or by Hirschfelder partitioning (the ACME multipoles of Ritchie (1985)). Other *ab initio* methods are defined by the basis set, such as the Cumulative Atomic Multipole Moments (CAMM) of Sokalski and Sawaryn (1987), the multipoles of Pullman and Perahia (1978) or Vigné-Maeder and Claverie (1988) and the Distributed Multipole Analysis (DMA) of Stone and Alderton (1985). However, in contrast to atomic charges, the choice of partitioning method is not critical, provided the order of multipoles and the number of sites are sufficient to represent accurately the molecular charge density. Spackman (1986) showed that various partitioning schemes gave very similar results for the HF dimer, provided the expansion was taken up to at least atomic quadrupoles. The atomic quadrupoles result from the contributions to the charge density involving two p orbitals, such as π and lone pair density, and so are essential for a reasonable description of molecular charge distributions. The atomic multipole series usually converge rapidly after the quadrupole moment, as the octupole, hexadecapole, etc., only describe d orbital contributions and refine the description of the bonding electron density. Since there are $2l + 1$ independent components of a multipole of rank l, (e.g. three for dipole, five for quadrupole, etc.) and it is the atomic site symmetry that determines whether any are zero, these multipole models involve many parameters. However, one advantage of these schemes is that the electrostatic property required (e.g. the potential in a given region outside the molecule) can be evaluated for the full expansion, and then the effect of simplifying the model by removing small components or higher multipoles can be assessed. In practice this is often easier than the procedure of fitting atomic multipole models directly to the electrostatic potential, an extension of the potential derived charge method which has been investigated by Williams (1988).

Thus, the main limitation to the accuracy of distributed multipole electrostatic models is the accuracy of the *ab initio* wave functions from which they were derived. For organic molecules containing up to about 30 atoms, SCF calculations can now be performed routinely, using direct methods. However, the omission of electron correlation leads to a significant overestimate of the electrostatic forces. The commonly used SCF 6-31G** wave function overestimates molecular dipoles by 12–15% according to Ryan (1994), although the commonly used corrective scaling factor of 0.91 was derived by Cox and Williams (1981) from eight molecules where the overestimate ranged from 4.5% for CH_3CN to 28.5% for NH_3. Scaling the atomic multipoles by the ratio of the dipole moment magnitudes does indeed give a good prediction of the changes in the electrostatic potential around *N*-acetylalanine *N'*-methylamide resulting from the inclusion of the effects of electron correlation and variations in the basis set (for split-valence and better quality) (Price *et al.*, 1992), which produces a corrective scaling factor of 0.9 for the SCF 6-31G** wave function. Until high quality correlated wave functions become routine for organic molecules, the wave function, or errors in scaling corrections, will always be a limitation on the accuracy of the electrostatic model.

Alternatively, atomic multipoles may be derived from high quality, low temperature X-ray data. Such experimental multipoles already include the effect of the polarization of the molecular charge density by the specific crystal environment. X-ray derived atomic multipole moments were used by Berkovitch-Yellin and Leiserowitz to study the role of electrostatic forces in the crystal packing motifs of amides (1980) and carboxylic acids (1982). They showed that the anisotropic atomic multipole moments made significant contributions to the energies of various packing motifs in amides, with the monopole–monopole energy being 60% or less of the total electrostatic energy.

The importance of the anisotropic multipoles (dipole, quadrupole, etc.) will naturally depend on the charge distribution in the molecule concerned. The errors introduced into the predicted electrostatic field around pyrimidine and uracil from the use of Mulliken analysis, rather than a full DMA representation of the wave function, are very significant (of similar magnitude to the actual field) in the regions around the lone pair and π electron density (Price and Richards, 1991). Although anisotropic multipoles are required to describe accurately the electrostatic forces arising from such non-spherical features in the atomic charge distribution, since electrostatic terms are long range, the electrostatic properties around a given functional group will also be strongly affected by the neighbouring functional groups, as is apparent in Plate X, which shows the electrostatic potential 1 Å from the van der Waals surface of cytosine. Also shown is the contribution to the potential from the dipole, quadrupole, octupole and hexadecapole moments on the atoms, which vary the potential in a complex fashion, making the potential around the carbonyl oxygen considerably less negative and increasing it most above the HC=CH part of the aromatic ring. The relative magnitude of these effects, and the difficulty of modelling the electrostatic forces in the van der Waals contact region by atomic charges, suggest that atomic multipole models should be used when simulating all but the simplest molecular crystal structures.

4.2.2 Implementation of atomic multipoles in a crystal structure relaxation program

The fundamental difference between an isotropic and an anisotropic atom–atom model is that the intermolecular forces between two atoms are no longer central and a torque is generated. Just as molecular orbital theory predicts where the lone pair or π electron density is relative to the atomic nuclei, the orientations of the atomic dipoles, etc., are fixed relative to the intramolecular bond. Thus, atomic dipoles will try to align with the local field, producing a torque which tries to move the intramolecular bond. Thus, as molecules rotate and translate during energy minimization, the atomic multipole components change, according to the rotation properties of tensors, relative to the global (crystallographic) coordinate system, whilst remaining fixed relative to molecule fixed axes. The easiest mathematical description is different for spherical or Cartesian definitions of the multipole tensors, although the conversion between the two types of multipole is straightforward.

For spherical tensors, the multipole moments are defined relative to the local axis system of the rigid molecule and so the energy of interaction between two multipole moments on different atoms is a function of the atom–atom separation **R** and some of the scalar products between this vector **R** and the unit local axis vectors on each atom ($\mathbf{x}_1,\mathbf{y}_1,\mathbf{z}_1,\mathbf{x}_2,\mathbf{y}_2,\mathbf{z}_2$). For example, the interaction between the out-of-plane component of the quadrupole tensor (represented by Q_{20} or Θ_{zz}) on site 1 and the x component of dipole ($Q_{11c} = \mu_x$) on site 2 has the form:

$$U_{\text{estatic}} = (4\pi\varepsilon_0)^{-1} Q^1_{20} Q^2_{11c} R^{-7} \tfrac{3}{2} [-5(\mathbf{z}_1 \cdot \mathbf{R})^2(\mathbf{x}_2 \cdot \mathbf{R}) + 2R^2(\mathbf{z}_1 \cdot \mathbf{R})(\mathbf{z}_1 \cdot \mathbf{x}_2) + R^2(\mathbf{x}_2 \cdot \mathbf{R})]$$

The explicit formulae for all terms in the multipole expansion up to R^{-5}, which includes the quadrupole–quadrupole, octupole–dipole and hexadecapole–charge terms, have been published (Price *et al.*, 1984; Stone, 1991). The chain-rule type formalism for the associated forces, torques and second derivatives has also been established, along with the derivatives with respect to the strain matrix which defines the unit cell shape, by Willock *et al.* (1995), with related analyses by Popelier and Stone (1994).

This formalism has been used to adapt THBREL, a crystal structure relaxation program which has been widely used in modelling inorganic solids, to model rigid molecules whose electrostatic interactions are described by a DMA model (Willock *et al.*, 1995). The program uses the Cartesian coordinates of the centres of mass of each molecule, rotations around the molecule fixed axes and the six strain matrix elements as variables, so the relaxation of the crystal structure to a minimum in the lattice energy is not restricted to specific space groups.

4.3 Using distributed multipoles to model molecular crystal structures

Going beyond the isotropic atom–atom model, by introducing accurate, anisotropic electrostatic models into crystal structure modelling, required considerable effort in the development of suitable programs such as DMAREL. Is the effort only useful for highly detailed modelling of individual systems, or will it make a significant difference to our ability to model the room temperature crystal structures of polar and hydrogen bonded molecules?

The use of a DMA electrostatic model plus an empirical isotropic atom–atom *6-exp* potential has already been tested (Coombes *et al.*, 1996) for around 40 compounds, ranging from nucleic acid bases to nitrobenzenes and nitroanilines. Thus, the molecules contain nitrogen atoms within amide, amine and nitro groups and within aromatic rings. Some of the polar molecules exhibit N–H···O=C or N–H···N intermolecular hydrogen bonding. The group includes molecules of interest to the pharmaceutical, energetic materials and non-linear optics industries.

The molecules were kept rigid at the experimental X-ray structure, except that the C–H and N–H bond lengths were standardized to 1.08 Å and 1.01 Å, respectively. The electrostatic model used was the DMA of a 6–31G** SCF wave function, with all multipoles multiplied by a factor of 0.9 to model crudely the effect of electron correlation on the molecular charge distribution. An initial repulsion-dispersion model was constructed from earlier studies. Combining rules were assumed, so that the repulsion-dispersion interaction between an atom i of type ι and atom k of type κ in another molecule, separated by R_{ik} is given by

$$U_{ik} = a_\iota a_\kappa \exp\left(-(b_\iota + b_\kappa)R_{ik}\right) - \frac{c_\iota c_\kappa}{R_{ik}^6}$$

The parameters of Williams and Cox (1984), derived from fitting the azahydrocarbons, were used for C, N and all H atoms bonded to C atoms. The O parameters were taken from fits to oxohydrocarbons (Cox *et al.*, 1981). The parameters of Williams *et al.* were the most suitable for this simple comparison, as they had been carefully derived in conjunction with an electrostatic model, whereas those of Gavezzotti and Filippini (1994) had absorbed the electrostatic contribution. However, it should be noted that they were not derived for any molecule containing both N and O atoms, nor any markedly polar or hydrogen bonded system. The parameters for polar hydrogen atoms H_p(–N) were derived (Coombes, 1996) from a fit to an *ab initio* surface for formamide/formaldehyde (Mitchell and Price, 1990).

Some sample results for relaxations from the room temperature experimental structure are given in Tables 11.2 and 11.3 (under the heading SDMA). This shows that, for many molecules, this simple scheme gives a structure prediction within a few percent in the lattice parameters, which is about as good as can be expected from using a static minimization method and such grossly transferred repulsion-dispersion parameters. For the vast majority of the structures, the results are at least of the quality that is normally considered acceptable when fitting parameters to a limited range of molecules. For example, the amide potentials of Hagler *et al.* (1974) predict the structure of urea with an r.m.s. error in the cell edges of 1.2%, in contrast to 1.6% (Table 11.2), whereas for formamide this charge based potential scheme produces an r.m.s. error of 4.8% in contrast to the DMA based potential which has an error of only 2.3%. In some cases, these potentials succeed far better than published isotropic

Table 11.2 Prediction of molecular crystal structures using distributed multipole electrostatic models.

	Urea, CH_4N_2O UREAXX09		*p*-Benzoquinone, $C_6H_4O_2$, BNZQUI		Purine, $C_5H_4N_4$, PURINE		3,5-Diamino-2,4,6-trinitrobenzamide, $C_7H_6N_6O_7$, BICWEP	
	(Guth *et al.*, 1980)		(Trotter, 1960)		(Watson *et al.*, 1965)		(Ammon and Bhattacharjee, 1982)	
	Expt	Error	Expt	Error	Expt	Error	Expt	Error
a(Å)	5.645	−0.075	7.055	−0.057	15.553	−0.214	16.557	−0.082
b(Å)	5.645	−0.075	6.795	0.006	9.374	0.186	10.245	−0.130
c(Å)	4.704	0.091	5.767	−0.117	3.664	0.000	5.940	−0.007
β(°)	90.0	0.0	101.47	−2.67	90.0	0.0	91.840	−0.923
Δx(Å)		0.046		0.083		0.168		0.109
$\Delta\theta$(°)		0.000		6.843		7.533		0.598
r.m.s. % error		1.559		1.262		1.394		0.791
U_s (kJ/mol)	−77.506	−78.321	−71.602	−73.969	−101.545	−106.721	−166.220	−167.692
ΔH_{sub} (kJ/mol)	−98.6		−62.8					

Errors (P(calc)-P(expt)) in the predicted molecular crystal structures, calculated by minimizing the static lattice energy, starting from the experimental structure, for a model potential which includes a distributed multipole electrostatic model. The electrostatic term uses a DMA of a 6–31G** SCF wave function, with all multipoles scaled by a factor of 0.9. The repulsion-dispersion potentials are taken from the literature (see text). The r.m.s. % error is calculated over the three cell edge lengths. U_s is the calculated lattice energy, given at both the experimental and relaxed crystal structures. This can be compared with the experimental heat of sublimation ΔH_{sub} (Chickos, 1987), where available.

Table 11.3 Sensitivity of molecular crystal structure predictions to the electrostatic model.

	Cytosine, $C_4H_5N_3O$, CYTSIN01 (McClure and Craven, 1973)				0-Dinitrobenzene, $C_6H_4N_2O_4$, ZZZFYW01 (Herbstein and Kapon, 1990)				m-Dinitrobenzene, $C_6H_4N_2O_4$, DNBENZ10 (Trotter and Williston, 1966)			
	Expt	SDMA	DMA	CHAR	Expt	SDMA	DMA	CHAR	Expt	SDMA	DMA	CHAR
a(Å)	13.044	−0.069	−0.348	−0.345	7.945	−0.003	−0.027	−0.213	13.257	−0.152	−0.184	−0.618
b(Å)	9.496	−0.003	−0.062	−0.254	12.975	−0.040	−0.048	−0.474	14.048	−0.143	−0.201	−5.160
c(Å)	3.814	0.042	0.103	0.250	7.421	−0.359	−0.374	0.139	3.806	−0.021	−0.014	2.251
β(°)	90.0	0.0	0.0	0.0	111.88	−2.054	−1.961	−1.785	90.0	0.0	0.0	0.0
Δx(Å)		0.216	0.261	0.337		0.158	0.155	0.321		0.157	0.179	3.482
$\Delta\theta$(°)		2.569	4.805	5.856		3.931	3.892	5.266		3.945	3.953	38.848
r.m.s. % error		0.711	2.219	4.368		2.795	2.922	2.831		0.942	1.171	40.292
U_{estat}(kJ/mol)	−105.744	−101.579	−135.280	−143.730	−32.317	−33.057	−41.374	−69.910	−21.494	−21.996	−27.872	−94.411
U_s(kJ/mol)	−128.486	−131.753	−156.530	−170.335	−98.061	−100.678	−108.484	−129.518	−88.484	−90.064	−95.292	−149.572
ΔH_{sub}(kJ/mol)		−155.0					−87.9				−87.0	

Errors (P(calc)-P(expt)) in the relaxed crystal structures using a fixed repulsion-dispersion potential (as in Table 11.2) and various electrostatic models derived from a DMA of a 6–31G** SCF wave function: DMA, full multipoles up to hexadecapole; SDMA, all multipoles scaled by 0.9; CHAR, just the charge component of the DMA. The r.m.s. % errors were calculated over the three cell lengths. The electrostatic contribution, U_{estat}, to the total lattice energy, U_s, is given at both the experimental (for the SDMA model) and relaxed crystal structures. This can be compared with the experimental heat of sublimation ΔH_{sub} (Chickos 1987), where available.

atom–atom potentials, for example, for *p*-benzoquinone, Table 11.3 predicts an r.m.s. error of 1.3% in the cell lengths, in contrast to 4.1% by Cox *et al.* (1981). However, for most of the organic molecules that we have investigated, there are no published crystal structure modelling results.

A test of the role of the electrostatic interaction in determining crystal structures is given in Table 11.3. Here the energy relaxation results are given first for the above model, second for the same model without the scaling to allow for electron correlation and third for a model which excludes anisotropic multipoles. The results clearly demonstrate that the predicted structure can be very sensitive to the electrostatic model, but that this sensitivity varies. The structure of *ortho*-dinitrobenzene is equally well calculated by all three electrostatic models, despite the significant variation in the electrostatic contribution to the lattice energy. In contrast, the anisotropic multipoles play a major role in stabilizing the observed structure of *meta*-dinitrobenzene, which is well predicted by the DMA model, but minimizes to a qualitatively different structure when the atomic dipoles, quadrupoles and higher order multipoles are removed. The contribution to the electrostatic forces from the anisotropic multipoles (illustrated in Plate X) also plays an important role in determining the relaxed crystal structure of cytosine, which is also sensitive to the scaling of the electrostatic forces, since allowing for electron correlation brings the predicted structure of cytosine well within the target accuracy. In contrast, the predicted structures of the dinitrobenzenes are relatively insensitive to the scaling of the distributed multipole electrostatic model. The scaling has a more significant effect on the predicted lattice energies, and is in line with the relative contribution of the electrostatic contribution to the total energy. The lattice energies can be compared with the experimental heats of sublimation, though the approximations in this comparison and the errors associated with the experimental measurement led Pertsin and Kitaigorodsky (1987) to conclude that discrepancies of up to 3–4 kcal/mol (13–17 kJ/mol) should not cause any concern when judging a potential. Thus the sparse available heat of sublimation data is not very useful in assessing the model potentials.

The structures discovered so far that are poorly modelled by this scheme are those where a small change in the relative tilt of the conjugated rings makes a significant difference to the lattice parameters, although there is a very small energy change associated with these changes in tilt angle. In some cases, such as benzene, this softness has been experimentally observed in the thermal expansion (Table 11.1). Such structures are sensitive to the electrostatic model, to a degree that a crude scaling factor is inadequate; and they will also be highly sensitive to other terms in the potential.

However, with the exception of these unusually sensitive structures, the accurate electrostatic model does very well in predicting the relative orientation in hydrogen bonding and polar molecules, despite the crudeness of the repulsion-dispersion potential. Attempts to improve the predictions by optimizing the repulsion potential parameters have not yet been successful, as reasonable changes in the parameters tend to make only small differences to the predicted structures, with some being improved while others deteriorate.

4.3.1 Anisotropy in other terms in the potential

The other terms in the potential can clearly be improved. The assumption that nitrogen atoms in different bonding environments, with different short-range inductive effects, have the same repulsion-dispersion parameters is clearly an approximation. The differences in their atomic charge distributions are included in the electrostatic model. For example, in the azabenzenes, the nitrogen charge component in a DMA is approximately twice as large when the aromatic nitrogen is bonded to two carbon atoms (C–N–C) than when one neighbour is another electronegative nitrogen atom (C–N–N) (Price and Stone, 1983). Such a difference in the charge associated with an atom must also affect its repulsion-dispersion interactions. Similarly, although allowing different parameters for different hybridization states (e.g. sp^2 and sp^3 nitrogens) within the isotropic model may be an improvement, as implemented in AMBER (Weiner *et al.*, 1984), it would be more realistic to introduce anisotropy into the repulsion. This may well be needed for close N···N interactions, as both our model and that of Williams and Weller (1983) produce large errors ($\Delta b \approx 0.5$ Å) in the predicted structure of *s*-tetrazine, despite including the electrostatic effects of the lone pair density; but an anisotropic nitrogen repulsion model produces a reasonable description (Price and Stone, 1984).

Evidence that anisotropy in the repulsive wall can be important for some atoms follows from the plots of effective van der Waals radius as a function of the angle between the intermolecular contact vector and the C–X intramolecular bond, derived from the Cambridge Structural Database by Nyburg and Faerman (1985). For chlorine, bromine, iodine and sulphur, it is very evident that two atoms can approach each other more closely for linear (head-on) contacts than when the contact is side on. The common occurrence of 'short' Cl···Cl contacts, varying from 3.25 Å to the accepted spherical van der Waals diameter of 3.52 Å, has caused considerable debate. However, detailed calculations of the various contributions to the intermolecular forces (Price *et al.*, 1994) show quite clearly that the exchange-repulsion does vary significantly with the relative orientation of the Cl atoms in contact. Thus, an anisotropic repulsion potential would be needed to reproduce a series of crystal structures which included all of the Cl···Cl contacts over the range. Indeed, Wheeler and Colson (1976) showed that it is impossible to get an isotropic atom–atom potential which can satisfactorily model all three phases of *p*-dichlorobenzene, because the shortest Cl···Cl van der Waals contacts are 3.73, 3.38 and 3.79 Å for the α, β and γ phases, respectively. No spherical repulsive wall can extrapolate correctly to other orientations when the repulsion is significantly anisotropic. Nevertheless, although anisotropic repulsion may be necessary to reproduce some short van der Waals contact distances, Hsu and Williams (1980) have shown that it is certainly possible to get reasonable cell dimension predictions for many perchlorohydrocarbons without the inclusion of this feature.

Further improvements in our model potentials and simulation methods will therefore undoubtedly increase the detailed accuracy of molecular crystal structure predictions and will be required for crystal structures that correspond to weakly defined minima. However, for a routine transferable scheme, the addition of a realistic *ab initio* based electrostatic model clearly improves the range of molecules where a minimum in the lattice energy is close to the observed structure. The use of a theoretically derived, rather than an empirical potential, also increases confidence in the extrapolation of the potential to regions sampled in hypothetical crystal structures.

5 DEVELOPMENTS IN CRYSTAL STRUCTURE PREDICTION METHODS

The second requirement for genuine crystal structure prediction is a method of generating hypothetical molecular crystal structures which can be used as starting points for energy minimizations. We wish such a method to generate sufficient starting points that all low energy structures which might be possible polymorphs would be found. The problem of exploring a complex energy surface with multiple minima has been discussed in Chapter 5, which emphasized the role of techniques based on simulating annealing. Here, we focus on systematic search methods which are particularly appropriate for molecular crystals and which are becoming practicable because of the increase in computer power.

All the methods discussed here are designed for organic molecules and are only capable of generating the most common packing symmetries; moreover, they only consider structures where there is only one independent molecule in the unit cell and only certain space groups. They should, nevertheless, be able to predict the vast majority of the organic crystal structures in the Cambridge Structural Database.

The program MOLPAK (Holden *et al.*, 1993) generates starting points for energy minimization by considering the packing density of the hypothetical structure. The program can currently handle the triclinic, $Z = 2$ monoclinic and the primitive $Z = 4$ monoclinic and orthorhombic space groups, which account for about 70% of the known C, H, N, O, F containing crystal structures with one molecule in the asymmetric unit cell. For each space group, molecules are brought into contact (as defined by a crude repulsion model) with a central molecule in the directions and relative orientations of the coordination sphere corresponding to the space group. The packing volume is then evaluated as the central molecule is systematically rotated within the constraints of the space group symmetry. The most dense structures are then refined by energy minimization, using the program WMIN. For several molecules containing nitro groups on an aromatic or saturated hydrocarbon framework, some with other O, N or F substituents, this search always predicted a low energy structure with the same lattice dimensions as were obtained by minimization from the experimental geometry. However, a search covering the most important $P\bar{1}$, $P2_1$, $P2_1/c$ and $P2_12_12_1$ space groups often found other minima within 1 kcal/mol of the lattice energy of the observed structure and in some cases the energy was below. This was mainly attributed to inadequacies in the effective repulsion-dispersion *6-exp* potential parameters that Holden *et al.* (1993) had developed for the WMIN refinement stage in the procedure, but could indicate unknown polymorphism. The spectacular success in predicting the structure of the local minimum nearest to the experimental structure was achieved using the molecular structure found in the crystal as a test probe. For a further set of molecules, the test probe structure was determined using AM1 semi-empirical calculations to optimize the geometry. In this case, the disagreement in the cell parameters varied from minor ($<$ 0.1 Å) up to 0.5 Å, and is clearly a function of molecular flexibility as well as the molecular packing. The authors identified the ability to use conformationally flexible probes, and refinement of the intermolecular potentials used in the final lattice energy refinement as the key areas for improving this very promising method to give an effective crystal structure prediction program.

An alternative method has been proposed by Gavezzotti (1991), which selects promising starting points on the basis of the interaction energy of clusters of two to four molecules, rather than packing density. Clusters containing one or two of the symmetry operators (translation, inversion, glide or screw axis) are constructed and their geometries varied to find promising structures. This assumes that at least one of the clusters containing a screw, inversion or glide (or translation for $P1$) must have a substantial intermolecular attraction, so that it survives in the solid. Nuclei with low interaction energies are then built into full three-dimensional crystal structures, by a systematic variation of the polar coordinates of each vector, within ranges defined by the statistically observed relationships between the molecular and cell dimensions. The combinations of operators for the space groups $P1$, $P\bar{1}$, $P2_1$, Pc, $P2_1/c$ and $P2_12_12_1$ are already considered by the program PROMET (Gavezzotti, 1991), but other operators or combinations of operators found in the less common space groups could be added, so that the approach could access the space groups that account for more than 80% of crystal structures in the Cambridge Structural Database. The translational search is performed by first building up a string, then a layer, and then the three-dimensional structure. The new structures which fall within a statistically plausible range of packing efficiency and cohesive energy are then refined, first by a steepest descent and then using PCK83 (Williams, 1983), using a highly refined set of potential energy parameters and charges for hydrocarbons.

The method was moderately successful for the trial hydrocarbon structures. It predicted the correct crystal structures of some hydrocarbons and for others showed that many crystal structures with small energy differences are possible. In a few cases the observed crystal structure could not be found. However, it would be premature to contrast these results with those of MOLPAK when the potential energy surfaces of the two types of molecule are so different and when the final optimization process with PCK83 was observed to be sensitive to the starting point and optimization parameters. Flat potential surfaces with a multitude of adjacent minima are particularly likely for hydrocarbons.

The prediction of the optimum packings in one or two dimensions, on the basis of close packing and interaction energy, also appears promising as a route into molecular crystal structures. Scaringe (1990) has developed a technique for predicting layer structures, based on the observation that only seven of the eighty crystallographic layer groups will allow the close packing of arbitrary shaped molecules with one symmetry independent molecule per unit cell. He estimated that perhaps 94% of observed crystal structures can be analysed in terms of layers with one of these seven symmetries. With only seven groups, it is possible to systematically search for low energy, close-packed structures by varying the orientational angles, as allowed by the symmetry, to give a three-dimensional potential surface. The molecular centre is confined to the plane, but the tilt of the molecule is optimized. The method was tested on a dozen conjugated hydrocarbons (ranging from naphthalene to dimethylbenzanthracene) and molecules that have a distinctive three-dimensional shape, such as helicenes or a bicyclic molecule. For most molecules, a predicted layer structure was in close agreement with that found experimentally within the crystal, with variations of only a few tenths of an Angstrom in the lattice constants, a few hundredths in the displacements and up to 10° in the orientational angles. Often there were several low energy layer structures within a few kcal/mol of the global minimum, and it was not always the global minimum found in the search that was in best agreement with the experimental layer. The worst results were for the helicenes, where the molecular mass centre was displaced from the layer plane by 0.5 Å or more in the experimental structure, which was not allowed in the search.

A key result from these layer structure predictions is that, even when the interlayer interactions are up to half the strength of the intralayer interactions, this is insufficient to markedly perturb the ideal structure of the free layer. This suggests that a crystal structure prediction method based on building up layers may be successful for molecules that are not obviously stratified. Indeed, the work of Perlstein (1992, 1994) suggests that even one-dimensional structures may be sufficiently well preserved within the crystal to allow his Monte-Carlo based method of predicting one-dimensional aggregates to be developed into a method of predicting three-dimensional crystal structures. The aggregate within the crystal is defined by the nearest neighbours which have the highest interactions energy with the central molecule. Only four types of one-dimensional aggregate commonly occur in molecular crystals. These aggregates consist of lines of molecules related by a translation, screw, glide or inversion operation. Perlstein (1994) showed that of 60 aggregate structures selected at random from the Cambridge Structural Database, 53 had structures that were close to a local energy minimum for the isolated chain, and this local minimum was less than 6 kcal/mol above the global minimum. Commonly, the global minimum energy aggregate had surface cavities that could not be filled by parts of the molecule.

There are therefore several methods of reducing the problem of generating hypothetical crystal structures so that it is within current computational capabilities and remains applicable to the majority of crystal structures. Although they vary in the assumption behind the initial search, either seeking high density (MOLPAK), favourable interaction energy for a nucleus (PROMET), or formation of one-dimensional or two-dimensional aggregates, all these approaches use the close packing principle and lattice energy calculations. These schemes are all promising and their relative efficiency or complementarity has yet to be assessed. This may well depend on the types of molecule and interaction present. MOLPAK (Holden *et al.*, 1993) was specifically designed for the prediction of energetic materials, where density is an important parameter, so a search defined by density will be most effective for practical applications. The evaluation of the interaction energy of a cluster may well be more successful for polar or hydrogen bonded molecules, where the variation in packing energy with orientation will be more dramatic and discriminating than it was for hydrocarbons, the first published examples of the use of PROMET. However, the user adaptation of the method could be important. For example, for molecules capable of forming multiply hydrogen bonded dimers or chains, clusters providing nuclei for the chain structure must not be ignored in a PROMET analysis, even though they are not as energetically favoured as the dimer. In such cases, building up structures on the basis of the one- or two-dimensional components should be very effective.

These methods will all generate hypothetical crystal structures, which need to be subject to a full lattice energy minimization before comparison with experiment. The method used for energy minimization may also be important. For example, the use of a minimization program which is not constrained by the space group symmetry might lead to the location of minima of different symmetry from the starting point. However, the problem of locating all the minima will be very dependent on the nature of the potential energy surface. When there is a very flat potential energy surface with multiple adjacent minima, any minimization procedure will have problems and the use of static methods is inappropriate. However, even these preliminary results suggest that many molecules will have more than one low energy static crystal structure. Indeed, a systematic search for possible crystal structures of the six hexopyranoses within the $P2_12_12_1$ space group produced of the order of a thousand possible structures for each molecule within 10 kcal/mol (van Eijck *et al.*, 1995). Such conformationally flexible sugars may be unusual in the number of low energy structures, which could be related to the notoriously poor crystallization propensity of the sugars. Nevertheless, it is clear that observed structures will often not be significantly more stable than many hypothetical structures. Considerably more research is required before we will genuinely understand and predict which crystalline structures can be formed by a given molecule.

6 OUTLOOK

There have been significant recent advances in both the model potentials that can be used in molecular crystal structure prediction and new methods of searching for hypothetical structures. We may expect that these will come together over the next decade, so that the prediction of a molecular crystal structure from first principles becomes possible. This should lead to an understanding of polymorphism.

ACKNOWLEDGEMENTS

I would like to thank Mr D. S. Coombes for carrying out the calculations for Tables 11.2 and 11.3, and the EPSRC and The Wellcome Foundation Limited for financial support for the crystal structure prediction work.

REFERENCES

Allen, F.H. and Kennard, O. (1993) *Chem. Des. Automation News*, **8**, 1, 31.
Allen, F.H., Davies, J.E., Galloy, J.J., Johnson, O., Kennard, O., Macrae, C.F., Mitchell, E.M., Mitchell, G.F., Smith, J.M. and Watson, D.G. (1991) *J. Chem. Inf. Comput. Sci.*, **31**, 187.
Allinger N.L. and Yan, L. (1993) *J. Am. Chem. Soc.*, **115**, 11 918.
Ammon, H.L. and Bhattacharjee, S.K. (1982) *Acta Crystallogr., B*, **38**, 2083.
Belsky, V.K. and Zorkii, P.M. (1977) *Acta Crystallogr., A*, **33**, 1004.
Berkovitch-Yellin, Z. and Leiserowitz, L. (1980) *J. Am. Chem. Soc.*, **102**, 7677.
Berkovitch-Yellin, Z. and Leiserowitz, L. (1982) *J. Am. Chem. Soc.*, **104**, 4052.
Buckingham, A.D. and Fowler, P.W. (1985) *Canad. J. Chem.*, **63**, 2018.
Chickos, J.S. (1987) In *Molecular Structure and Energetics*, Vol. 2 (eds J.F. Leibman and A. Greenberg), VCH, New York; and personal communication of updated tables.
Coombes, D.S. (1996), *Phil. Mag.*, **73**, 117.
Coombes, D.S., Price, S.L., Willock, D.J. and Leslie, M. (1996) *J. Phys. Chem.*, **100**, 7352.
Cox, S.R. and Williams, D.E. (1981) *J. Comput. Chem.*, **2**, 304.
Cox, S.R., Hsu, L.-Y. and Williams, D.E. (1981) *Acta Crystallogr., A*, **37**, 293.
Dykstra, C.E. (1993) *Chem. Rev.*, **93**, 2339.
English, C.A. and Venables, J.A. (1974) *Proc. Roy. Soc., A*, **340**, 57.
Filippini, G. and Gavezzotti, A. (1993) *Acta Crystallogr., B*, **49**, 868.
Gavezzotti, A. (1989) *J. Am. Chem. Soc.*, **111**, 1835.
Gavezzotti, A. (1991), *J. Am. Chem. Soc.*, **113**, 4622.

Gavezzotti, A. and Filippini, G. (1994) *J. Phys. Chem.*, **98**, 4831.
Guth, H., Heger, G., Klein, S., Treutmann, W. and Scheringer, C. (1980) *Z. Kristallogr.*, **153**, 237.
Hagler, A.T., Huler, E. and Lifson, S. (1974) *J. Am. Chem. Soc.*, **96**, 5327.
Hagler, A.T., Lifson, S. and Dauber, P. (1979) *J. Am. Chem. Soc.*, **101**, 5122.
Herbstein, F.H. and Kapon, M. (1990) *Acta Crystallogr., B*, **46**, 567.
Holden, J.R., Du, Z. and Ammon, H.L. (1993) *J. Comput. Chem.*, **14**, 422.
Hsu, L.-Y. and Williams, D.E. (1980) *Acta Crystallogr., A*, **36**, 277.
Kitaigorodsky, A.I. (1973) *Molecular Crystals and Molecules,* Academic Press, New York.
Laidig, K.E. (1993) *J. Phys. Chem.*, **97**, 12760.
Lifson, S., Hagler, A.T. and Dauber, P. (1979) *J. Amer. Chem. Soc.*, **101**, 5111.
Maitland, G.C., Rigby, M., Smith, E.B. and Wakenham, W.A. (1981) *Intermolecular Forces,* Clarendon, Oxford.
McClure, R.J. and Craven, B.M. (1973) *Acta Crystallogr., B*, **29**, 1234.
Mighell, A.D., Himes, V.L. and Rodgers, J.R. (1983) *Acta Crystallogr., A*, **39**, 737.
Mitchell, J.B.O. and Price, S.L. (1990) *J. Comput. Chem.*, **11**, 1217.
Nyburg, S.C. and Faerman, C.H. (1985) *Acta Crystallogr., B*, **41**, 274.
Perlstein, J. (1992) *J. Am. Chem. Soc.*, **114**, 1955.
Perlstein, J. (1994) *J. Am. Chem. Soc.*, **116**, 455.
Pertsin, A.J. and Kitaigorodsky, A.I. (1987) *The Atom–Atom Potential Method,* Springer Series in Chemical Physics 43, Springer-Verlag, Berlin.
Popelier, P.L.A. and Stone, A.J. (1994) *Mol. Phys.*, **82**, 411.
Price, S.L. (1985) *Chem. Phys. Letts.*, **114**, 359.
Price, S.L. (1987) *Mol. Phys.*, **62**, 45.
Price, S.L. (1996) *Phil. Mag.*, **73**, 95.
Price, S.L. and Richards, N.G.J. (1991) *J. Comput. Aided Mol. Des.*, **5**, 41.
Price, S.L. and Stone, A.J. (1983) *Chem. Phys. Letts.*, **98**, 419.
Price, S.L. and Stone, A.J. (1984) *Mol. Phys.*, **51**, 569.
Price, S.L., Stone, A.J. and Alderton, M. (1984) *Mol. Phys.*, **52**, 987.
Price, S.L., Andrews, J.S., Murray, C.W. and Amos, R.D. (1992) *J. Am. Chem. Soc.*, **114**, 8268.
Price, S.L., Stone, A.J., Lucas, J., Rowland, R.S. and Thornley, A.E. (1994) *J. Am. Chem. Soc.*, **116**, 4910.
Pullman, A. and Perahia, D. (1978) *Theor. Chim. Acta.*, **48**, 29.
Ritchie, J.P. (1985) *J. Am. Chem. Soc.*, **107**, 1829.
Ritchie, J.P., Kober, E.M. and Copenhaver, A.S. (1993) *Proc. 10th International Symposium on Detonation*, Office of Naval Research.
Ryan, M.D. (1994) In *Modelling the Hydrogen Bond,* ACS Symposium Series, **569**, p. 36.
Scaringe, R.P. (1990) In *Electron Crystallography of Organic Molecules* (eds J.R. Fryer and D.L. Dorset), Kluwer, Dordrecht.
Spackman, M.A. (1986) *J. Chem. Phys.*, **85**, 6587.
Sokalski, W.A. and Sawaryn, A. (1987) *J. Chem. Phys.*, **87**, 526.
Stone, A.J. (1991) In *Theoretical Models of Chemical Bonding* (ed. Z.B. Maksić), Springer, Berlin.
Stone, A.J. and Alderton, M. (1985) *Mol. Phys.*, **56**, 1047.
Trotter, J. (1960) *Acta Crystallogr.*, **13**, 86.
Trotter, J. and Williston, C.S. (1966) *Acta Crystallogr.*, **21**, 285.
van Eijck, B.P., Mooij, W.T.M. and Kroon, J. (1995) *Acta Crystallogr., B*, **51**, 99.
Vigné-Maeder, F. and Claverie, P. (1988) *J. Chem. Phys.*, **88**, 4934.
Watson, D.G., Sweet, R.M. and Marsh, R.E. (1965) *Acta Crystallogr.*, **19**, 573.

Weiner, S.J., Kollman, P.A., Case, D.A., Singh, U.C., Ghio, C., Alagona, G., Profeta, S. and Weiner, P. (1984) *J. Am. Chem. Soc.*, **106**, 765.
Weiner, S.J., Kollman, P.A., Nguyen, D.T. and Case, D.A. (1986) *J. Comput. Chem.*, **7**, 230.
Wheeler, G.L. and Colson, S.D. (1976) *J. Chem. Phys.*, **65**, 1227.
Wiberg, K.B. and Rablen, P.R. (1993) *J. Comput. Chem.*, **14**, 1504.
Williams, D.E. (1983) PCK83, Quantum Chemistry Program Exchange, Indiana University, Bloomington, Indiana, Program 548.
Williams, D.E. (1988) *J. Comput. Chem.*, **9**, 745.
Williams, D.E. and Cox, S.R. (1984) *Acta Crystallogr., B*, **40**, 404.
Williams, D.E. and Starr, T.L. (1977) *Comput. Chem.*, **1**, 173.
Williams, D.E. and Weller, R.R. (1983) *J. Am. Chem. Soc.*, **105**, 4143.
Willock, D.J., Leslie, M., Price, S.L. and Catlow, C.R.A. (1995) *J. Comput. Chem.*, **16**, 628.

12

Amorphous Solids

B. Vessal

1 INTRODUCTION

In this chapter we show how atomistic computer simulation methods can yield unique insights into the structural properties of amorphous solids. The range and scope of the materials that can currently be studied by modelling techniques is illustrated by recent results, predominantly on glassy materials (defined below), although a brief discussion of other classes of amorphous materials is given towards the end of the chapter.

Amorphous solids are compounds without long range order and periodicity (Cusack, 1987). Glasses are a subset of amorphous solids which show a distinct glass transition temperature. Amorphous materials present one of the greatest challenges to the applications of modelling techniques. Experimental methods cannot at present yield accurate structures at the atomic level: they give averaged structures as in the Radial Distribution Functions (RDFs) provided by diffraction measurements, or very local information as in Extended X-ray Absorption Fine Structure (EXAFS) and Nuclear Magnetic Resonance (NMR) experiments; knowledge on intermediate level order is sparse. Computer modelling techniques have, however, the opportunity of yielding detailed models for the structures of amorphous materials at the short and intermediate level.

Several methods have been developed for generating models of the structure of amorphous materials at the atomic level. These include random network models (which may be made by hand or, as discussed later, generated on a computer), energy minimization (or relaxation) methods, Monte Carlo (MC), Molecular Dynamics (MD), Reverse Monte Carlo (RMC), and Quantum Mechanical (QM) methods. The aim of this chapter is to discuss the generation of models for amorphous solids mainly using the most widely used and successful MD and RMC techniques, giving examples from the literature.

The first MD simulation of an amorphous system was reported by Rahman *et al.* (1976). They prepared a 500-particle amorphous Lennard–Jones system by rapidly quenching a liquid configuration. The structure generated was similar to

COMPUTER MODELLING IN INORGANIC CRYSTALLOGRAPHY
ISBN 0-12-164135-X

a random closed packing of hard spheres. Shortly after their pioneering work Woodcock *et al.* (1976) reported the first MD simulation of several vitreous ionic systems namely KCl, BeF_2, $ZnCl_2$, and SiO_2. These two seminal papers set the stage for studies in later years of several amorphous systems by MD, which we will briefly describe in the following pages.

2 TECHNIQUES

We discuss first the basic aspects of the simulation procedure: the energy evaluation using summation methods and the interatomic potentials employed in some of the more widely studied systems. Details of the simulation techniques employed are then reviewed.

2.1 Coulomb sums

Most simulations of amorphous materials have described the structure in terms of a large simulation box to which periodic boundary conditions are applied. The periodicity is, of course, artificial, but if the simulation box is sufficiently large the effect of imposing periodicity should be unimportant. Given this approximation we may use the standard Ewald procedure (Chapter 4) for undertaking the Coulomb summation.

We note, however, in the present context that as discussed in Chapter 4, there are two alternative techniques to the Ewald sum method for evaluating the long range Coulomb interactions. One is the Particle–Particle/Particle–Mesh method (PPPM) (Eastwood *et al.*, 1980) and the other is the Cell Multipole Method (CMM) (Greengard and Rokhlin, 1987). The computational cost for both PPPM and CMM scale as N, the number of particles, while for the Ewald sum the cost scales as $N^{3/2}$ (Fincham, 1994). Of the two alternative techniques, the CMM is gaining more popularity mainly because it is applicable to non-periodic and inhomogeneous systems as well and it is more amenable to parallelization. CMM is slower than the Ewald sum for small systems but it is faster for very large systems. However, it is not certain yet at which value of N the crossover occurs. Values between 300 and 30 000 have been quoted (Fincham, 1994).

Short range potentials are normally handled by standard real-space summation with a specified cut-off. The types of potential commonly employed are discussed in the next section, as in many cases they differ from those used in modelling crystalline structures. These differences are partially due to the greater variety of coordination numbers and bond lengths which places extra demands on the potential functions.

2.2 Potential forms

One of the most widely used potential function forms in the MD simulation of glass structures is a special version of the standard Born–Mayer form (Chapter 3) which has been used in many studies of ionic crystals and which attempts to relate the potential to properties of the individual interacting ions. Known as the Born–Mayer–Huggins (BMH) potential, the functional form is as follows:

$$E_{ij}(r) = Z_i Z_j e^2/r + (1 + Z_i/n_i + Z_j/n_j)\, b \exp[(\sigma_i + \sigma_j - r_{ij})/\rho] \tag{1}$$

where Z is the charge on the different ions i,j, e is the charge on a proton, n is the number of outer shell electrons, σ is a distance parameter characteristic of the ionic radius, and b and ρ are constants. Another widely used (and again older) function is the Pauling potential:

$$E_{ij}(r) = Z_i Z_j e^2/r_{ij}[1 + \text{sgn}\,(Z_i Z_j)\,((s_i + s_j)/r_{ij})^n] \tag{2}$$

where Z_i,Z_j and s_i,s_j are the effective charges and radii of atoms i and j and n is a measure of the hardness of the repulsion. The 'sgn' function is used to keep the non-coulombic part of the potential always positive.

A widely used model, which goes beyond the pair potential approximation, is that due to Keating (1966) which has been used mainly for the relaxation of random network models (Keating, 1966). It is in concept a 'molecular mechanics' like potential as it consists of a bond stretching and bond bending term (Gaskell, 1991):

$$E = E_s + E_b \tag{3}$$

$$E_s = 3\alpha/16r_0^2 \sum_{i,j} (r_{ij}^2 - r_0^2)^2 \tag{4}$$

$$E_b = 3\beta/8r_0^2 \sum_{i,j>k} (r_{ij}r_{ik} - r_0^2 \cos\theta_0)^2 \tag{5}$$

where α and β are stretching and bending force constants, r_0 and r_{ij} are the equilibrium and actual internuclear distances between atoms i and j and θ_0 is the equilibrium angle jik.

Recent work has employed more sophisticated potentials including the following.

2.2.1 Vessal–Amini–Leslie–Catlow potential

Vessal *et al.* (1989, 1993) have used a four-range Buckingham potential to model the short range interactions between different ions. The different components of the potential are as follows:

$$E_{ij} = A_{ij} \exp(-r_{ij}/\rho_{ij}) \qquad r_{ij} < r_1 \tag{6}$$

$$E_{ij} = \sum_{m=0}^{5} A_m r_{ij}^m \qquad r_1 < r_{ij} < r_2 \tag{7}$$

$$E_{ij} = \sum_{m=0}^{3} B_m r_{ij}^m \qquad r_2 < r_{ij} < r_3 \tag{8}$$

$$E_{ij} = -C_{ij}/r_{ij}^6 \qquad r_3 < r_{ij} < r_c \tag{9}$$

where r_c is the short range cut-off.

The function is splined at r_1, r_2, and r_3 so that the energy, and first and second derivatives, are continuous. The function has a minimum at r_2.

Two different forms for the three-body interactions have been used together with the above two-body potential. The original form models the interaction between three atoms i, j, and k according the the following formula:

$$E_{ijk} = \tfrac{1}{4} A_{ijk}(B_{ijk})^2 \exp(-r_{ij}/\rho_1) \exp(-r_{ik}/\rho_2) \tag{10}$$

where

$$A_{ijk} = k_{ijk}/2(\theta_0 - \pi)^2 \tag{11}$$

$$B_{ijk} = (\theta_0 - \pi)^2 - (\theta - \pi)^2 \tag{12}$$

and where k_{ijk} is the three-body spring constant, θ_0 is the equilibrium bond angle and θ is the calculated bond angle.

The following alternative form of the three-body potential has also been used to improve the description of the Si–O–Si bond angle distribution (BAD):

$$E_{ijk} = k_{ijk}[\theta^n(\theta - \theta_0)^2(\theta + \theta_0 - 2\pi)^2 - n/2\pi^{n-1}(\theta - \theta_0)^2(\pi - \theta_0)^3] \tag{13}$$

2.2.2 Stillinger–Weber potential

This potential form has been widely used in simulating amorphous semiconductors (Stillinger and Weber, 1985):

$$E = E_2 + E_3 \tag{14}$$

where

$$E_2 = A(B/r_{ij}p - 1/r_{ij}q) \exp(1/(r_{ij} - \alpha)) \tag{15}$$

and

$$E_3 = h(r_{ij}, r_{ik}, \theta_{jik}) + h(r^{ji}, r_{jk}, \theta_{ijk}) + h(r_{ki}, r_{kj}, \theta_{ikj}) \tag{16}$$

where α is the potential cut-off and θ_{jik} is the angle between r_{ij} and r_{ik} and

$$h(r_{ij}, r_{ik}, \theta_{jik}) = \lambda \exp[\gamma/(r_{ij} - \alpha) + \gamma/(r_{ik} - \alpha)] (\cos\theta_{jik} - \cos\theta_0)^2 \tag{17}$$

The above (and other) types of potential function have been implemented in simulation studies of amorphous materials.

2.3 Techniques

Both Monte Carlo (MC) and Molecular Dynamics (MD) techniques (Chapters 1, 3 and 5) have been used in modelling glass structures. Detailed aspects of the implementation for glassy systems will be considered below. The Reverse Monte Carlo (RMC) technique discussed in Chapter 6 has also been applied to construct atomistic models of glass structures based on X-ray and neutron scattering data. Details of applications to glass structures will be given towards the end of the chapter. The next section focuses on the approaches used in 'preparing' glass structures using simulation techniques.

2.4 Methods of preparation

Simulated glass structures are prepared computationally in the same way as real glasses. Models of crystalline systems are melted; the melts are then quenched, freezing the structure into a disordered glassy phase. The main problem with this technique is in the time scale of the melting and more particularly of the quench, since with current computational capabilities, these are several orders of magnitude greater than the rates achieved experimentally. As a consequence, simulated glass structures have very high fictive temperatures. Nevertheless, the information obtained on glass structures is of great value.

Although most of the amorphous materials modelled by MD have been prepared by melting either a crystalline or random structure and quenching the resulting melt to generate the appropriate glassy structure, other methods of preparation have also been used such as pressure induced amorphization, defect induced amorphization, and radiation induced amorphization. Examples will be considered below.

3 APPLICATIONS

We concentrate in this section on the application of MD techniques to the structure of glasses, although later sections consider other amorphous materials and techniques, principally RMC.

3.1 Molecular dynamics simulations of glass structures

3.1.1 Oxide glasses

(a) Silica The first MD study of the structure of vitreous silica was reported by Woodcock *et al.* (1976). They used full ionic charges on the oxygens and silicons and a BMH two-body term. This simulation succeeded in generating an

amorphous structure in which almost all of the silicons were four-coordinated. The Si–O first nearest neighbours were at the correct distance from each other, but the O–O distances were too short and the Si–Si distances were too long. There was no discussion of the Bond Angle Distribution (BAD), but in general early MD simulations of silica using only two-body interactions gave a broad O–Si–O BAD and the peak in the Si–O–Si BAD was at larger angles than that suggested by experiment.

Vessal *et al.* (1989, 1993) undertook a simulation of vitreous silica at constant volume starting from a molten silica configuration with the experimental density of vitreous silica at 7000 K. A cubic simulation box with periodic boundary conditions was used containing 216 Si^{4+} and 432 O^{2-} ions. The interatomic potentials included both two- and three-body terms and were of the type described in the previous section. The total time for annealing of vitreous silica was 36 ps. A time-step of 1 fs was employed. The RDFs and bond angle distributions (BADs) that are obtained for the vitreous system are shown in Figs. 12.1 and 12.2, respectively.

It is generally found that MD studies of glassy materials result in a glass structure with a high fictive temperature owing to the rapid cooling rates employed; glass transition temperatures are typically high by a factor of 2. Nevertheless the structures generated accord well with experiment. Thus the total RDF obtained for vitreous silica using the potential model derived by Vessal *et al.* (1993) is compared with neutron diffraction results in Fig. 12.3, and shows good agreement with the experimental results obtained from neutron scattering data by Grimley *et al.* (1990). Indeed, Wright (1993) has compared their model to other models and finds it even better than the Bell and Dean model (Bell and Dean, 1972).

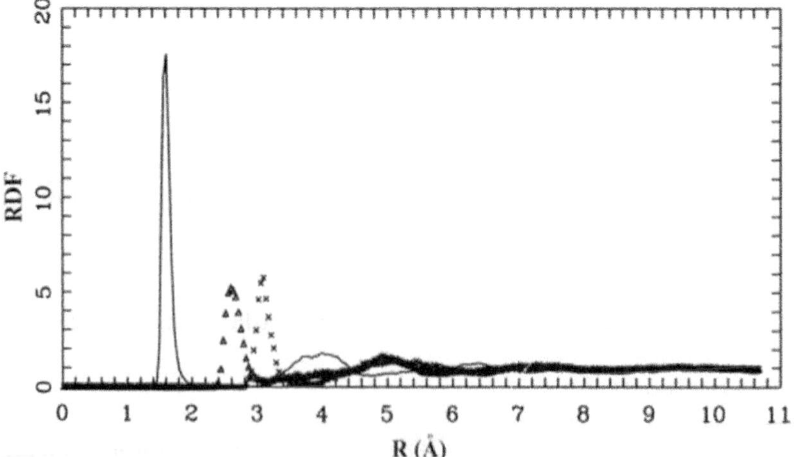

Figure 12.1 RDF for vitreous silica from MD simulation: ×××, Si–Si; ——, Si–O; △△△, O–O.

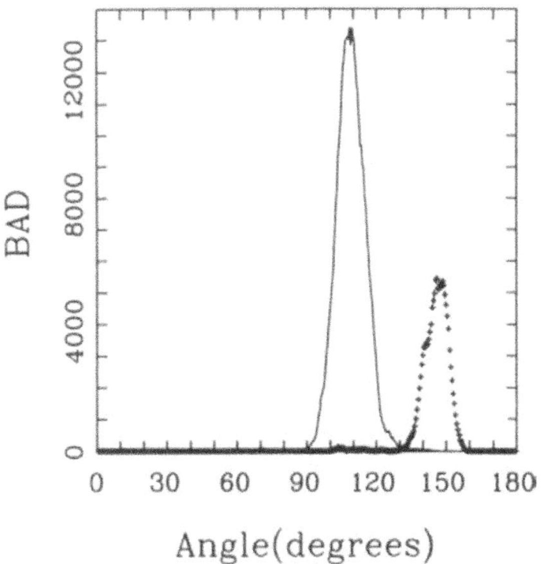

Figure 12.2 BAD for vitreous silica from MD simulation: ——, O–Si–O; +++, Si–O–Si.

One of the major advantages of simulated structures is that they allow us to explore intermediate range order, in particular ring statistics. Figure 12.4 shows the ring analysis for the vitreous silica structure. There is only one three-membered ring in the whole structure and the maximum in the figure is for six-membered rings, in agreement with experiment.

One of the useful methods of characterizing an amorphous structure is by the use of Voronoi polyhedra. A Voronoi polyhedron is the analog of the Wigner–Seitz cell applied to amorphous solids or liquids. The Voronoi polyhedron surrounding any given atom is the smallest polyhedron formed by the planes that perpendicularly bisect the lines joining the atom to its neighbours. The value of this concept is that the statistical distribution of quantities such as polyhedron volumes, number of faces per polyhedron, and the number of edges per face constitutes an economical description of the amorphous structure. The results are presented in Figs. 12.5–12.8. The quantity plotted in Fig. 12.7 is a measure of the asphericity of the polyhedra defined as

$$\omega = (S/4\pi)/(3V/4\pi)^{2/3} \tag{18}$$

where S is the surface area and V is the volume. Obviously ω has a value of unity for a sphere and a value greater than unity for aspherical objects. The long tail in the volume plot is due to edge effects.

(b) Mixed sodium and rubidium silicates Vessal *et al.* (1992) have reported simulations of a series of alkali silicates. The glass compositions varied from $Na_2Si_2O_5$ to $Rb_2Si_2O_5$, including a wide range of mixed-alkali compositions.

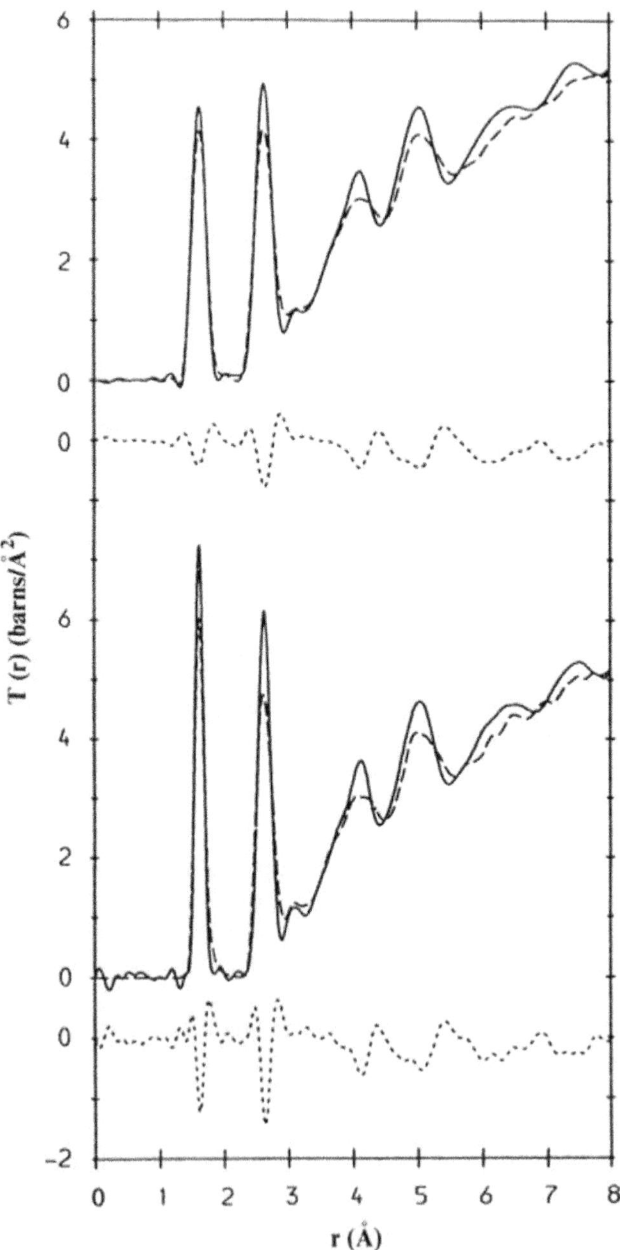

Figure 12.3 Experimental (——), calculated (– – –), and difference curve (- - - -) RDF for silica glass. Lower curves: comparison with experimental data of Grimley *et al.* (1990); upper curves: comparison with the more recent data of Wright *et al.* (1992).

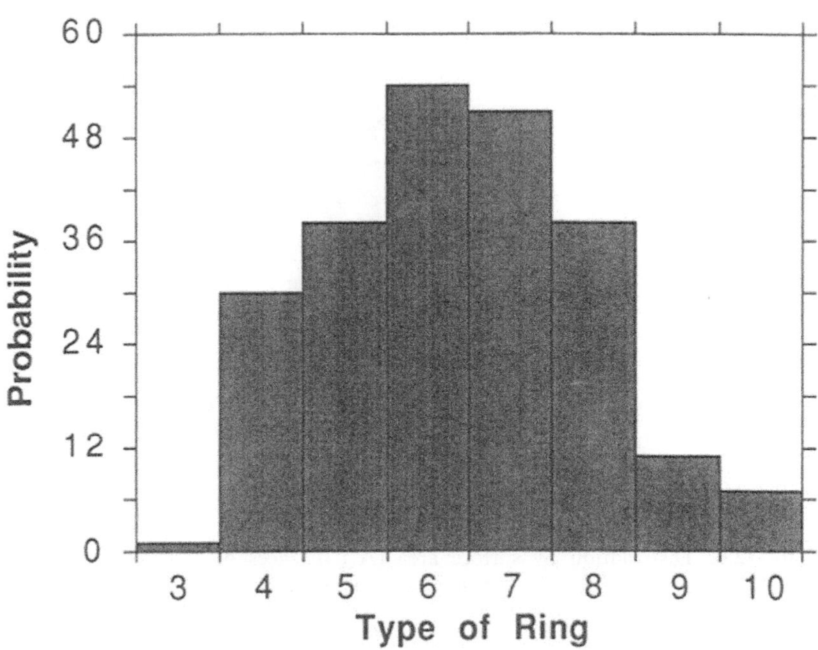

Figure 12.4 Ring analysis for the computer generated vitreous silica.

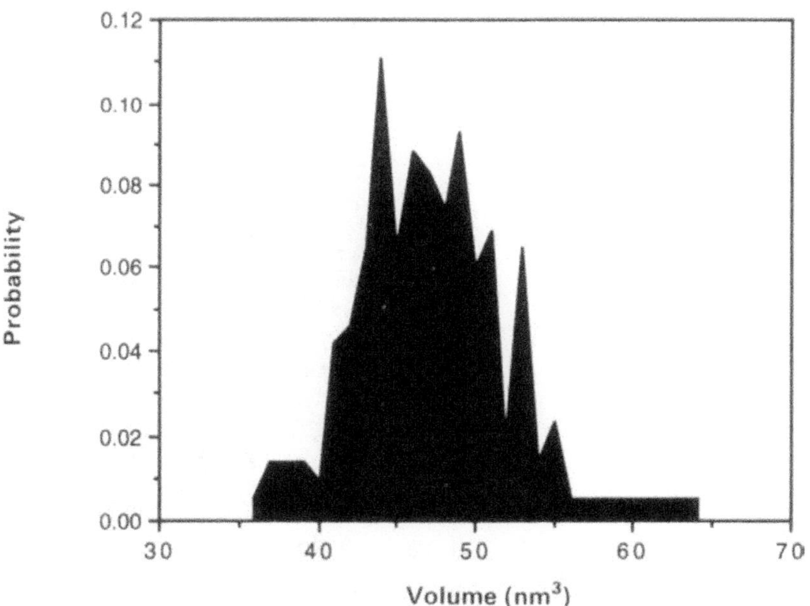

Figure 12.5 Distribution by volume (Å^3) of Voronoi polyhedra for simulated vitreous silica.

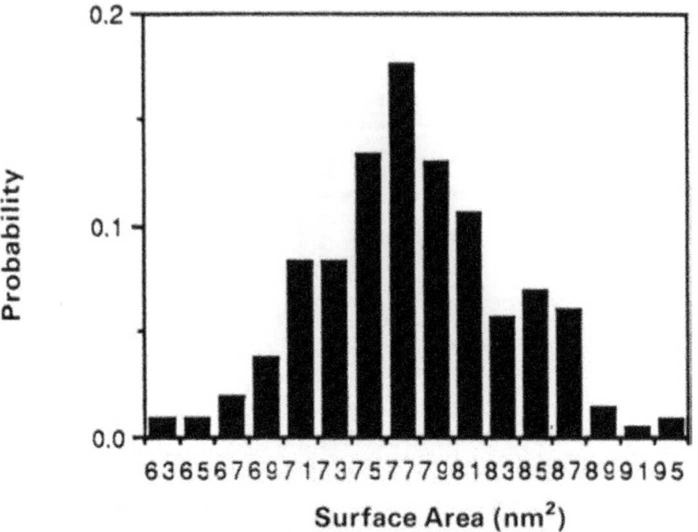

Figure 12.6 Distribution by surface area (Å^2) of faces of Voronoi polyhedra for simulated viteous silica.

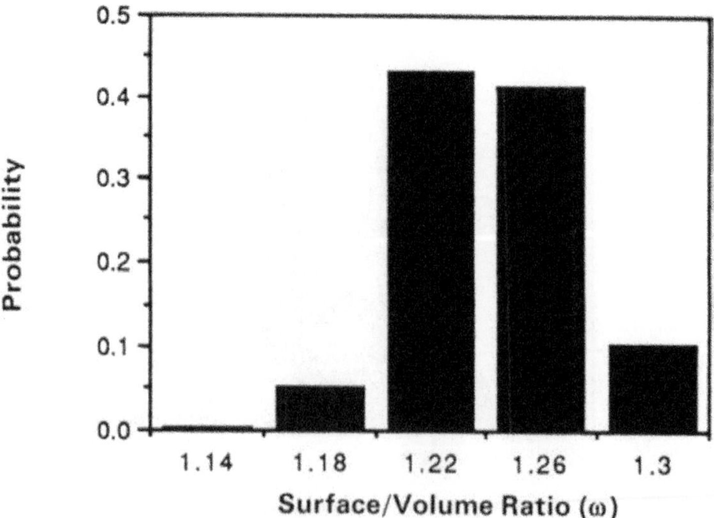

Figure 12.7 Surface to volume ratio for Voronoi tessellation for simulated vitreous silica.

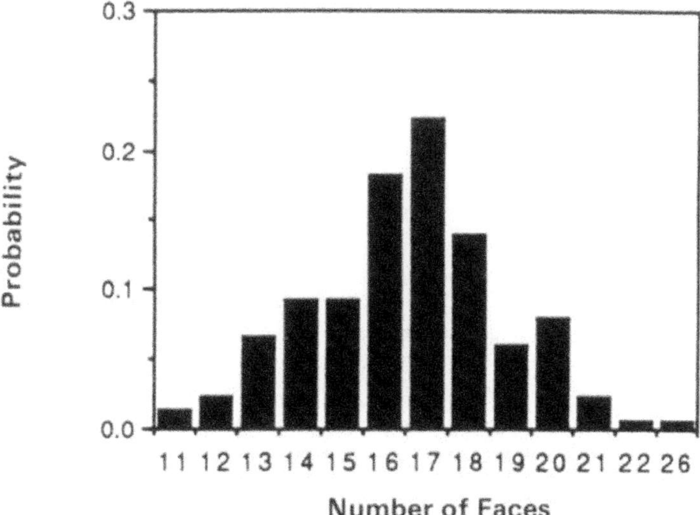

Figure 12.8 Distribution of number of faces of Voronoi polyhedra for simulated vitreous silica.

The partial RDFs for the simulated $NaRbSi_2O_5$ glass are shown in Fig. 12.9. The sharpest features belong to the principal intranetwork correlations: Si–O, O–O, and Si–Si, which are similar to the equivalent features found in vitreous silica with the exception that the Si–O peak at 1.61 Å is split into two contributions at 1.51 and 1.62 Å. Similar split peaks occur in the crystalline lithium and sodium disilicates. Compared to α-quartz, the bridging oxygen distance is increased in these layer structured silicates and the non-bridging oxygen distance decreased, distorting the SiO_4 tetrahedra and increasing the mean Si–O distance. The splitting of the Si–O coordination sphere for the simulated glass structure decreases with sodium content, disappearing for $Na_2Si_2O_5$ glass. Distortion of SiO_4 tetrahedra in $Na_2Si_2O_5$ glass has in fact not been detected in Si EXAFS (Greaves *et al.*, 1990), in agreement with the present results. Undistorted SiO_4 tetrahedra were also observed from Si EXAFS in $Li_2Si_2O_5$ and $K_2Si_2O_5$ glasses, which points to the absence in these glasses of steric constraints present in crystalline structures. The splitting when Rb is the major alkali, is evident in Fig. 12.9. Similar behaviour has been found in simulations of $(K, Cs)_2Si_2O_5$ glasses, where Si–O distances are split for Cs-rich compositions but not for $K_2Si_2O_5$ glass. Recent EXAFS measurements for silicon in single-alkali silicate glasses reveal commensurate increases in the average Si–O distances (Greaves *et al.*, 1990).

The BADs for network atoms in the simulated $Rb_2Si_2O_5$ glass are given in Fig. 12.10 and are typical of structures for the other compositions studied. O–Si–O correlations peak at the tetrahedral angle of 109.5°, which might be

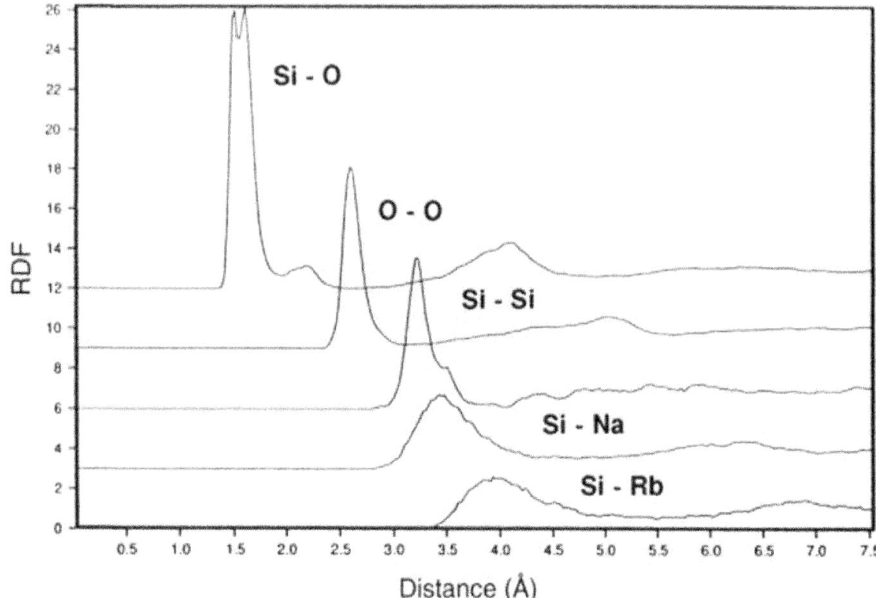

Figure 12.9 Partial RDFs for MD simulated $NaRbSi_2O_5$ glass.

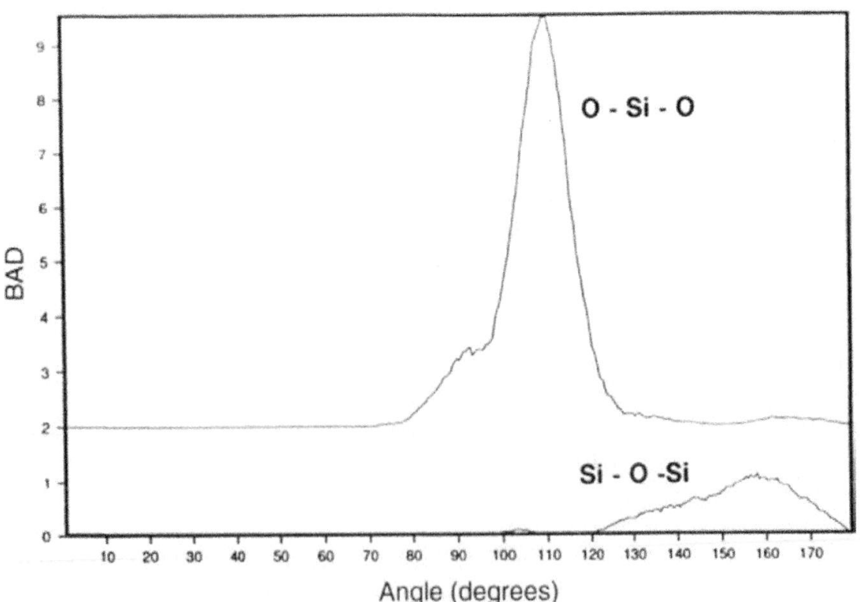

Figure 12.10 BADs for simulated $Rb_2Si_2O_5$ glass.

expected in view of the use of three-body terms in the potential. The distribution of Si–O–Si bond angles was not constrained and is much broader and asymmetric with a maximum close to 160°. Despite the depolymerizing effect of non-bridging oxygens, the short range geometry of the network in modified glasses appears to be close to that of vitreous silica.

The BADs relating to the modifier component in the $Rb_2Si_2O_5$ glass structure are rather broader than in the network distributions shown in Fig. 12.10, commensurate with more weakly bound units. What is particularly interesting is the fact that both the Si–O–Rb and O–Rb–O distributions peak at smaller angles than the Si–O–Si or even the O–Si–O distributions. The maximum for the Si–O–Rb correlations corresponds to an angle close to 100°. Calculations for Na-containing glasses give similar results, with distributions peaking at an angle of 95°. These angles suggest the non-bridging oxygens connecting alkalis and silicons have a coordination number substantially higher than the twofold-coordinated bridging oxygens connecting pairs of silicons in the network. The average coordination number of all types of oxygen in these structures is close to 3, indicating that a third of these, viz. the non-bridging oxygens have a coordination number of approximately 5. The O–Rb–O BAD, whilst this is broader than the Si–O–Rb distribution, still has its maximum around 100°, indicating a geometry for the alkali intermediate between tetrahedral and octahedral.

The distribution of coordination numbers for Rb sites in the single and mixed glass structures is shown in Fig. 12.11. This peaks at 5 for Rb-rich glasses, broadening a little and shifting to slightly higher values for Na-rich glasses. The results of these calculations are consistent with EXAFS measurements obtained for K and Cs in $(K, Cs)_2Si_2O_5$ glasses (Greaves *et al.*, 1990). Coordination numbers of between 5 and 6 were obtained with oxygen and alkali bond angles of between 92° and 108° for both single- and mixed-alkali glasses. It is worth noting that the angles obtained from EXAFS were deduced from the alkali–oxygen and alkali–alkali/silicon distances. The corresponding simulated correlations for Na–O, Rb–O, Na–Si/M and Rb–Si/M can be seen in Figs. 12.9 and 12.12. It is evident from the breadths and positions of these partial RDFs that they serve to smear out structure in the total RDF beyond 3.5 to 4 Å. This underlines the difficulty of obtaining selective information on alkali environments in mixed alkali glasses from X-ray or neutron diffraction data compared to the utility of element selective probes like EXAFS.

The contributions of the modifier components to the RDF of the simulated $NaRbSi_2O_5$ glass are shown in Fig. 12.12. The cation–cation correlations responsible for the small oxygen bond angles discussed above can be clearly distinguished. Na–Na distances peak initially at 3.1 Å and Rb–Rb distances close to 3.9 Å. This separation relates mainly to the different nearest neighbour distances: 2.4 Å for Na–O compared to 3.0 Å for Rb–O. In the total RDF measured by diffraction techniques these distances fall either side of the strong O–Si–O peak at 2.65 Å and can only be accurately measured by EXAFS. In

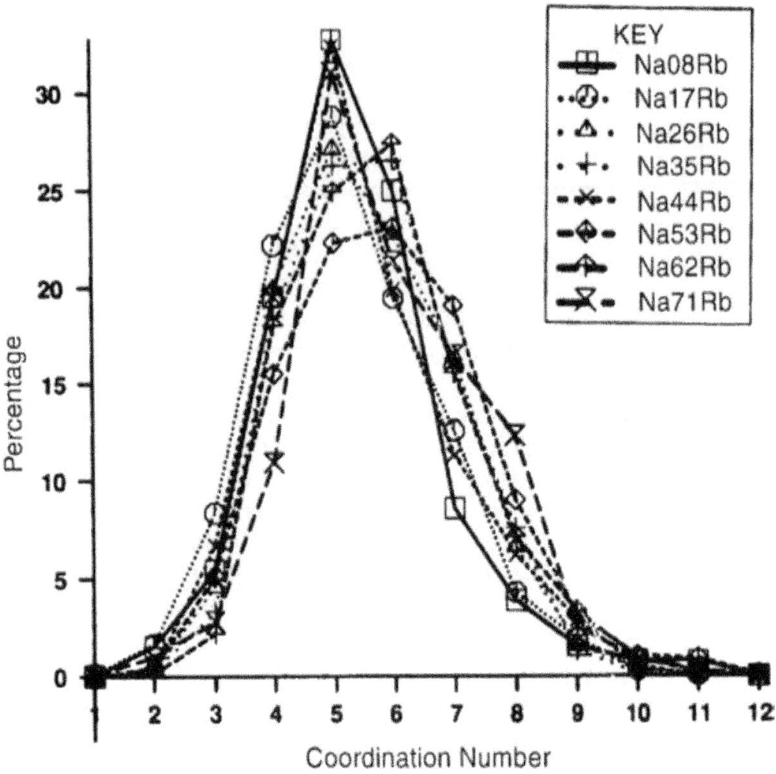

Figure 12.11 Distribution of coordination numbers of Rb sites for simulated $Na_xRb_{1-x}SiO_{2.5}$. Na08Rb means that we have 0 Na and 8 Rb atoms in the starting crystalline unit cell for the simulation of the alkalisilicate glass etc.

earlier experiments on $Na_2Si_2O_5$ glass an Na–O separation of 2.3 Å was recorded (Greaves *et al.*, 1981) and recent measurements on $Rb_2Si_2O_5$ glass (Greaves *et al.*, 1991) gave an Rb–O distance of 2.8 Å. The results of both these studies agree reasonably closely with the simulated distances obtained in the present study and indicate the accuracy of the potentials used for Na and Rb.

Two further important results are contained in Fig. 12.12. Note first how the Rb–O partial RDF contains further structure beyond the nearest neighbour peak at 2.9 Å – notably a second peak at 5.4 Å. The Rb–Rb distribution also exhibits a further peak, this time around 7.5 Å. In a similar way the Na–O partial RDF peaks at 2.4 Å and 4.8 Å. Moreover, Na–Na distances occur at 6.5 Å as well as at 3.0 Å. The identification of first (O), second (M), third (O), and fourth (M) nearest neighbours around each and every alkali demonstrates that modifying oxides, although they are minor components in the glass, do indeed form clusters within the simulated networks as envisaged in the modified random network model (Greaves, 1985). Now in mixed-alkali silicate glasses it

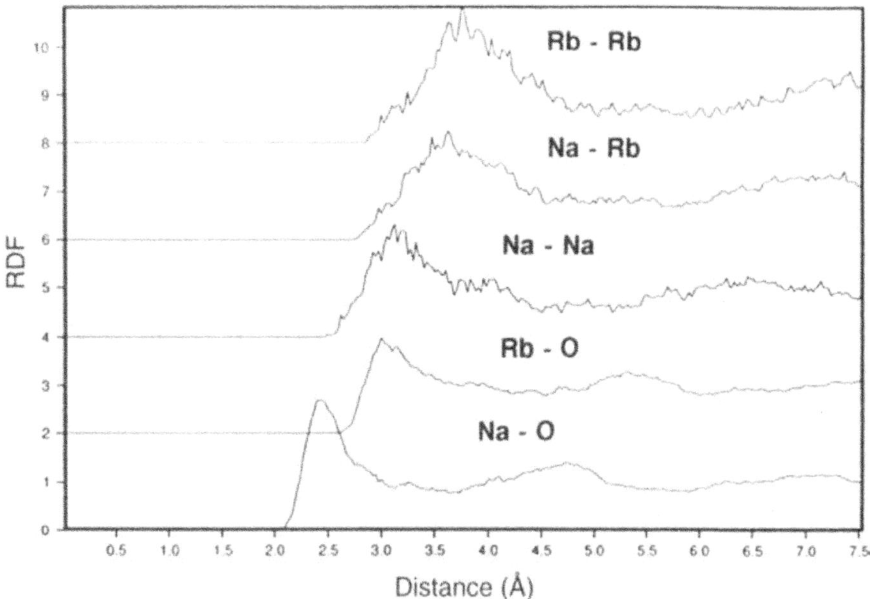

Figure 12.12 Partial RDFs for simulated glass $NaRbSi_2O_5$.

is also apparent from Fig. 12.12 that the two alkalis are *intimately* mixed. The Na–Rb partial RDF has similar features to the Na–Na and Rb–Rb distributions. The intensities are of similar magnitude and the first and second peaks fall approximately midway between the peaks for single alkali pairs, demonstrating that within the modified regions in the glass structure the alkalis are randomly distributed. The same conclusions were drawn, albeit tentatively, from studying the EXAFS of $(K, Cs)_2Si_2O_5$ glasses (Greaves *et al.*, 1990) where comparative disorder between large and small alkalis was detected. Indeed in the simulated glasses a small increase in nearest neighbour disorder around Rb has been found as Na is added to the composition. The opposite is true for the Na oxygen shell which becomes slightly better ordered as a result of added Rb.

It is well known that the ionic conductivity of alkali glasses falls markedly when more than one alkali is present (Day, 1976). This phenomenon known as the mixed alkali effect is primarily related to an increase in the Arrhenius activation energy of the ionic conductivity. Vessal *et al.* (1992) have suggested that although ionic mobility is expected to be higher in the microsegregated alkali-rich regions than in the surrounding silicate network, the stochastic mixing of alkalis nevertheless impedes the hopping process of a given alkali ion. This is manifested as an increase in the average activation energy for hopping and results in a lowering of the total ionic conductivity.

Table 12.1 Experimental and simulation details for mixed sodium and potassium silicates (ρ^0 is given in units of 10^{-2} atoms/Å^2).

	Experiment			Simulation					
Sample	χ_{Na_2O}	χ_{K_2O}	ρ^0	Na	K	Si	O	ρ^0	R_χ(%)
N5	0.446	–	7.343	114	–	128	313	7.362	12.7
NK	0.229	0.235	6.837	54	54	117	288	6.85	12.9
K	–	0.514	6.165	–	106	102	257	6.168	14.1

(c) Mixed sodium and potassium silicates Wright *et al.* (1993) have undertaken a combined neutron diffraction and MD study of the structure of a sodium, a potassium and a mixed sodium and potassium silicate glass. The compositions of the glasses discussed here are summarized in Table 12.1. The starting structures were generated from that of crystalline $Na_2Si_2O_5$, the appropriate number of atoms being removed and added, to obtain the closest agreement with the experimental composition.

Figure 12.13 compares the correlation functions obtained from the molecular dynamics simulations of the three glasses to the neutron diffraction data, after weighting the individual components $t_{ij}(r)$ with the appropriate neutron scattering length products and convolution with the peak function $P(r)$; the corresponding R_χ factors are given in Table 12.1. A quantitative measure of agreement between simulation and experiment is R_χ defined as:

$$R_\chi = (\Sigma_i[T_{exp}(r_i) - T_{sim}(r_i)]^2/\Sigma_i T^2_{exp}(r_i))^{1/2} \quad (19)$$

where $T_{exp}(r)$ is the experimental total correlation function while $T_{sim}(r)$ is the simulated total correlation function. The dotted lines in Fig. 12.13 represent the difference between the simulations and experiment and indicate that the major discrepancy occurs in the region of the intratetrahedral Si–O and O–O peaks, mainly due to the fact that these distances in the simulation are slightly shorter than those in the real material. Unlike simulations using two-body potentials, the width of the intratetrahedral O–O peak is in good agreement with that from the diffraction experiments. However, the bond length distribution is too broad and at least part of the reason for this can be seen from Fig. 12.14. The top curves show results based on the use of a single distance to fit the first peak in the experimental correlation function for vitreous K_2O–SiO_2, which yields a mean bond length of 1.625 ± 0.005 Å with an r.m.s. variation of 0.064 ± 0.005 Å. A fit was also performed using separate Si–O distances for the bridging and for the non-bridging oxygen atoms (centre curves) and the resulting unbroadened Si–O peak compared with the same function from the molecular dynamics simulation (bottom curves). It can be seen that the latter is partly resolved into a doublet, with the contribution from the non-bridging oxygen atoms occurring at lower distances compared with that from the bridging oxygen atoms, whereas the

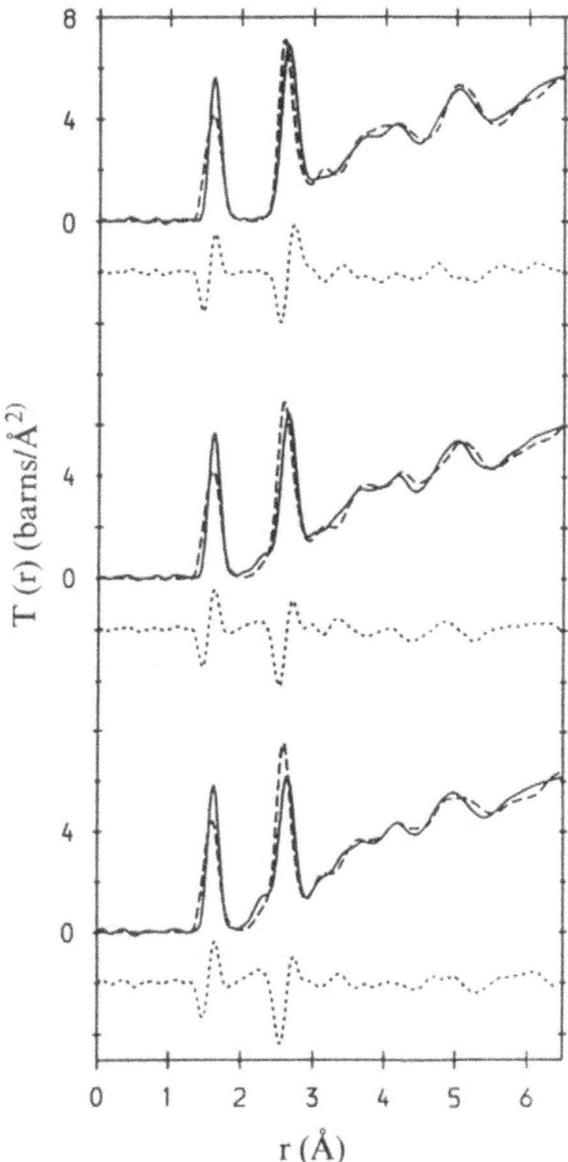

Figure 12.13 Real space correlation function for K_2O–SiO_2 (top), Na_2O–K_2O–SiO_2 (middle), and Na_2O–SiO_2 (bottom). Solid lines: neutron diffraction; dashed lines: molecular dynamics; dotted lines: difference curves (experiment − simulation). T(r) is the total correlation function.

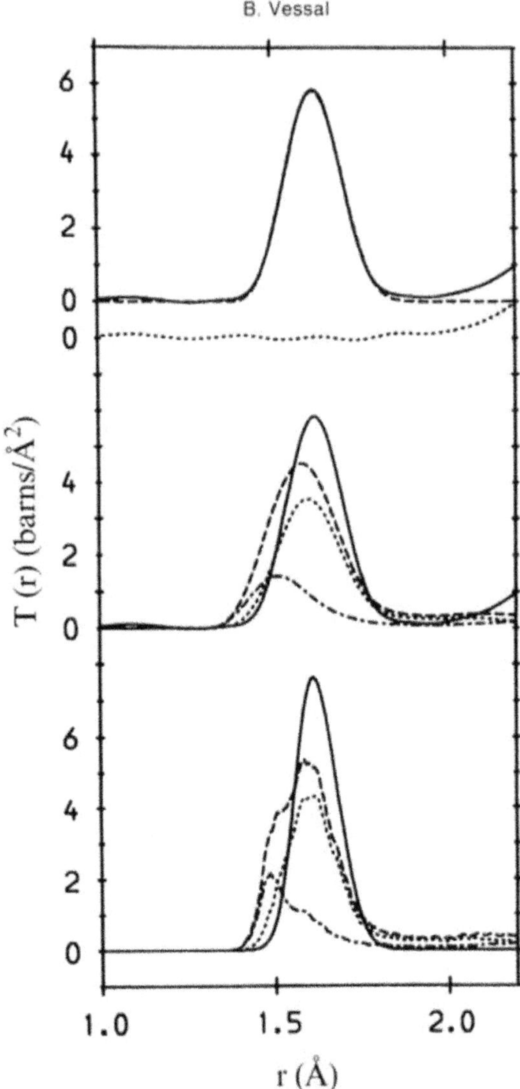

Figure 12.14 Comparison of Si–O bond length distribution from simulation for Na_2O–SiO_2 glass to that from neutron diffraction data. Top: fit to first (Si–O) peak in neutron correlation function (——, experiment; - - - -, fit; · · · · ·, residual); middle: comparison of resolution broadened peak for simulation (– – –) to experiment (——) showing separate contributions from non-bridging (–· –·–) and bridging (· · · · ·) oxygen atoms; bottom: comparison of unbroadened distributions for simulation to that generated from neutron peak fit parameters (key as middle graph).

former is a much narrower asymmetric peak. Thus the difference in the Si–O bond length for the bridging and non-bridging atoms is too large in the simulation. The fitting procedure based on two distances in the peak yields mean bond lengths for the non-bridging and bridging oxygen atoms of 1.596 and

1.639 Å, respectively, giving a difference of 0.043 Å. The corresponding quantities for crystalline $Na_2Si_2O_5$ are 1.578, 1.630 and 0.052 Å, which are within the uncertainties of the fit.

In the experimental correlation function for the soda-silica glass, the Na–O interactions give rise to a well-defined shoulder on the low r side of the first O–O peak, but this is much less pronounced in the simulation (cf. Fig. 12.13) due to the shorter O–O distance and the Na–O distances which are conversely slightly too long. Molecular graphics representations of the structures generated by recent simulations of alkali silicate glasses, such as that in Plate XI, strongly suggest that the alkali ions have a non-uniform distribution and hence it is important to try to investigate their local environment. This may be achieved using a modified version of the difference technique (Wright *et al.*, 1991) as shown by the top two curves in Fig. 12.15. Interpretation of the simple difference curve (dashed line), obtained by subtracting the correlation function for pure vitreous SiO_2 from that for $xNa_2O.SiO_2$, is complicated by the structure arising from the increase in the intratetrahedral Si–O and O–O distances which occurs on adding Na_2O to SiO_2. It is therefore necessary to

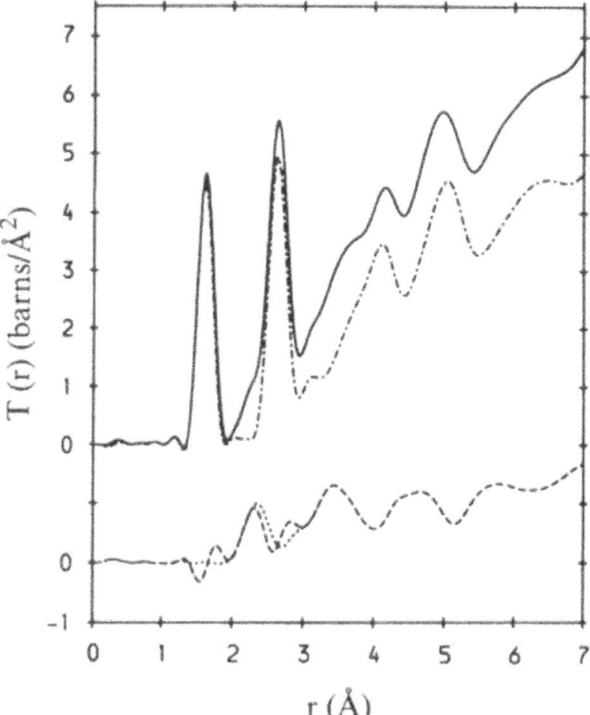

Figure 12.15 Analysis of first neighbour Na–O distribution for the Na_2SiO_2 glass using modified difference technique (Wright *et al.*, 1991). From the top: Na_2O–SiO_2 (——) and SiO_2 (–· –·–); difference (- - - -) and modified difference (· · · · ·) correlation functions.

work with the modified difference curve (dotted line), generated from the peak fit residuals, as discussed in detail by Wright *et al.* (1991). The latter suggests that the first Na–O distance distribution is asymmetric and this is supported by the Na–O contribution from molecular dynamics simulation which is compared to the modified difference function in the low curves of Fig. 12.15. A similar asymmetry is seen in the K–O component of the K_2O–SiO_2 simulation and in both M–O distributions for the mixed-alkali glass (where M is the alkali atom).

The disorder inherent to the vitreous state means that not all of the oxygen atoms in the first neighbour shell around the sodium ions will be non-bridging and, similarly, that the cavities in which some of the Na^+ ions reside will not be sufficiently regular for them to be at the ideal distance from every oxygen atom. In this situation, the positively charged Na^+ ions will tend to adopt the optimum configuration with respect to the more flexible negatively charged non-bridging oxygen atoms rather than the more constrained bridging oxygen atoms. Thus it might be expected that the Na–O distribution will comprise an approximately symmetric peak at the non-bridging oxygen atom distance plus a high r tail arising from the bridging oxygen atoms, and that fitting an asymmetric peak to the modified difference function would reflect mainly the non-bridging oxygen atoms. This is consistent with the Na(O) coordination number 3.0 ± 0.5 extracted from the fit, since in an oversized cavity the Na^+ ion will normally only be 'in contact' with three oxygen neighbours. The coordination number is also lower than would be expected in comparison to crystalline α-$Na_2O.2SiO_2$ where each Na^+ ion has five oxygen neighbours, four in the range 2.290–2.386 Å (mean Na–O distance 2.347 Å, cf. 2.338 Å from the fit) and the fifth at 2.600 Å. Interestingly, however, the single non-bridging oxygen atom in the first coordination shell of the crystal is not the one at 2.600 Å but that at 2.386 Å.

The study clearly demonstrates the powerful role which molecular dynamics can play in interpreting diffraction data for multicomponent glasses. This is particularly true for systems, such as Na_2O–SiO_2, where it is not possible to use techniques such as isotopic substitution to investigate the individual component correlation functions.

(d) Titanium and zirconium wadeites The system $K_2(Ti,Zr)Si_3O_9$ (Ti- and Zr-wadeite) provides an interesting test of hypotheses regarding nucleation and crystallization, because TiO_2 and ZrO_2 are two of the oxides most commonly used as nucleating agents. Since these oxides are major components of the wadeite system it would be expected that these compositions would be effectively nucleated and undergo bulk crystallization. However, as shown by Dickinson (1989) Ti-wadeite does not behave as expected; instead, it surface crystallizes. Zr-wadeite, on the other hand, nucleates homogeneously and does not surface crystallize. Based on Raman and EXAFS spectra of corresponding crystals and glasses, Dickinson (1988, 1989) proposed that these differences are the result of structural dissimilarity between the crystal and melt in the case of

Ti-wadeite, but structural similarity in the case of Zr-wadeite. These differences/similarities influence the energetics of nucleation, with the result that it is more difficult to form a critical nucleus of Ti-wadeite. Vessal and Dickinson (1994) have undertaken both constant volume and constant pressure molecular dynamics simulations to test this hypothesis.

Molecular graphics representations of the Ti-wadeite crystal and the corresponding glass simulated at constant volume are shown in Plates XII and XIII. It is evident from Plate XII that crystalline Ti-wadeite contains three-membered rings of SiO_4 tetrahedra connected together by TiO_6 octahedra, and the potassium atoms occupy interstitial positions. When the structure of the glass is analysed (Plate XIII) only one three-membered ring of SiO_4 tetrahedra can be found. This is in accord with the Raman experiments reported by Dickinson (1989).

The first near neighbour peaks in the RDF of both the Ti–O and Zr–O pairs for crystalline and glassy Ti- and Zr-wadeites are compared in Figs 12.16 and 12.17. The Zr–O first neighbour distance for crystalline Zr-wadeite is at 2.03 Å while that for glassy Zr-wadeite is at 1.98 Å. However the Ti–O first neighbour distance shows remarkable differences between the crystalline and the glassy state. The Ti–O RDF for the crystalline phase of Ti-wadeite has a first sharp peak at 1.94 Å while that of the corresponding glass shows evidence of a bimodal distribution of Ti–O bond lengths at 1.75 and 1.81 Å.

The Zr-wadeite crystal and glass RDF comparison is in excellent agreement

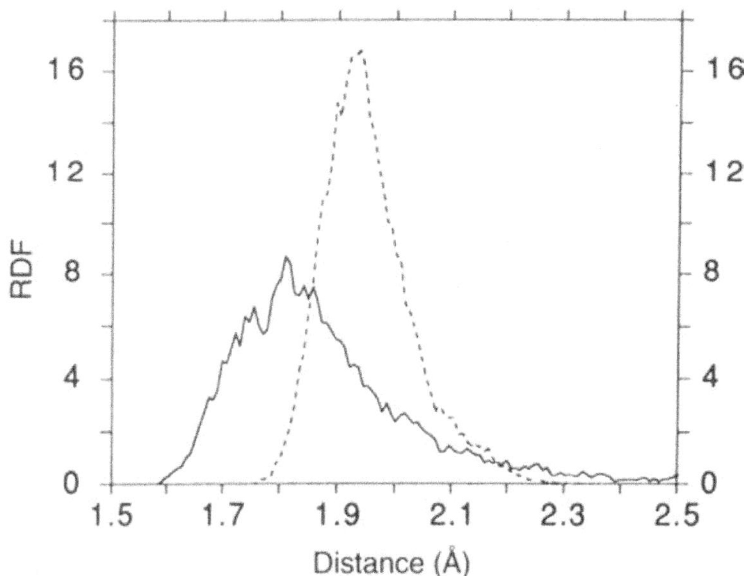

Figure 12.16 Ti–O RDF for Ti-wadeite crystal (dashed curve) and glass (full curve).

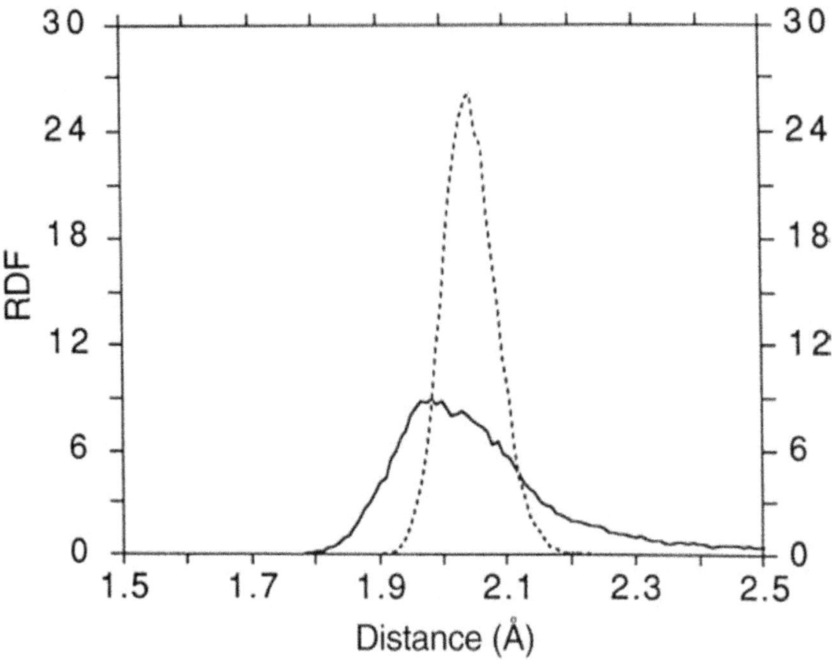

Figure 12.17 Zr–O RDF for Zr-wadeite crystal (dashed curve) and simulated glass (full curve).

with EXAFS measurements (Dickinson, 1988). The MD results for the Ti-wadeite system, however, do not agree as well with EXAFS measurements, which are best fit by two shells with one Ti–O distance at 1.67 Å and the other at 1.97 Å. Nevertheless, considering that the Ti–O potentials have not been optimized, the results are very encouraging. Moreover, the simulations are definitely consistent overall with there being strong similarity between crystal and glass in the Zr-wadeite system and a very strong dissimilarity in the Ti-wadeite system.

(e) B_2O_3 The most accurate MD study of B_2O_3 has been reported by Takada (1993). He has derived an accurate potential which contains terms that are dependent on coordination number. In his simulations the strength of the B–O bond increases when boron changes from tetrahedral to trigonal planar coordination. He had developed two models, one with 25% of boron atoms involved in boroxol rings (i.e. rings containing three B and three O atoms) and the other with 42% of boron atoms in boroxol rings compared to a value of 40% suggested by the experimental lower limit (Johnson and Wright, 1982). In his first model 99% of borons are threefold coordinated and 1% are twofold coordinated while his second model has 94% threefold and 6% fourfold coordinated borons.

3.1.2 Chalcogenide glasses

(a) $GeSe_2$ Vashishta *et al.* (1988, 1989a,b) have developed a potential model for simulating the structure of $GeSe_2$ glass. Their potential model consists of partial charges on the atoms, a Pauling type repulsive term and an r^{-4} term representing very crudely the effects of the polarizability of the selenium atoms. In their later study (1989b) they have also included Stillinger–Weber type three-body interactions in the potential. With this potential they are able to simulate all the features in the structure factor ($S(Q)$) including the First Sharp Diffraction Peak (FSDP). They find that the origin of the FSDP stems from Ge–Ge, and Ge–Se correlations between 4 and 8 Å and they propose that the anomalous decrease in its height on cooling is due to frustration enhanced by increased density.

(b) $SiSe_2$ Antonio *et al.* (1988) have simulated the structure of $SiSe_2$ using a potential model which is similar in form to that used for $GeSe_2$. They find that most of the $SiSe_4$ tetrahedra are edge sharing from the analysis of different BADs.

(c) B_2S_3 Recently an MD study on vitreous B_2S_3 was reported by Balasubramanian and Rao (1994). They used a 500 atom system and constant volume MD to generate a simulated structure which comprises a planar chain forming four-membered B_2S_2 rings (or edge sharing BS_3 triangles). The potential they developed is a partial charge model plus an exponential repulsive term and an r^{-4} term. They also included a Stillinger–Weber type three-body term, and obtained 76% three-coordinated borons and 24% two-coordinated borons, while 82% of sulfurs are two-coordinated and 18% have only one near neighbour. The nearest neighbour distances for B–B are at 2.05 Å (1.95), B–S at 1.95 (1.86), and S–S at 3.35 (3.18 Å). The values quoted in brackets are from neutron diffraction results. The B–S–B angles peak at 90° while S–B–S angles have a peak at 122°.

3.1.3 Fluoride glasses

Inoue and MacFarlane (1987) have carried out MD simulations on the ZrF_4–BaF_2–$BaCl_2$–MCl_2 (M = Mg, Ca, Sr) system. They have investigated the relationship between the glass forming tendency and the distribution of Cl^- ions among the cation coordination spheres. They found that in these glassy systems localization of halide ions around some cations triggers crystallization.

The glass formation tendency of compositions in the ZrF_4–BaF_2 and ZrF_4–BaF_2–GdF_3 (ZBG) systems has been studied by MacFarlane *et al.* (1988). They defined a stabilization parameter (S) which takes the weighted average of the degree of localization of cation pairs. They found that S passes through a significant minimum (indicating the most random arrangement) in the glass forming region of each of a series of constant GdF_3 cuts of the ZBG phase diagram.

Simmons *et al.* (1988) have examined the structure of amorphous ZrF_4–BaF_2 binary by means of MD. They studied the effect of the addition of Ba^{2+} ions on the fluorozirconate cage structure. Beginning with the ZrF_4 side of the binary, they found that Zr^{4+} ions are primarily surrounded by eight F^- ions. They also discovered that the addition of Ba^{2+} ions has little effect on the cage structure surrounding the Zr^{4+} ions and appeared to find interstitial positions with high F^- coordination between the cages.

Uhlherr *et al.* (1990) have performed MD studies on ZrF_4–BaF_2–MF–M′F glasses (M, M′ = Li, Na, Cs). Addition of alkali metal fluorides to fluorozirconate glasses causes the ionic conductivity to drop to a minimum at around 20 mole% alkali fluoride, then to rise rapidly on increasing the alkali content. The common explanation for this behaviour is that these glasses are fluoride ion conductors below the 20 mole% limit, and alkali ion conductors above it. The simulations determined the hole size distribution in the mixed-alkali glasses and indicated that in both Li/Na and Na/Cs glasses there is a pronounced reduction in the number of larger holes on alkali mixing (holes of >1.0 Å in the Li/Na case and holes of >1.1 Å in the Na/Cs case), and a corresponding increase in the number of smaller holes. They proposed that a fluoride ion utilizes holes of radius ~ 1.24 Å (based on an estimate of the radius of a fluoride ion of this value from the F–F RDF) in its migration and that the reduction in the concentration of such vacancies is responsible for the observed mixed-alkali effect.

Moore *et al.* (1991) used MD to probe the structure of CdF_2–LiF–AlF_3–PbF_2 (CLAP) glasses and the effect of oxide addition to these systems. They found that the average coordination numbers for Cd, Li, Al, and Pb are 7.8, 5.5, 6.2, and 9.6, respectively, in the undoped glass. The oxide ions are found to be predominantly located near Al and Pb in the doped glass. They also found that each oxide ion replaces a fluoride ion in the Al and Cd coordination spheres but substitutes for two fluoride ions when coordinated to Pb.

Uhlherr *et al.* (1991) have undertaken MD studies of ZrF_4–BaF_2–LnF_3 (Ln = Y, La–Lu) glasses. Their results show that the rare-earth elements La–Pm exhibit ninefold coordination while for Y and Sm–Lu both eight- and nine-coordinated polyhedra are observed. They found that in the latter case, the $[LnF_9]^{6-}$ units, each containing a ‘long’ bond, become less common with increasing atomic number across the lanthanide series, resulting in a decrease in the observed mean Ln–F coordination number from Sm to Lu. The structural role of the lanthanide ion is found to be intermediate between the strong network former, such as Zr, and a modifier species. They proposed that the lanthanide components stabilize fluorozirconate glasses by increasing the mean Zr–F coordination number.

Uhlherr and MacFarlane (1993) have studied ZrF_4–BaF_2–NaF glasses. They have found that these glasses consist of corner and edge sharing ZrF_8

polyhedra. Compositions that form glasses under experimental conditions are found to be characterized by a bridging/non-bridging F^- ion ratio in the range 1:1 to 3:1, with at least 10% of the anions acting as edge sharing linkages, and a negligible fraction of free anions.

3.2 Melting

All the simulated glasses discussed above were prepared by the melt–quench procedure. It is therefore appropriate to consider the extent to which the simulation reproduces the melting of the original crystal.

3.2.1 Melting of SiO_2

Vessal *et al.* (1989) have studied melting of β-cristobalite at constant pressure. One of the incentives for their study has been provided by the uncertainty and controversy surrounding the behaviour of SiO_2 on melting. Water and silica behave similarly in many respects. Both liquid water and the silicon dioxide melt exhibit density maxima in their density–temperature curves (Angell and Kanno, 1976). Both show an increase in the particle mobility that accompanies an increase in density under pressure (Angell *et al.*, 1982). It has been shown recently that ice I, α-quartz, and coesite amorphize under pressure (Mishima *et al.*, 1984; Hemley *et al.*, 1988); Vessal *et al.* (1989) have observed an increase in density on melting of β-cristobalite. This implies a negative Clapeyron slope, and to investigate this phenomenon further they have undertaken a simulation at a pressure of 50 kbar where they have observed a decrease in the melting point, reaffirming their earlier observation.

Figure 12.18 shows the plot of internal energy, U, against the temperature during the heating and cooling of the system. β-Cristobalite is heated from A to B, melts between B and C, and the melt is heated from C to D. It is then cooled down from D to F. A clear first order transition occurs between B and C from which the melting point and the latent heat can be obtained. The specific heats for solid and melt are obtained from the slope of the AB and DE curves. During the cooling of the system from D, recrystallization does not occur, and the points from C to E represent the supercooled liquid. At a temperature of around one third of the melting point, the specific heat of the system is observed to decrease from the liquid to the crystal value (slope of lines DE and EF). This behaviour is characteristic of glass formation and we identify this temperature as the glass transition temperature of the system.

The volume of the system, V, is calculated at the end of each heating or cooling cycle. Figure 12.19 shows variations in the volume with the temperature. The points A to F represent the same stages in the heating/cooling cycle as those defined in Fig. 12.18. Melting of the system can be seen quite clearly in the

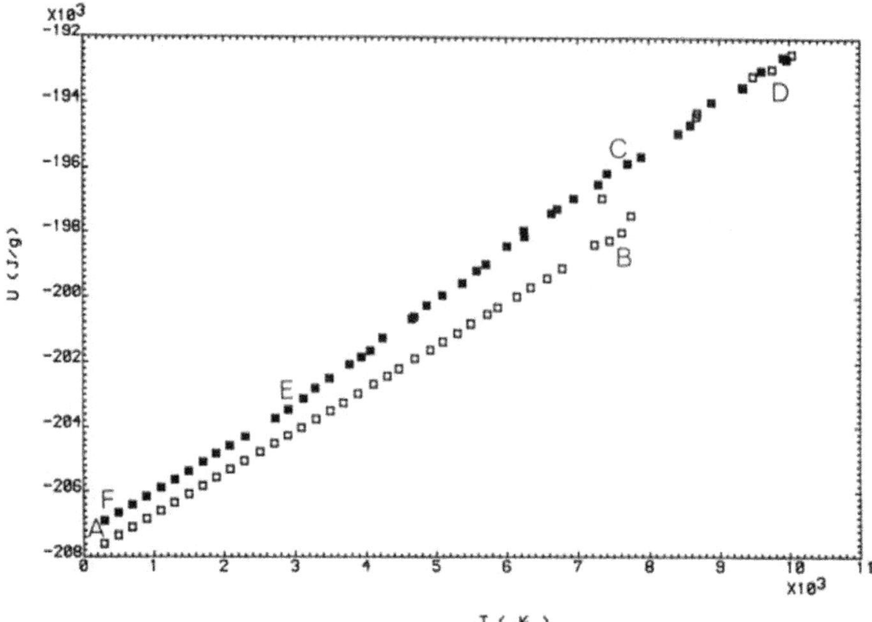

Figure 12.18 Plot of total energy against temperature for simulation of silica. White squares show points during heating of system while black squares are points during cooling process. See text for details of A–F.

figure where the volume of the system decreases by about 20%. The results reveal an interesting isomorphism with the behaviour of water.

Although there are reports suggesting that the density of the silica melt is lower than those of its crystalline forms (Bacon *et al.*, 1960), the current estimates do suggest an isomorphism with water, i.e. a negative Clapeyron slope which is due to the higher density of silica melt relative to its crystalline forms (Hemley *et al.* (1988), and references therein). These simulations clearly also suggest that the earlier experiments were inaccurate. As a further test Vessal *et al.* (1989) have run the present simulation at a pressure of 50 kbar and they have observed a decrease in the melting point of β-cristobalite, demonstrating that the behaviour of the simulation shows full internal consistency.

3.3 Pressure-induced amorphization

After the pioneering work of Mishima *et al.* (1984) on the amorphization of ice I under pressure, there has been considerable interest in this new form of amorphization. Recently Hemley *et al.* (1988) have observed that both α-quartz and coesite amorphize under pressure.

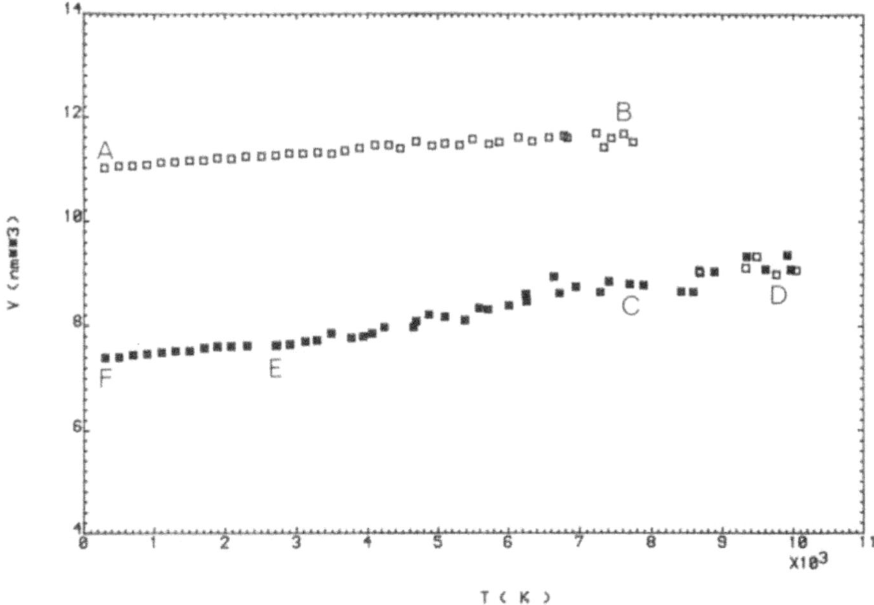

Figure 12.19 Plot of volume against temperature for simulation of silica. White squares show points during heating of system while black squares are points during cooling process. See text for details of A–F.

The first MD study of pressure-induced solid state amorphization was undertaken by Tse and Klein (1987). They showed that at 80 K ice Ih undergoes a transition to a high density amorphous form at around 13 kbar.

Handa *et al.* (1990) have used MD to study pressure-induced phase transitions in ethylene oxide and krypton hydrates. The MD results suggest that under pressure the water molecules in the hydrates collapse around the guest molecules, and the repulsive forces between the guest and the water molecules are mainly responsible for the reversible transition to the original structure when the pressure is released.

3.3.1 Amorphization of berlinite

Vessal (1991) has studied the simulation of amorphization of crystalline $AlPO_4$ berlinite using the constant pressure MD method. When the system is depressurized gradually, amorphous $AlPO_4$ is observed to recrystallize back to the crystalline phase. These observations are in agreement with recent experiments (Kruger and Jeanloz, 1990). The simulation was undertaken with a pressure interval of 1 GPa between two consecutive cycles. During pressurization, the behaviour of the system was studied to detect a phase transition. After amorphization, the amorphous material was pressurized further to see if there

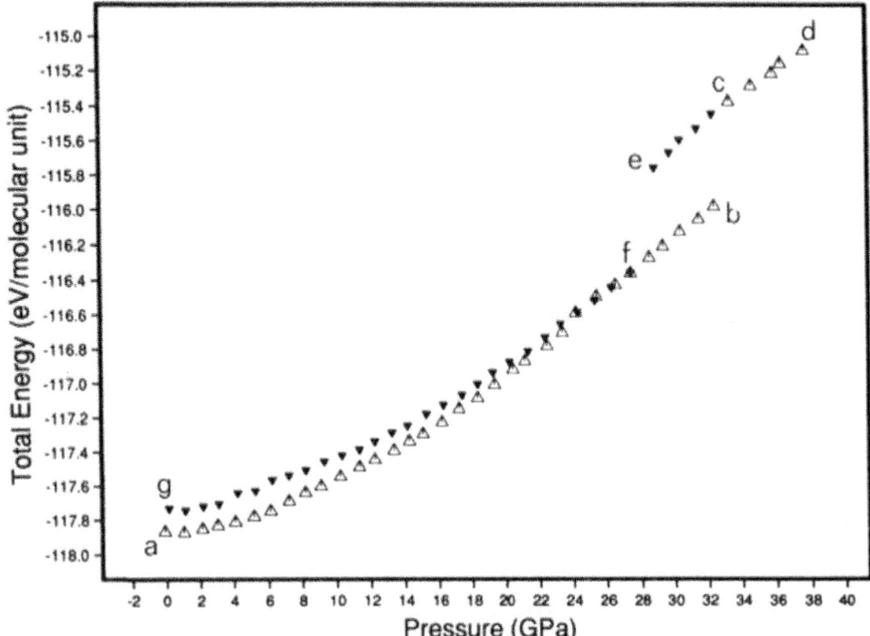

Figure 12.20 Plot of total energy against pressure for simulation of $AlPO_4$. White triangles show points during pressurization of system while black triangles are points during depressurization process. See text for detail of a–g.

was any change in its structure. Then the process of depressurization was undertaken with a pressure interval of −1 GPa between two consecutive cycles.

The internal energy is plotted against the pressure in Fig. 12.20. The crystal is pressurized from a to b and amorphizes between b and c; the amorphous material is pressurized from c to d. The system is then depressurized from c to g. The amorphous material 'recrystallizes' between e and f. A clear first order transition occurs between points b and c from which the transition pressure and the heat of amorphization $L_a = U(c) - U(b)$ are obtained. The observed heat of amorphization is 58 kJ/mol. During the depressurization of the system from c to g another first order phase transition occurs between points e and f from which the transition pressure and the heat of recrystallization $L_a = U(f) - U(e)$ are obtained. The observed heat of recrystallization is −58 kJ/mol.

It can be seen from Fig. 12.20 that berlinite amorphizes at a pressure of 33 GPa and amorphous aluminium phosphate recrystallizes at a pressure of 27 GPa. The observed phase transitions are in accord with the experimental results (Kruger and Jeanloz, 1990), but the transition pressures are different from experiment (15 ± 3 GPa and 5 GPa, respectively, with a hysteresis of about 13–15 GPa). In this study the hysteresis is about 6 GPa. This discrepancy could be attributed to several factors. First, the fact that a perfect crystal is

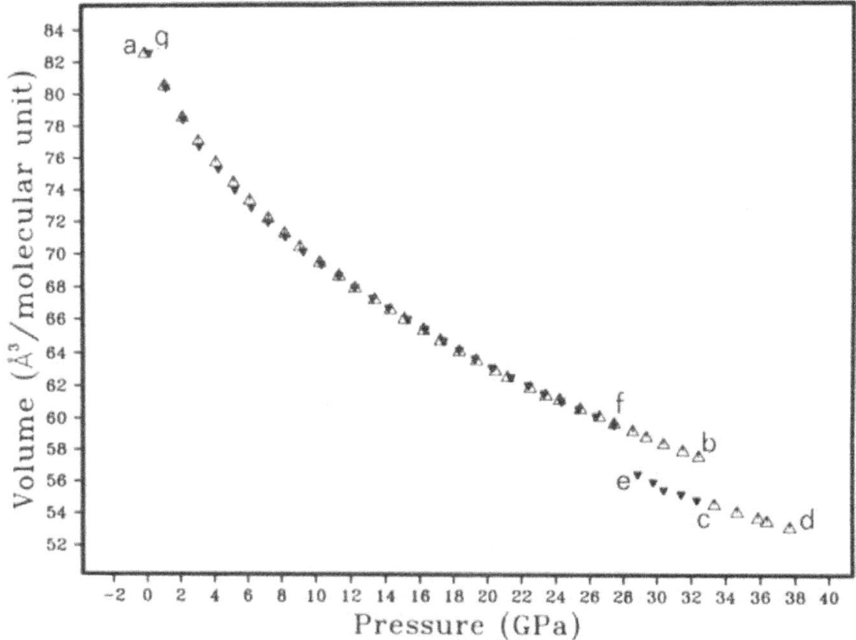

Figure 12.21 Plot of volume against pressure for simulation of $AlPO_4$. White triangles show points during pressurization of system while black triangles are points during depressurization process. See text for details of a–g.

being simulated and the absence of defects cause the system to be amorphized at a much higher pressure than expected. Second, the effect of the periodic boundary conditions and the lack of any free surfaces will also lead to high transition pressures, as we would expect that solid state amorphization should start at the surface of the crystal as in the melting process. Third, the timescale of the simulations is short and it is expected that if the system is pressurized more slowly, the transition will be observed at a lower pressure.

The volume of the system, V, is calculated at the end of each pressurization or depressurization cycle. Figure 12.21 shows variations in the volume with pressure. The points a–g are as in Fig. 12.20. Amorphization of the system is clearly apparent from the figure when the volume of the system decreases by about 5%. During the depressurization, the system expands upon the recrystallization of the amorphous materials, with a volume increase of about 5%.

Apart from the U–P and V–P curves there are several other quantities that can be used to show the amorphization and recrystallization transitions. The accumulated values of the Mean Square Displacements (MSDs) of the different types of atom in the system over the last 800 steps of each cycle during the pressurization process are plotted in Fig. 12.22. It can be seen from this figure that the MSDs of all three types of atom peak at a pressure of 33 GPa which is the crystalline to amorphous transition pressure. Another interesting feature of

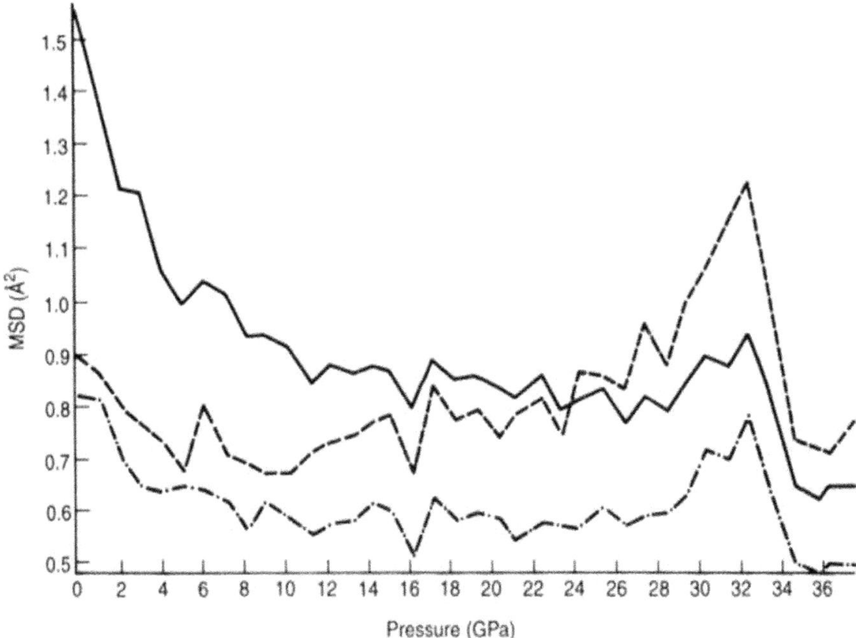

Figure 12.22 Mean square displacement of aluminium (- - - -), oxygen (——), and phosphorus (–· –·–) against pressure during simulated compression of $AlPO_4$.

this graph is that the MSDs of oxygens are higher than those of the aluminium atoms up to a pressure of 24 GPa. Above this pressure the behaviour is reversed with higher aluminium MSDs.

There are striking changes in the structure of $AlPO_4$ as it is subjected to mechanical stress and the effects of the crystalline–amorphous transition is clearly seen. The change of the Al–Al radial distribution function (RDF) with pressure is shown in Fig. 12.23. Generally, as crystalline berlinite is subjected to mechanical stress the peaks in the RDFs are pushed towards shorter distances, indicating a decrease in atomic distances under pressure. The reverse is, as expected, the case for the depressurization process as atomic distances increase when the pressure on the system is gradually removed. The effects of the phase transitions can be seen in all the calculated RDFs. During the pressurization process a sudden change at 33 GPa is observed. The sharp peaks become very broad and long range order is lost. This clearly indicates a transition from the crystalline to the amorphous state. The same is true with the depressurization process where an abrupt change is observed at 27 GPa, where the broad peaks become very sharp and long range order is regained (see Fig. 12.23). This change in the RDFs indicates 'recrystallization' from the amorphous to the crystalline state.

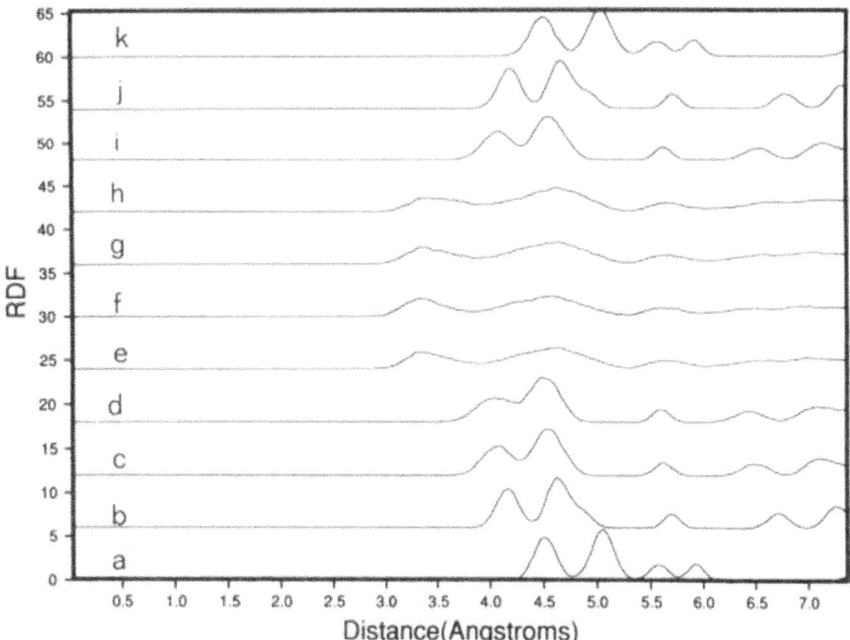

Figure 12.23 Al–Al RDF in simulated $AlPO_4$ at (a) 0 GPa, (b) 19 GPa, (c) 29 GPa, (d) 32 GPa, (e) 33 GPa, (f) 37 GPA, (g) 32 GPa, (h) 28 GPa, (i) 27 GPa, (j) 16 GPa, and (k) 0 GPa.

The four main peaks in the Al–Al RDF are located at 4.52, 5.07, 5.56, and 5.39 Å at zero pressure (see Fig. 12.23). As the pressure on the system is increased these peaks shift to shorter distances. The peak at 5.56 Å will disappear completely at a pressure of 24 GPa. At a pressure of 33 GPa the shape of the RDF changes drastically, which indicates the amorphization of berlinite, where three broad peaks at 3.36, 4.65, and 5.63 Å are observed in the Al–Al RDF at 33 GPa. During the decompression process the reverse trend is observed. At 28 GPa there is clearly a phase transition and the system becomes more ordered. The four Al–Al peaks that are observed for the 'recrystallized' material are at exactly the same positions as the peaks for the starting crystalline material.

The Al–O–Al BAD is depicted in Fig. 12.24. This graph shows the two transitions clearly, since the shape of the BAD changes drastically during both amorphization and recrystallization processes. A flat curve is observed at pressures below 24 GPa because at these pressures there is only one aluminium atom around each oxygen atom, with this BAD having two main peaks at 120° and 139°.

Turning now to the coordination numbers of alumimium, we find that these change substantially during the amorphization and recrystallization processes.

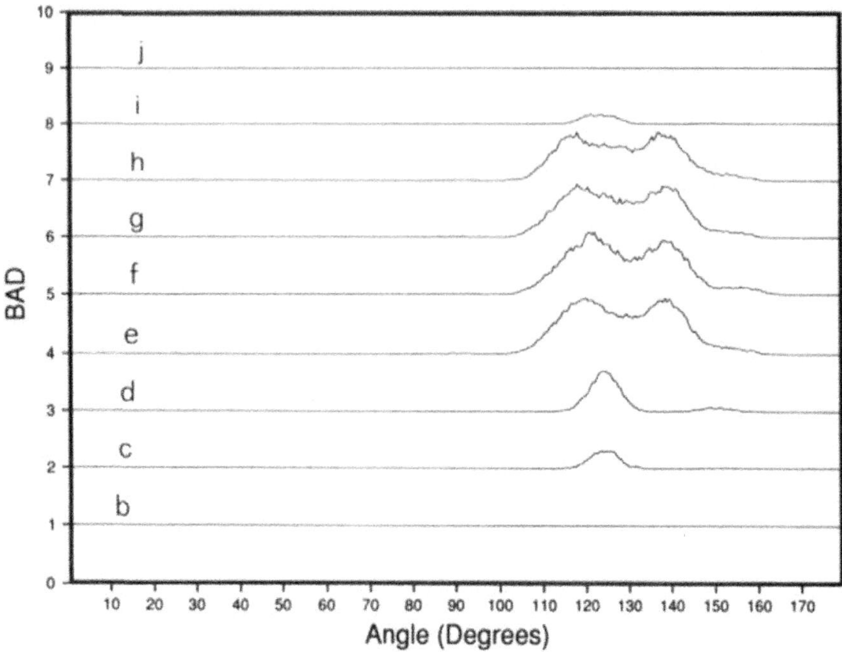

Figure 12.24 Al–O–Al BAD in simulated $AlPO_4$ at (a) 0 GPa, (b) 19 GPa, (c) 29 GPa, (d) 32 GPa, (e) 33 GPa, (f) 37 GPa, (g) 32 GPa, (h) 28 GPa, (i) 27 GPa and (j) 16 GPa.

In Fig. 12.25 the coordination numbers of aluminium atoms in the crystal at zero pressure, at a pressure of 15 GPa, and in the amorphous state are plotted. When the system is pressurized, the percentage of the four-coordinated aluminium atoms gradually decreases, while that of the five-coordinated aluminium atoms gradually increases until there is a sharp drop from 36% to just under 2% of four-coordinated aluminium atoms upon amorphization. The percentage of the five-coordinated aluminium atoms also drops from 49% to just under 37% upon amorphization. On the other hand, there is a sharp increase in the percentage of six-coordinated aluminium atoms from 15% to 51% upon amorphization. There are no seven-coordinated aluminium atoms in the crystalline state but upon amorphization at 33 GPa, 10% of aluminium atoms are seven-coordinated. Five- and six-coordinated aluminium atoms have been observed recently in glasses prepared at high pressures using NMR spectroscopy (Ohtani *et al.*, 1985; Fitzgerald *et al*,. 1989; Kohn *et al.*, 1991).

As noted above, upon depressurization the amorphous $AlPO_4$ recrystallizes. This is the point at which the percentage of four-coordinated aluminiums jumps from 2% to 59%, while the percentage of five-coordinated aluminiums decreases from 48% to 38% and that of the six-coordinated aluminium atoms decreases sharply from 38% to just 3%. The seven-coordinated species disappear altogether upon recrystallization. As the crystal is depressurized

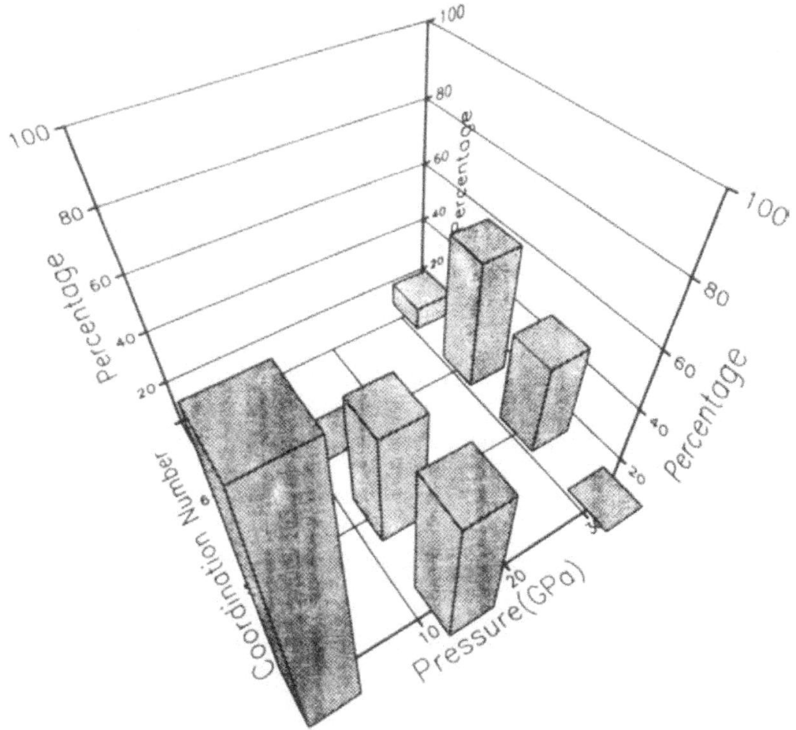

Figure 12.25 Plot of coordination number of Al versus pressure in simulated $AlPO_4$.

further the percentage of the five- and six-coordinated aluminiums decreases gradually until at zero pressure all of the aluminium atoms are four-coordinated.

During the whole pressurization and depressurization process all of the phosphorus atoms have a coordination number of four, and all of the oxygens are bonded to only one phosphorus atom. In the crystalline berlinite at zero pressure all of the oxygen atoms are bonded to just one aluminium atom. As the system is pressurized, oxygens that are bonded to two aluminiums are observed. At 29 GPa the percentage of oxygens coordinated to two aluminiums is 4%. Upon amorphization this percentage jumps from 10% at 32 GPa to 35% at 33 GPa. At 33 GPa, 2% of oxygens are coordinated to three aluminium atoms, while in the crystalline state there are none of these types of oxygen coordination. Also, upon recrystallization the percentage of oxygens that are coordinated to two aluminium atoms drops sharply from 31% to 1%, and all of the oxygens coordinated to three aluminium atoms disappear. At 24 GPa, all of the oxygen atoms are coordinated to just one aluminium.

In Plate XIV molecular graphics representations of the structure of berlinite at four different pressures are shown. As the pressure on the crystal is increased, the AlO_4 and the PO_4 tetrahedra become distorted, until at a pressure of 33 GPa an amorphous structure is formed, as we have noted.

Vessal (1991) has examined the connectivity of both the aluminium and the phosphorus atoms during the simulation. The PO_4 tetrahedra become distorted upon pressurization but the coordination number of the phosphorus atoms does not change. Thus the connectivity of the phosphorus atoms remains intact during the whole process. The behaviour is different with the aluminium atoms because of the weaker Al–O bond (compared to that of the P–O bond). Upon pressurization five- and six-coordinated aluminium atoms begin to form. When the pressure is released, perfect AlO_4 tetrahedra are formed again with the same connectivity as before the amorphization process. In other words, each aluminium is bonded to the same oxygens both before amorphization and after recrystallization. Therefore the amorphization is clearly displacive, in agreement with the original experimental high pressure data.

3.4 Defect-induced amorphization

Hsieh and Yip (1987) have simulated the radiation-induced amorphization in crystals by introducing interstitials in a model fcc lattice using constant stress MD. They have shown that the amorphization threshold depends both on the concentration and the rate of insertion of the defect. The amorphous structure that they obtained is very similar to the structure obtained by quenching of a liquid. They also found that in the defective crystals the elastic constants C_{11} and C_{44} decrease while C_{12} increases somewhat. For the amorphous system the same trend is observed but the effect is much more pronounced. The values of the elastic constants for the amorphous system are again close to those for the quenched liquid.

Massobrio *et al.* (1989, 1990) studied defect-induced amorphization in $NiZr_2$ using a constant number of particles, constant pressure and constant temperature (NPT) MD and tight-binding potentials, by randomly exchanging a number of Ni and Zr atoms. They observed a volume increase in the system. The change in volume is more pronounced as the degree of chemical disorder introduced increases. The amorphous structure is similar to a structure obtained from a quenched liquid.

Another example of elegant work is that of Sabochick and Lam (1991) who studied defect-induced amorphization of CuTi, $CuTi_2$, and Cu_4Ti_3 using an NPT MD method and embedded-atom potentials. They introduced the defects by exchanging Cu and Ti atoms randomly, by removing atoms from their lattice sites (i.e. creating vacancies), or inserting atoms at random positions (i.e. creating interstitials). Their results indicate that the presence of point defects (i.e. vacancies and interstitials) is essential in triggering a crystalline to amorphous transition.

3.5 Reverse Monte Carlo

The RMC method was extensively discussed in Chapter 6 and has been widely used in modelling amorphous systems. Indeed its earliest use dates back to the work of Averbach's group (Kaplow *et al.*, 1968; Rechtin and Averbach, 1974, 1975; Rechtin *et al.*, 1974; Renninger and Rechtin, 1974) on the structure of vitreous selenium and As–Se glasses. In their original study, Kaplow *et al.* (1968) used different lattices as their starting configuration. They took a more realistic approach in their later papers where they placed several restrictions on building the initial configuration. They took into account the known bond angles and lattic energies in building the starting structures. As–As distances less than 3.0 Å were kept to a minimum. Thermal Boltzmann probability factors ($\exp(-E/k_BT)$ were taken into account (Rechtin *et al.*, 1974), where E is the energy change due to the addition of an atom and is estimated from the lattice energies, k_B is the Boltzmann constant and T is absolute temperature.

In glassy Se the dominant configuration for Se atoms turned out to be slightly distorted Se_8 rings. In the mixed As–Se system, clusters of connected atoms appeared. The dominant feature is the tendency to form As_2Se_3-like structures consisting of bridging Se atoms connecting three-coordinated As atoms.

Recently the RMC method has been used for simulating the structures of vitreous silica and vitreous B_2O_3 starting from random configurations (McGreevy and Pusztai, 1988; Keen and McGreevy, 1990; Bionducci *et al.*, 1994). Although the structures produced using the RMC technique are in quantitative agreement with diffraction experiments, the resulting structures have too many undercoordinated atoms and are therefore in this sense unacceptable (Wright, 1994).

Vessal and Wright (manuscript in preparation) have started from a vitreous silica structure generated by MD (Vessal *et al.*, 1993) with 648 atoms in a cubic box of 21.397 Å on each side. The MD structure used is the closest MD model to the experiment and exhibits an R_χ factor of 9.1% calculated between 1 and 8 Å (Wright, 1993). They have attempted to refine the MD structure using RMC. This has resulted in a structure that is very close to neutron diffraction results (R_χ of 1.6% calculated between 1 and 8 Å – see Fig. 12.26). This structure, however, remains unsatisfactory because it yields an average coordination number of 3.74 for silicons. Other applications of RMC methods to silicate systems are given in Chapter 6.

4 CONCLUSIONS

Atomistic computer simulation techniques can now provide good structural models of a wide range of glasses. Such models accord well with available experimental data and give new insights into the fascinating structural features of glasses. As computers have become more powerful the size of the models that

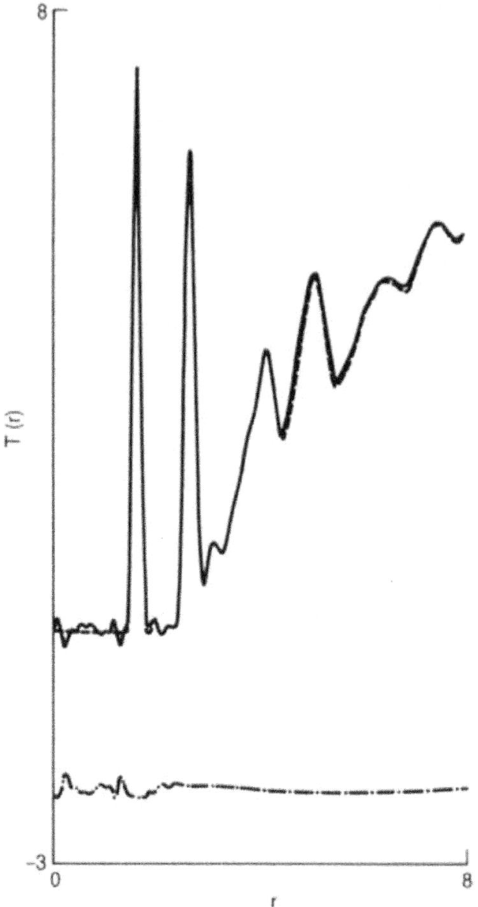

Figure 12.26 Experimental (Grimley *et al.*, 1990) (——), calculated (- - - -), and difference curve (–·–·–) RDF for silica glass.

can be generated has increased significantly. We may also anticipate that quantum mechanical techniques, which were previously too expensive for the study of glass structures, will become more practicable for the study of these materials.

ACKNOWLEDGEMENTS

I would like to acknowledge many useful discussions with M. Amini, M. Leslie, C.R.A. Catlow, D. Fincham, A.C. Wright, A.C. Hannon, J.E. Dickinson, G.N. Greaves, J.M. Newsam, D. Rigby, J.W. Adams, P.T. Marten and C.M. Freeman.

REFERENCES

Angell, C.A. and Kanno, H. (1976) *Science*, **193**, 1121.

Angell, C.A., Cheeseman, P.A. and Tamaddon, S. (1982) *Science*, **218**, 885.

Antonio, G.A., Kalia, R.K. and Vashishta, P. (1988) *J. Non-Cryst. Solids*, **106**, 305.

Bacon, J.F., Hasapis, A.A. and Wholley, J.W. (1960) *Phys. Chem. Glasses*, **1**, 90.

Balasubramanian, S. and Rao, K.J. (1994) *J. Phys. Chem.*, **98**, 9216.

Bell, R.J. and Dean, P. (1972) *Phil. Mag.*, **25**, 1381.

Bionducci, M., Buffa, F., Licheri, G., Musinu, A., Navarra, G. and Piccaluga, G. (1994) *J. Non-Cryst. Solids*, **177**, 137.

Cusak, N.E. (1987) *The Physics of Structurally Disordered Matter*, Adam Hilger, Bristol.

Day, D.E. (1976) *J. Non-Cryst. Solids*, **21**, 343.

Dickinson, J.E. (1988) *EOS Trans. Amer. Geophys. Union*, **69**, 501.

Dickinson, J.E. (1989) In *Proc. XV International Congress on Glass* (Lenigrad) (ed. O.V. Nazurin), Leningrad Nauka, Leningrad, p. 192.

Eastwood, J.W., Hockney, R.W. and Lawrence, D. (1980) *Comput. Phys. Commun.*, **19**, 215.

Fincham, D. (1994) *Mol. Simul.*, **13**, 1.

Fitzgerald, J.J., Dec, S.F. and Hamza, A.I. (1989) *Am. Mineral.*, **74**, 1405.

Gaskell, P.H. (1991) *Glasses and Amorphous Materials, Materials Science and Technology*, Vol. 9, (ed. J. Zarzycki), VCH, Weinheim, p. 175.

Greaves, G.N. (1985) *J. Non-Cryst. Solids*, **71**, 203.

Greaves, G.N., Fontaine, A., Raoux, P. and Gurman, S.J. (1981) *Nature*, **293**, 611.

Greaves, G.N., Catlow, C.R.A., Vessal, B., Charnock, J., Henderson, C.M.B., Zhu, R., Qiao, S., Wang, Y., Gurman, S.J. and Houde-Walter, S. (1990) *Inst. Phys. Conf. Ser.*, **111**, 411.

Greaves, G.N., Gurman, S.J., Catlow, C.R.A., Chadwick, A.V., Houde-Walter, S., Dobson, B.R. and Henderson, C.M.B. (1991) *Phil. Mag. A*, **64**, 1059.

Greengard, L. and Rokhlin, V. (1987) *J. Comput. Phys.*, **73**, 325.

Grimley, D.I., Wright, A.C. and Sinclair, R.N. (1990) *J. Non-Cryst. Solids*, **119**, 49.

Handa, Y.P., Tse, J.S., Klug, D.D. and Whalley, E. (1990) *J. Chem. Phys.*, **94**, 623.

Hemley, R.J., Jephcoat, A.P., Mao, H.K., Ming, L.C. and Manghnani, M.H. (1988) *Nature*, **334**, 52.

Hsieh, H. and Yip, S. (1987) *Phys. Rev. Lett.*, **59**, 2760.

Inoue, S. and MacFarlane, D.R. (1987) *J. Non-Cryst. Solids*, **95/96**, 585.

Johnson, P.A.V. and Wright, A.C. (1982) *J. Non-Cryst. Solids*, **50**, 281.

Kaplow, R., Rowe, T.A. and Averbach, B.L. (1968) *Phys. Rev.*, **168**, 1068.

Keating, P.N. (1966) *Phys. Rev.*, **145**, 637.

Keen, D.A. and McGreevy, R.L. (1990) *Nature*, **344**, 423.

Kohn, S.C., Dupree, R., Mortuza, M.G. and Henderson, C.M.B. (1991) *Am. Mineral.*, **76**, 309.

Kruger, M.B. and Jeanloz, R. (1990) *Science*, **249**, 647.

MacFarlane, D.R., Inoue, S., Browne, J.O. and Uhlherr, A. (1988) *Mater. Sci. Forum*, **32–33**, 125.

McGreevy, R.L. and Pusztai, L. (1988) *Mol. Simul.*, **1**, 359.

Massobrio, C., Pontikis, V. and Martin, G. (1989) *Phys. Rev. Lett.*, **62**, 1142.

Massobrio, C., Pontikis, V. and Martin, G. (1990) *Phys. Rev. B*, **41**, 10486.

Mishima, O., Calvert, L.D. and Whalley, E. (1984) *Nature*, **310**, 393.

Moore, L.J., MacFarlane, D.R. and Uhlherr, A. (1991) *Mater. Sci. Forum*, **67–68**, 413.

Ohtani, E., Taulelle, F. and Angell, C.A. (1985) *Nature*, **314**, 78.

Rahman, A., Mandell, M.J. and McTague, J.P. (1976) *J. Chem. Phys.*, **64**, 1564.

Rechtin, M.D. and Averbach, B.L. (1974) *Solid State Commun.*, **13**, 491.

Rechtin, M.D. and Averbach, M.D. (1975) *Phys. Stat. Sol. A*, **28**, 283.
Rechtin, M.D., Renninger, A.L. and Averbach, B.L. (1974) *J. Non-Cryst. Solids*, **15**, 74.
Renninger, A.L. and Rechtin, M.D. (1974) *J. Non-Cryst. Solids*, **16**, 1.
Sabochick, M.J. and Lam, N.Q. (1991) *Phys. Rev. B*, **43**, 5243.
Simmons, J.H., O'Rear, G., Swiler, P. and Wright, A.C. (1988) *J. Non-Cryst. Solids*, **106**, 325.
Stillinger, F.H. and Weber, T.A. (1985) *Phys. Rev. B*, **31**, 5262.
Takada, A. (1993) *Computer Modelling of Crystalline and Vitreous Boric Oxide*, PhD Thesis, University of London.
Tse, J.S. and Klein, M.L. (1987) *Phys. Rev. Lett.*, **58**, 1672.
Uhlherr, A. and MacFarlane, D.R. (1993) *J. Non-Cryst. Solids*, **161**, 98.
Uhlherr, A., MacFarlane, D.R. and Bastow, T.J. (1990) *J. Non-Cryst. Solids*, **123**, 42.
Uhlherr, A., MacFarlane, D.R., Moore, L.J. and Thomas, P.D. (1991) *Mat. Sci. Forum*, **67–68**, 431.
Vashishta, P., Kalia, R.K. and Ebbsjo, I. (1988) *J. Non-Cryst. Solids*, **106**, 301.
Vashishta, P., Kalia, R.K., Antonio, G.A. and Ebbsjo, I. (1989a) *Phys. Rev. Lett.*, **62**, 1651.
Vashishta, P., Kalia, R.K. and Ebbsjo, I. (1989b) *Phys. Rev. B*, **39**, 6034.
Vessal, B. (1991) *Trans. Am. Cryst. Assoc.*, **27**, 37.
Vessal, B. and Dickinson, J.E. (1994) *Mater. Res. Soc. Proc.*, **321**, 129.
Vessal, B., Amini, M., Fincham, D. and Catlow, C.R.A. (1989) *Phil. Mag. B*, **60**, 753.
Vessal, B., Greaves, G.N., Marten, P.T., Chadwick, A.V., Mole, R. and Houde-Walter, S. (1992) *Nature*, **356**, 504.
Vessal, B., Amini, M. and Catlow, C.R.A. (1993) *J. Non-Cryst. Solids*, **159**, 184.
Woodcock, L.V., Angell, C.A. and Cheeseman, P. (1976) *J. Chem. Phys.*, **65**, 1565.
Wright, A.C. (1993) *J. Non-Cryst. Solids*, **159**, 264.
Wright, A.C. (1994) *J. Non-Cryst. Solids*, **179**, 84.
Wright, A.C., Clare, A.G., Bachra, B., Sinclair, R.N., Hannon, A.C. and Vessal, B. (1991) *Trans. A.C.A.*, **27**, 239.
Wright, A.C., Bachra, B., Brunier, T.M., Sinclair, R.N., Gladden, L.F. and Portsmouth, R.L. (1992) *J. Non-Cryst. Solids*, **150**, 69.
Wright, A.C., Bachra, B., Vessal, B., Clare, A.G., Sinclair, R.N. and Hannon, A.C. (1993) In *Fundamentals of Glass Science and Technology*, ESG, Venice, p. 211.

Index